Methods in Molecular Biology

Series Editor
John M. Walker
School of Life and Medical Sciences
University of Hertfordshire
Hatfield, Hertfordshire, AL10 9AB, UK

For further volumes:
http://www.springer.com/series/7651

Xenopus

Methods and Protocols

Editor

Kris Vleminckx

Department of Biomedical Molecular Biology, Ghent University, Ghent, Belgium

Humana Press

Editor
Kris Vleminckx
Department of Biomedical Molecular Biology
Ghent University
Ghent, Belgium

ISSN 1064-3745 ISSN 1940-6029 (electronic)
Methods in Molecular Biology
ISBN 978-1-4939-8783-2 ISBN 978-1-4939-8784-9 (eBook)
https://doi.org/10.1007/978-1-4939-8784-9

Library of Congress Control Number: 2018952461

This Humana Press imprint is published by the registered company Springer Science+Business Media, LLC part of Springer Nature.
The registered company address is: 233 Spring Street, New York, NY 10013, U.S.A.

Preface

Xenopus Protocols: Methods for Functional Genomics and Disease Modeling

The last couple of years have seen unbiased and systematic analysis of gene mutations, genomic rearrangements, and RNA expression in clinical practice, for instance in patients afflicted by Mendelian disorders or cancer. While whole exome and whole genome sequencing (WES and WGS) allow detailed and fast identification of mutations and variants in the afflicted patients, it mostly remains a major challenge to identify the specific genetic defects that are causally linked to the disease. In addition, in order to better understand the molecular basis of the disease, functional models can be required. While some aspects of disease can be functionally addressed in cell culture, induced pluripotent stem cells, or organoids, more complex developmental or physiological processes require modeling in a living organism. Evidently, highly evolutionary conserved processes such as regulation of the cell cycle or RNA splicing can be addressed in organisms like yeast, *C. elegans*, or the fruit fly. However, evolutionary more derived biological process and the required presence of orthologous genes impose the use of higher organisms, with the mouse historically as an evident choice. However, highly accessible nonmammalian vertebrates that allow straightforward genome editing (e.g., via TALEN or CRISPR/Cas9) will be of increasing importance to pinpoint the genetic mutations and variations underlying specific diseases. While these models could be used as a triage system for the subsequent generation of a mammalian disease model, they can as well function themselves as a platform for basic research, novel therapeutic target identification, and early drug development.

Genome editing via guided nucleases such as TALEN and **CRISPR/Cas9** has created a true revolution in functional genomics and genetics. This is also the case for nonmammalian vertebrate model organisms where it is now possible to mutate specific genes (i.e., performing factual reverse genetics), rather than having to rely on random mutagenesis approaches or transient knock-down of gene expression (e.g., via RNAi or Morpholino injections). Especially for the *Xenopus* **and zebrafish** research communities, this has started an exciting new era of genome editing that creates unique opportunities for **modeling human disease**. Certainly *Xenopus tropicalis*, **which unlike** ***Xenopus laevis*** **and the zebrafish has a true diploid genome, has emerged as a powerful model**, extremely well positioned for modeling human disease. This book wants to serve as a comprehensive guidebook for using *Xenopus* in this newly emerged experimental landscape.

The first part of the book provides protocols for implementation of CRISPR/Cas9 and TALEN experiments in *Xenopus*, from the design stage up until the genotyping. In addition to the straightforward approaches to obtain knock-out animals, based on repair via nonhomologous end joining (NHEJ), also methods for the generation of knock-in animals are presented. Evidently, up until recently few *Xenopus* researches felt the need to grow up tadpoles and establish lines, since most of the manipulations were temporarily restricted to the embryonic and tadpole stages. However, the implementation of genome editing now imposes the need for efficient and reliable husbandry and animal care. I believe this is still a major hurdle for many *Xenopus* researchers, who were spoiled by the sturdiness of the

Xenopus laevis embryos and tadpoles. In truth, we need to approach the *Xenopus tropicalis* animals from a totally different perspective, i.e., as an animal requiring our daily attention. Two disease-related applications of CRISPR/Cas9-based genome editing are presented, i.e., in cancer modeling and in screening of congenital heart disease.

To avoid possible early adverse effects of genome modifications in the embryo, genetic manipulation in tadpole stages may be desirable. Hence protocols for electroporation are included, albeit not yet in the context of CRISPR/Cas9 or TALEN, although I believe it is only a matter of time before this will be applied to the *Xenopus* model.

The expansion of *Xenopus* research to late tadpole and post-metamorphic stages exposes a major hiatus. While making knock-outs is relatively simple and fast, most of the time and workload for many future *Xenopus* researchers will go into the phenotyping and they will be faced with a scarcity of methods for analysis, although they can often rely on adapted versions of protocols developed in the mouse or in zebrafish. The book describes methods for phenotyping at the organismal level, the cellular level, and at the level of the proteome. This also includes two exemplary protocols for establishing primary cell cultures from both larval and adult stages. While most of these methods are currently available for *Xenopus laevis*, they should be easily transferable to the *Xenopus tropicalis* model.

This book is a timely update and addition to previous editions in this Methods in Molecular Biology series and should continue to keep *Xenopus* at the frontline for biochemical, cell biological, and developmental studies. However, the easy and straightforward methods for genome editing should now in addition establish our beloved organism also as a solid and versatile preclinical disease model and will foster future collaborative interactions with clinical geneticists, cancer researchers, and other people working on human disease. There is a wide open field ahead of us that is inviting to be filled by current and future *Xenopus* researchers.

Ghent, Belgium ***Kris Vleminckx***

Contents

Preface . . . . *v*
Contributors . . . . *ix*

1 Husbandry, General Care, and Transportation of *Xenopus laevis* and *Xenopus tropicalis* 1
Sean McNamara, Marcin Wlizla, and Marko E. Horb

2 Generation and Care of *Xenopus laevis* and *Xenopus tropicalis* Embryos 19
Marcin Wlizla, Sean McNamara, and Marko E. Horb

3 Methods for CRISPR/Cas9 *Xenopus tropicalis* Tissue-Specific Multiplex Genome Engineering 33
Thomas Naert and Kris Vleminckx

4 Targeted Genome Engineering in *Xenopus* Using the Transcription Activator-Like Effector Nuclease (TALEN) Technology 55
Tom Van Nieuwenhuysen and Kris Vleminckx

5 Genotyping of CRISPR/Cas9 Genome Edited *Xenopus tropicalis* 67
Thomas Naert and Kris Vleminckx

6 BATCH-GE: Analysis of NGS Data for Genome Editing Assessment 83
Wouter Steyaert, Annekatrien Boel, Paul Coucke, and Andy Willaert

7 A Simple Knock-In System for *Xenopus* via Microhomology Mediated End Joining Repair 91
Ken-ich T. Suzuki, Yuto Sakane, Miyuki Suzuki, and Takashi Yamamoto

8 How to Generate Non-Mosaic CRISPR/Cas9 Mediated Knock-In and Mutations in F0 *Xenopus* Through the Host-Transfer Technique 105
Emmanuel Tadjuidje and Sang-Wook Cha

9 Targeted Electroporation in the CNS in *Xenopus* Embryos 119
Hovy Ho-Wai Wong and Christine E. Holt

10 Conditional Chemogenetic Ablation of Photoreceptor Cells in *Xenopus* Retina 133
Albert Chesneau, Odile Bronchain, and Muriel Perron

11 Cancer Models in *Xenopus tropicalis* by CRISPR/Cas9 Mediated Knockout of Tumor Suppressors 147
Thomas Naert and Kris Vleminckx

12 CRISPR/Cas9 F0 Screening of Congenital Heart Disease Genes in *Xenopus tropicalis* 163
Engin Deniz, Emily K. Mis, Maura Lane, and Mustafa K. Khokha

13 Quantitative Proteomics of *Xenopus* Embryos I, Sample Preparation 175
Meera Gupta, Matthew Sonnett, Lillia Ryazanova, Marc Presler, and Martin Wühr

14 Quantitative Proteomics for *Xenopus* Embryos II, Data Analysis 195
Matthew Sonnett, Meera Gupta, Thao Nguyen, and Martin Wühr

15 Dye Electroporation and Imaging of Calcium Signaling in *Xenopus* Nervous System 217
Lukas Weiss, Thomas Offner, Thomas Hassenklöver, and Ivan Manzini

16 X-FaCT: Xenopus-Fast Clearing Technique 233
Pierre Affaticati, Sébastien Le Mével, Arnim Jenett, Laurie Rivière, Elodie Machado, Bilal B. Mughal, and Jean-Baptiste Fini

17 Cell Cycle Analysis of the Embryonic Brain of Fluorescent Reporter *Xenopus tropicalis* by Flow Cytometry 243
Rivka Noelanders and Kris Vleminckx

18 Manipulating and Analyzing Cell Type Composition of the *Xenopus* Mucociliary Epidermis 251
Peter Walentek

19 Evaluating Blood Cell Populations in *Xenopus* Using Flow Cytometry and Differential Counts by Cytospin 265
Jacques Robert, Eva-Stina Edholm, and Francisco De Jesus Andino

20 Isolation and Culture of Amphibian (*Xenopus laevis*) Sub-Capsular Liver and Bone Marrow Cells 275
Amulya Yaparla and Leon Grayfer

21 Isolation and Primary Culture Methods of Adult and Larval Myogenic Cells from *Xenopus laevis*. 283
Kazi Taheruzzaman and Akio Nishikawa

Index *301*

Contributors

PIERRE AFFATICATI • *Tefor Core Facility, Paris-Saclay Institute of Neuroscience, CNRS, Université Paris-Saclay, Gif-sur-Yvette, France*

ANNEKATRIEN BOEL • *Center for Medical Genetics, Ghent University Hospital, Ghent, Belgium*

ODILE BRONCHAIN • *Paris-Saclay Institute of Neuroscience, CNRS, Univ Paris Sud, Université Paris-Saclay, Orsay Cedex, France*

SANG-WOOK CHA • *Cincinnati Children's Research Foundation, Cincinnati, OH, USA; Department of Biology and Agriculture, School of Natural Sciences, University of Central Missouri, Warrensburg, MO, USA*

ALBERT CHESNEAU • *Paris-Saclay Institute of Neuroscience, CNRS, Univ Paris Sud, Université Paris-Saclay, Orsay Cedex, France*

PAUL COUCKE • *Center for Medical Genetics, Ghent University Hospital, Ghent, Belgium*

FRANCISCO DE JESUS ANDINO • *Department of Microbiology and Immunology, University of Rochester Medical Center, Rochester, NY, USA*

ENGIN DENIZ • *Pediatric Genomics Discovery Program, Section of Pediatric Critical Care, Department of Pediatrics, Yale University School of Medicine, New Haven, CT, USA*

EVA-STINA EDHOLM • *Department of Microbiology and Immunology, University of Rochester Medical Center, Rochester, NY, USA; The Norwegian College of Fishery Science, University of Tromsø, Tromsø, Norway*

JEAN-BAPTISTE FINI • *Evolution des Régulations Endocriniennes, Département « Adaptation du Vivant », UMR 7221 MNHN/CNRS, Sorbonne Universités, Paris, France*

LEON GRAYFER • *Department of Biological Sciences, George Washington University, Washington, DC, USA*

MEERA GUPTA • *Department of Molecular Biology and Lewis-Sigler Institute for Integrative Genomics, Princeton University, Princeton, NJ, USA; Department of Chemical and Biological Engineering, Princeton University, Princeton, NJ, USA*

THOMAS HASSENKLÖVER • *Department of Animal Physiology and Molecular Biomedicine, Institute of Animal Physiology, Justus-Liebig-University Giessen, Giessen, Germany*

CHRISTINE E. HOLT • *Department of Physiology, Development and Neuroscience, University of Cambridge, Cambridge, UK*

MARKO E. HORB • *National Xenopus Resource, Marine Biological Laboratory, Woods Hole, MA, USA*

ARNIM JENETT • *Tefor Core Facility, Paris-Saclay Institute of Neuroscience, CNRS, Université Paris-Saclay, Gif-sur-Yvette, France*

MUSTAFA K. KHOKHA • *Pediatric Genomics Discovery Program, Section of Pediatric Critical Care, Department of Pediatrics, Yale University School of Medicine, New Haven, CT, USA; Department of Genetics, Yale University School of Medicine, New Haven, CT, USA*

MAURA LANE • *Pediatric Genomics Discovery Program, Section of Pediatric Critical Care, Department of Pediatrics, Yale University School of Medicine, New Haven, CT, USA*
SÉBASTIEN LE MÉVEL • *Evolution des Régulations Endocriniennes, Département « Adaptation du Vivant », UMR 7221 MNHN/CNRS, Sorbonne Universités, Paris, France*
ELODIE MACHADO • *Tefor Core Facility, Paris-Saclay Institute of Neuroscience, CNRS, Université Paris-Saclay, Gif-sur-Yvette, France*
IVAN MANZINI • *Department of Animal Physiology and Molecular Biomedicine, Institute of Animal Physiology, Justus-Liebig-University Giessen, Giessen, Germany; Center for Nanoscale Microscopy and Molecular Physiology of the Brain, University of Göttingen, Göttingen, Germany*
SEAN MCNAMARA • *National Xenopus Resource, Marine Biological Laboratory, Woods Hole, MA, USA*
EMILY K. MIS • *Pediatric Genomics Discovery Program, Section of Pediatric Critical Care, Department of Pediatrics, Yale University School of Medicine, New Haven, CT, USA*
BILAL B. MUGHAL • *Evolution des Régulations Endocriniennes, Département « Adaptation du Vivant », UMR 7221 MNHN/CNRS, Sorbonne Universités, Paris, France*
THOMAS NAERT • *Department of Biomedical Molecular Biology, Ghent University, Ghent, Belgium; Cancer Research Institute Ghent, Ghent, Belgium*
THAO NGUYEN • *Department of Molecular Biology and Lewis-Sigler Institute for Integrative Genomics, Princeton University, Princeton, NJ, USA; Department of Chemical and Biological Engineering, Princeton University, Princeton, NJ, USA*
AKIO NISHIKAWA • *Department of Biological Science, Faculty of Life and Environmental Science, Shimane University, Matsue, Shimane, Japan*
RIVKA NOELANDERS • *Department of Biomedical Molecular Biology, Ghent University, Ghent, Belgium; Cancer Research Institute Ghent, Ghent, Belgium*
THOMAS OFFNER • *Department of Animal Physiology and Molecular Biomedicine, Institute of Animal Physiology, Justus-Liebig-University Giessen, Giessen, Germany; Center for Nanoscale Microscopy and Molecular Physiology of the Brain, University of Göttingen, Göttingen, Germany*
MURIEL PERRON • *Paris-Saclay Institute of Neuroscience, CNRS, Univ Paris Sud, Université Paris-Saclay, Orsay Cedex, France*
MARC PRESLER • *Department of Systems Biology, Harvard Medical School, Boston, MA, USA*
LAURIE RIVIÈRE • *Tefor Core Facility, Paris-Saclay Institute of Neuroscience, CNRS, Université Paris-Saclay, Gif-sur-Yvette, France*
JACQUES ROBERT • *Department of Microbiology and Immunology, University of Rochester Medical Center, Rochester, NY, USA*
LILLIA RYAZANOVA • *Department of Molecular Biology and Lewis-Sigler Institute for Integrative Genomics, Princeton University, Princeton, NJ, USA*
YUTO SAKANE • *Graduate School of Science, Hiroshima University, Hiroshima, Japan*
MATTHEW SONNETT • *Department of Molecular Biology and Lewis-Sigler Institute for Integrative Genomics, Princeton University, Princeton, NJ, USA*
WOUTER STEYAERT • *Center for Medical Genetics, Ghent University Hospital, Ghent, Belgium*
KEN-ICH T. SUZUKI • *Graduate School of Science, Hiroshima University, Hiroshima, Japan*

MIYUKI SUZUKI • *Graduate School of Science, Hiroshima University, Hiroshima, Japan*
EMMANUEL TADJUIDJE • *Department of Biological Sciences, Alabama State University, Montgomery, AL, USA*
KAZI TAHERUZZAMAN • *Department of Biological Science, Faculty of Life and Environmental Science, Shimane University, Matsue, Shimane, Japan*
TOM VAN NIEUWENHUYSEN • *Department of Biomedical Molecular Biology, Ghent University, Ghent, Belgium*
KRIS VLEMINCKX • *Department of Biomedical Molecular Biology, Ghent University, Ghent, Belgium; Cancer Research Institute Ghent, Ghent, Belgium; Center for Medical Genetics, Ghent University, Ghent, Belgium*
PETER WALENTEK • *Renal Division, Department of Medicine, University Freiburg Medical Center, Freiburg, Germany; Center for Biological Systems Analysis (ZBSA), University of Freiburg, Freiburg, Germany*
LUKAS WEISS • *Department of Animal Physiology and Molecular Biomedicine, Institute of Animal Physiology, Justus-Liebig-University Giessen, Giessen, Germany*
ANDY WILLAERT • *Center for Medical Genetics, Ghent University Hospital, Ghent, Belgium*
MARCIN WLIZLA • *National Xenopus Resource, Marine Biological Laboratory, Woods Hole, MA, USA*
HOVY HO-WAI WONG • *Department of Physiology, Development and Neuroscience, University of Cambridge, Cambridge, UK*
MARTIN WÜHR • *Department of Molecular Biology and Lewis-Sigler Institute for Integrative Genomics, Princeton University, Princeton, NJ, USA*
TAKASHI YAMAMOTO • *Graduate School of Science, Hiroshima University, Hiroshima, Japan*
AMULYA YAPARLA • *Department of Biological Sciences, George Washington University, Washington, DC, USA*

Chapter 1

Husbandry, General Care, and Transportation of *Xenopus laevis* and *Xenopus tropicalis*

Sean McNamara, Marcin Wlizla, and Marko E. Horb

Abstract

Maintenance of optimal conditions such as water parameters, diet, and feeding is essential to a healthy *Xenopus laevis* and *Xenopus tropicalis* colony and thus to the productivity of the lab. Our prior husbandry experience as well as the rapid growth of the National *Xenopus* Resource has given us a unique insight into identifying and implementing these optimal parameters into our husbandry operations. Here, we discuss our standard operating procedures that will be of use to both new and established *Xenopus* facilities.

Key words *Xenopus laevis*, *Xenopus tropicalis*, Aquatic recirculating systems, Husbandry

1 Introduction

A number of reasons make both *Xenopus laevis* and *Xenopus tropicalis* powerful systems for the study of developmental, cellular, and molecular biology as well as biomedical research modeling human disease [1, 2]. Many of these reasons are practical in nature and include a closed lifecycle with a relative ease of maintaining the animals in the lab, the ability to generate large quantities of equivalent research material all year round through the use of gonadotropic hormone injection to induce ovulation at any time of the year, and robust and relatively rapid development of early embryonic cultures [3]. These practical reasons, in particular, are dependent on proper husbandry practices that are essential for maintaining a colony of healthy frogs capable of providing high quality eggs and embryos necessary for driving productivity of the lab forward.

One of the goals of the National *Xenopus* Resource (NXR) is to serve as a central repository and distribution center for a diverse number of *Xenopus* stocks of high interest to the research community [4]. Since the establishment of the NXR in 2010 we have grown our colony from zero frogs to now house over 3000 *X. tropicalis* and 5000 *X. laevis*. This rapid growth as well as the

Kris Vleminckx (ed.), *Xenopus: Methods and Protocols*, Methods in Molecular Biology, vol. 1865,
https://doi.org/10.1007/978-1-4939-8784-9_1,

necessity to quickly replenish animal lines that we distribute, has motivated us to optimize *Xenopus* husbandry protocols for all life history stages, thus assuring rapid and robust growth from fertilization to sexual maturity.

Many researchers starting their own labs have limited understanding of the steps necessary for successfully establishing a new colony and frequently this task is delegated to institutional animal care facilities that may similarly lack the necessary expertise. Our goal in this chapter is to provide our protocols and experiences in establishing a new colony and maintaining it over the long term. We discuss a number of topics including recirculating *Xenopus* housing modules, optimal water parameters, diets and feeding schedules, and animal distribution.

2 Materials

2.1 Animals

Wild type outbred *X. laevis* and *X. tropicalis* can be purchased from a variety of commercial vendors, including Nasco, *Xenopus* One, and *Xenopus* Express. These companies are an excellent resource for obtaining large quantities of outbred *Xenopus* and they work closely with each researcher to provide animals at all stages. More specialized strains and lines of *Xenopus*, such as the inbred *X. laevis* J strain and the Nigerian *X. tropicalis* strain, are available for purchase through regional stock centers, including the National *Xenopus* Resource (NXR, RRID:SCR_013731, http://www.mbl.edu/xenopus), the European *Xenopus* Resource Centre (EXRC, RRID:SCR_007164, https://xenopusresource.org/), the *Xenopus laevis* Research Resource for Immunobiology (XLRRI, https://www.urmc.rochester.edu/microbiology-immunology/xenopus-laevis.aspx) and the *Xenopus tropicalis* National BioResource Project of Japan (http://home.hiroshima-u.ac.jp/amphibia/xenobiores_en/iweb_en/Top.html). In addition, these stock centers distribute transgenic and mutant lines for both *X. laevis* and *X. tropicalis.* Detailed information about the various suppliers, including contact information, can be found online at Xenbase (RRID:SCR_003280, http://www.xenbase.org/other/obtain.do).

2.2 Animal Housing

There are three main options for housing and maintaining *Xenopus* animals: recirculating, flow through, and static fill and dump systems. System choice will depend on the needs of the lab and available funds. Over the long term, we have found that recirculating aquatic systems are the best option for research laboratory sized operations. First, modern recirculating systems automatically measure and control a number of water parameters including temperature, pH, conductivity, and filter inflow and outflow pressures. This aids in maintaining the system water at peak quality for the housed animals, which is not possible with fill and dump systems. Second, although flow through systems also measure and regulate the same parameters, they are considerably less economical. In a flow through configuration a large volume of conditioned water is replaced daily, whereas in a recirculating system typically only 10% of its total volume is exchanged in a 24-h period. Another choice that can be made during the system selection is the size of the tanks that will house the animals. This choice will depend on the species, age, number of animals in the colony, and the expected frequency of their use; whichever size is chosen, we recommend the use of transparent plastic tanks that permit for easy visual inspection of the animals inside. The following companies provide both stand-alone and multi-rack recirculating frog systems and have extensive experience in helping research labs selecting the most appropriate configurations for housing animals (*see* **Note 1**).

1. Aquaneering, Inc., San Diego, CA, USA.
2. Aquatic Enterprises, Inc., Seattle, WA, USA.
3. Aqua Schwarz, GmBH, Göttingen, Germany.
4. Iwaki Aquatic Systems and Services, Holliston, MA, USA.
5. Tecniplast USA, West Chester, PA, USA.

2.3 Food

1. Sera Micron Growth Food (sera North America, Montgomeryville, PA, USA).
2. Brine Shrimp Flake (Brine Shrimp Direct, Ogden, UT, USA).
3. Nasco Frog Brittle (Powder) (Nasco, Fort Atkins, WI, USA).
4. Nasco Frog Brittle (Small Nuggets) for Post-Metamorphic *Xenopus* (Nasco, Fort Atkins, WI, USA).
5. Nasco Frog Brittle (Large Nuggets) for Adult *Xenopus* (Nasco, Fort Atkins, WI, USA).
6. Bio Vita Fry 1.2 mm Pellet (Bio-Oregon, Longview, WA, USA).
7. Bio Vita Fry 2.0 mm Pellet (Bio-Oregon, Longview, WA, USA).
8. BioTrout 4.0 mm Pellet (Bio-Oregon, Longview, WA, USA).

2.4 Water Quality Maintenance and Testing Kits

1. Sea salt used to control conductivity, we use Reef Salt from Seachem Laboratories (Madison, GA, USA).
2. Biocarbonate to buffer and regulate pH (ProLine brand, Pentair AES, Apopka, FL, USA).
3. Nitrifying bacteria (ProLine brand, Pentair AES, Apopka, FL, USA).
4. pH Aquarium Fresh Water Test Kit, pH range 6.0–7.6 (API, Chalfont, PA, USA).

5. Ammonia Fresh Water and Salt Water Test Kit (API, Chalfont, PA, USA).
6. Nitrite NO_2 Fresh Water and Salt Water Test Kit (API, Chalfont, PA, USA).
7. Nitrate NO_3 Fresh Water and Salt Water Test Kit (API, Chalfont, PA, USA).

2.5 Cleaning and Sterilization Reagents

1. Bleach (6.0% NaClO). Available from supermarket cleaning supplies isle. Dilute with pure water to make a 10% bleach solution (final concentration of NaClO is 0.6%), and use to sterilize tanks, tank accessories, frog nets, etc.
2. Ethanol. Source may depend on local regulations. Purchase 190 proof (95%) and dilute to 70% with pure water.
3. Dechlorinator ($Na_2S_2O_3$). We use ProLine brand (Pentair AES, Apopka, FL, USA), which contains both the anhydrous (CAS# 7772-98-7) and pentahydrate salt (CAS# 10102-17-7) molecules. It can be acquired from other sources and whether the anhydrous or hydrate salt is used is not important. 1.6–2.6 ppm of $Na_2S_2O_3$ per 1 ppm of chlorine is typically sufficient to dechlorinate water.
4. Virkon Aquatic (DuPont, Wilmington, DE, USA).

2.6 Anesthesia and Euthanasia Reagents

1. Ethyl 4-aminobenzoate, 98%; aka. Benzocaine (CAS# 94-09-7): make 10% stock in 95% Ethanol and store at room temperature.
2. Tricaine Methanesulfonate; aka. Tricaine-S, aka. MS 222 (CAS# 886-86-2).

3 Methods

3.1 Housing Animals

At the National *Xenopus* Resource, we house our adult, juvenile, and tadpole *X. laevis* and *X. tropicalis* in modular recirculating aquatic systems, and use both individual and multi-rack configurations (Fig. 1a). To ensure high water quality, these recirculating systems incorporate several filtration steps, including biological, mechanical, activated carbon, and UV sterilization. Listed below we outline some guidelines that we have found help maintain the systems and keep animals healthy and productive.

1. For adult animals, we recommend the use of so called flood-and-flush tanks. These are self-cleaning tanks that are designed for automatic clearing of undissolved solids, such as leftover food and excretion products. Use of these tanks minimizes the time and effort required to net out the solids and in addition maintains cleaner water.

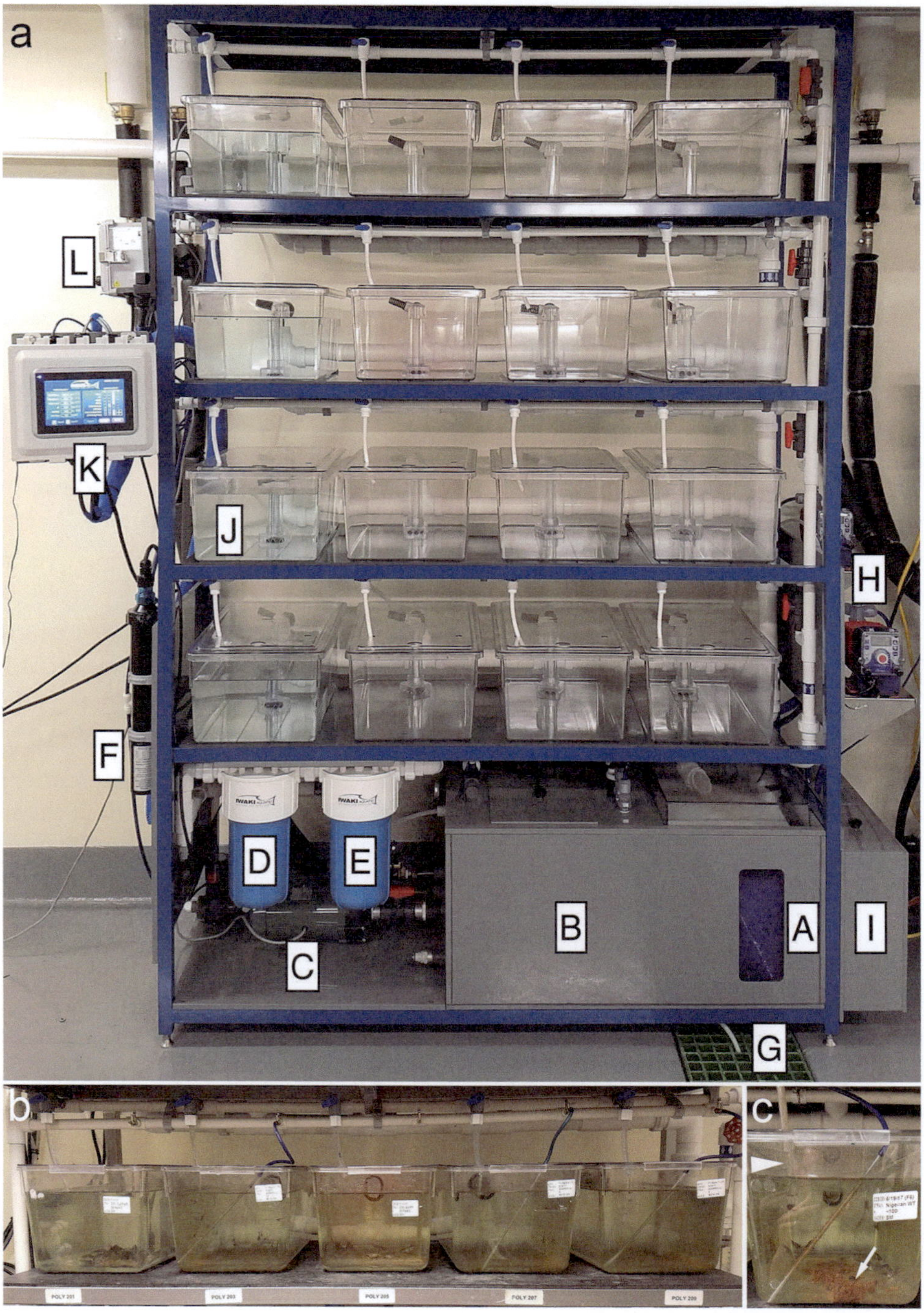

Fig. 1 Individual recirculating system configuration. (**a**) Example of an individual recirculating aquatic system designed by Iwaki Aquatic, the ITS-X model. This system holds 16 × 23 L flood-and-flush tanks, and has the capacity to heat and chill the water. It has four filtration processes and digital user interface display of water quality levels. Specific aspects of the system are labeled. A-biofilter and aeration, B-water storage, submersible heater, and chilling coil, C-water pump, D-mechanical filter, E-carbon filter, F-UV lamp/filter, G-water effluent, H-pH/conductivity dosing pumps, I-pH/conductivity dosing reservoirs, J-23 L flood-and-flush tank, K-human machine interface, L-programmable logic controller with pH/Conductivity probes. (**b**) Nursery tank Fig. 1 (continued) row. One row of 16 L nursery tanks shown as examples; the tanks have air lines added for aeration of water when growing tadpoles. (**c**) Single nursery tank. Arrowhead—incoming flow of water to the tank is much less than for adult frogs. Arrow—uneaten food and detritus settles to the bottom of the tank and provides tadpoles with additional food and places to hide

2. We do not recommend using the flood and flush tanks for growing tadpoles as this will result in flushing the animals out of the system. Tadpoles and froglets should be kept in tanks with only an overflow bulkhead (Fig. 1b).
3. For growing tadpoles, we recommend that the tanks have air lines or bubblers installed to help oxygenate the water. Care should be taken however, as too fast a bubbling rate can actually have a negative impact on the tadpoles by increasing the total dissolved gas pressure. We generally keep the flow rate at not more than 10 × 3–5 mm bubbles per second in a 16 L tank (Fig. 1b, c).
4. System readings should be checked and recorded daily to spot any potential developing problems; this is in addition to the automatic monitoring that is done by the systems.
5. Animal health and wellbeing should be confirmed via a daily visual inspection.
6. 10% of the system water should be exchanged daily; most systems will change the water slowly over the course of 24 h.
7. The animals should be kept in a 12 h light/12 h dark cycle, though this can be flexible, as others use 14 h light/10 h dark.
8. Animals can be provided with environmental enrichment; however, this should be kept to a minimum to limit the surface area available for algal growth. For *X. laevis* we recommend PVC piping cut lengthwise in half, of a length sufficient to allow the animals to hide under. Cut surfaces should be smoothened out to eliminate sharp edges that can injure the frogs. For *X. tropicalis* we recommend artificial aquarium foam lotus leaves that permit the animals to both hide underneath as well as rest on top.
9. The tank flow rate will vary with tank size and should be kept between 1.0–3.5 LPM for smaller to larger tanks respectively. Ideal flow should be high but without agitating the water.
10. To limit algal growth and detritus accumulation tanks should be spot cleaned as needed using aquarium scrubbing pads (*see* **Note 2**).
11. Tanks that are especially soiled should be replaced with clean tanks, and themselves disassembled and thoroughly cleaned. For cleaning, the tank and the lid should be scrubbed and rinsed with RO (Reverse Osmosis) water, sprayed with 10% bleach and allowed to stand for 30 min to 1 h, rinsed with RO water again, sprayed with 70% ethanol and allowed to stand for 30 min to 1 h, finally rinsed once more with RO water, allowed to dry, and put away.
12. Tank accessory parts, cleaning utensils, nets, and enrichment items should be cleaned following a similar process. Rinse in

RO water and scrub if necessary, let soak for 1 h in 10% bleach, let soak another hour in RO water with 10 g/L $Na_2S_2O_3$, soak for 10 min in 70% ethanol, rinse with RO water, air dry and put away.

3.1.1 Establishing a New Recirculating Aquatic System

When setting up a new recirculating aquatic system, the biofilter must be established to ensure that the ammonia produced by frog excretion is converted to nontoxic byproducts. Nitrifying bacteria are used in recirculating systems to create the ammonia/nitrogen cycle [5]. Ammonia (NH_3/NH_4^+) introduced from frog and food waste is oxidized by *Nitrosomonas* bacteria creating the byproduct nitrite (NO_2^-). High levels of nitrite are also harmful to frogs and it is further oxidized by *Nitrobacter* bacteria into nontoxic nitrate (NO_3^-). A startup container of bacteria can be purchased from aquatic system companies to seed the system.

1. Biomedia, on which the bacteria grow, should be rinsed before being placed into the biofilter.
2. On the first day, run one tank at the end of each row and place around 5–10 sentinel frogs total in the system. The UV lamp and water effluent exchange on the recirculating system should be OFF and no carbon included.
3. The next day, shut the system off except for the air pump to keep the biomedia aerated. Then, add the correct dosage of nitrifying bacteria to the biofilter, which depends on the amount of water in the system.
4. Allow bacteria to mix with the biomedia for 1–2 h then turn the system back on, leaving the UV lamp and water effluent OFF and no carbon. Keep the filter pad and mechanical cartridge filter in the system.
5. Measure the NH_3/NH_4^+, NO_2^-, and NO_3^- levels daily. In a couple of days to a few weeks the NH_3/NH_4^+ levels will reach above 1–2 ppm. At this time, turn on the UV and water effluent, add carbon and start up 1–2 more tanks.
6. Continue to measure the water quality every 1–2 days. NO_2^- levels should then increase followed by NO_3^-. This process can take a couple of weeks to 2 months.
7. Once both NH_3/NH_4^+ and NO_2^- levels reach below 0.5 ppm, the biofilter has been established and is ready to receive more frogs.

3.1.2 System Sterilization

At times, it may be necessary to sterilize an entire frog housing system before introducing any animals into it. The following protocol can be used to recover a system following a disease outbreak, when repurposing it for use from one species to another, or prior to restarting a used or old system after an extended period of inactivity.

1. If the system is still in use shut it down completely turning off the water pump, the dosing pumps, any biomedia agitators and air bubblers, and unplugging pH and conductivity probes.
2. Remove all biomedia and either sterilize it by soaking in a 10% bleach solution for at least 1 day or discard it.
3. Clean the sump by netting out any detritus. Do this several times letting the detritus settle between each time, then scrub the inside of the sump and any sump parts.
4. Remove all filtration from the system including filter pads, filter cartridges, and carbon filters. Leave the UV bulb in place.
5. Disassemble the frog tanks. Scrub the tanks and the lids if necessary, then rinse with water, spray with 10% bleach and let it stand for 1 h, rinse with water, spray with 70% ethanol and let it dry for 1 h, rinse again with RO water, and finally allow it to dry completely. Tank accessory parts should be scrubbed and rinsed, soaked 1 h in 10% bleach, soaked 1 h in a 10 g/L solution of $Na_2S_2O_3$ in RO water, soaked 10 min in 70% ethanol, rinsed with RO water, and air dried.
6. Drain 75% of the sump volume.
7. Reassemble the system by putting back all the tanks as well as other inserts. For the carbon cartridge leave the carbon out.
8. Start the water pump letting the sump and all the tanks fill to normal capacity.
9. Add bleach to the sump. The final concentration of NaClO running through the system should be 0.06%.
10. Adjust the effluent rate to its lowest setting but not less than 1% volume exchange per day.
11. After 2–3 days increase the effluent rate to its maximum setting, preferably at least 50% volume exchange per day.
12. After 7 days begin testing the system pH daily. Once the system pH matches that of the influent water add Virkon Aquatic to the system. The amount of Virkon required is 3.2 g per 1 m^2 of estimated internal surface area of the tanks and sump.
13. Return effluent to 1% exchange rate. Virkon will foam reaching and sterilizing surfaces in the system that water does not contact. At this point the system should be closely monitored as it is possible that the foam may overflow. To limit the risk of the overflow, water agitation in the tanks and sump should be reduced to a minimum and excessive foam should be sprayed with 70% ethanol to break up the bubbles.
14. After 7 days change the effluent to 50% exchange rate to flush Virkon out of the system.
15. Observe the system daily. Virkon should be mostly gone once no foam can be seen. Let the system run for 2–3 more days to make sure Virkon is completely flushed out.

16. Replace all the rubber tubing.
17. At this point the system is sterile and can be started up again using the procedure described in Subheading 3.1.1.

3.2 Water Parameters and Animal Density

The optimal water parameters and the density at which animals are kept vary both by species and by the life history stage. In general, the animals can tolerate a range of values, with the average presumably being the optimal. In Table 1 we list the optimal values at which we keep our animals with the range of tolerance provided in parentheses where appropriate. All values correspond to animals over 2 weeks old; care of younger animals is described in Chapter 2. All water is derived from RO Filtration. The systems are supplemented with Reef Salt and sodium bicarbonate to regulate conductivity and pH respectively. In mature systems water quality should be checked on a weekly basis. The readings for both *X. laevis* and *X. tropicalis* systems should be: NH_3/NH_4^+ = 0 ppm (0–0.5 ppm), NO_2^- < 1 ppm, NO_3^- < 40 ppm, alkalinity >40 ppm.

3.3 Feeding

To ensure rapid growth and maturation of animals, as well as maintenance of healthy and productive adults, it is essential to have a regular feeding schedule and use the appropriate food. The regimen described here has been developed in our facility and works well for animals kept in recirculating systems. The amount and frequency of feeding varies mainly by size and at early stages of development is identical for both *X. laevis* and *X. tropicalis.*

Table 1
Water quality parameters for *Xenopus*

	Age	Water temperature (°C)	pH	Conductivity (μS)	Density
X. laevis	Tadpole/Froglet	24 (23–25)	7.4 (7.2–7.6)	1000 (900–1100)	2–4 per 1 L
	Juvenile[a]	22 (20–24)	7.8 (7.0–8.5)	1600 (1200–1800)	1 per 1–2 L
	Adult	20 (18–20)	7.8 (7.0–8.5)	1600 (1200–1800)	1 per 3 L
X. tropicalis	Tadpole/Froglet	27 (26–27.5)	7.0 (6.8–7.2)	1000 (900–1100)	2–4 per 1 L
	Juvenile[a]	26 (25–27)	7.0 (6.8–7.2)	1200 (1000–1300)	1 per 1 L
	Adult	25 (24–26)	7.0 (6.8–7.2)	1200 (1000–1300)	1 per 1 L

[a]8–12 month old *X. laevis* and 5–8 month old *X. tropicalis*

3.3.1 Feeding Premetamorphic Tadpoles

1. Begin feeding when animals are older than stage 45. Feed them daily with sera Micron Growth Food while keeping them in a static container [6]. The amount used will depend on the size of the container in which the tadpoles are being grown and the number of tadpoles. A good rule of thumb is to add enough feed to turn the culture just slightly greenish. The following day the culture should not be green and if the tadpoles are feeding, strands of green digestion and excretion products should be present on the bottom of the container.
2. Sera Micron can be sprinkled on the top of the culture surface as most of it will eventually sink; however, some of it will invariably remain on the surface and over time settle on the sides of the container. To prevent settling on sides we recommend premixing the sera Micron with 0.1× MMR first, either in a weight boat by stirring it in with a Pasteur pipette or in a conical tube by vortexing, then adding this mixture to the tadpole culture.
3. When growing the tadpoles in the static tanks, large clutches of tadpoles will get dirty with time. Several steps can be taken to keep them relatively clean before the tadpoles reach a size large enough to be put in a recirculating system nursery tank. A pipette can be used to aspirate excretion products and settled food from the bottom of the container, more 0.1× MMR can be added to the culture, or a 50% volume exchange can be done with fresh 0.1× MMR. When exchanging 50% volume it is important that the MMR solution being added is at the same temperature as the one in which the tadpoles are growing.
4. At approximately 2 weeks of age tadpoles are transferred over to a nursery tank in a recirculating aquatic system. Once in the recirculating system the feeding schedule changes slightly, and tadpoles now should be fed twice daily, first in the early morning and second in the late afternoon. Just prior to feeding, the water flow in the tank should be decreased and kept at this lower flow rate for 1–2 h; shutting off the water is not recommended during feeding as this affects the stability of water temperature. Food is measured into a 50 mL conical tube, resuspended with system water, and the whole mix is added to the tadpole tank. Initially only sera Micron is used, but as the animals grow their diet is supplemented with a 1:1 mix of brine shrimp flakes and Nasco frog brittle powder ground together to a powder like consistency (Table 2). Care should be taken to not overfeed. The feeding regimen described in Table 2 works well for 16 L nursery tank, but can be adjusted to work with tanks of different sizes.
5. As the tadpoles are typically not grown in the self-cleaning flood-and-flush systems, food detritus will accumulate on the bottom of the tank over time (Fig. 1c). It is not essential to

Table 2
Feeding regimen for *Xenopus*

Age	Regimen	Food
Days 1–4 (In System)	2× Daily	0.1 g Sera Micron
Days 5–6	2× Daily	0.25 g Sera Micron
Days 7–11	2× Daily	0.4 g Sera Micron
Days 12–20	2× Daily	0.6 g Sera Micron
Days 21–27	2× Daily	0.4 g Sera Micron + 0.2 g Brine Shrimp Brittle Mix
Days 28–34	2× Daily	0.4 g Sera Micron + 0.4 g Brine Shrimp Brittle Mix
Days 35+	2× Daily	0.6 g Sera Micron + 0.6 g Brine Shrimp Brittle Mix
Post-Metamorhpic	2× Daily	1:1 ratio1.2 mm Bio Vita Fry Trout PelletsNasco Post-Metamorphic Brittle/ Powder
Juvenile froglets[a]	5× Week 10–12 pellets/frog	1:1 ratio 2 mm Bio Vita Fry Trout PelletsNasco Post-Metamorphic Brittle
Adult *X. tropicalis*	3× Week 8–10 pellets/frog	1:1 ratio 2 mm Bio Vita Fry Trout PelletsNasco Post-Metamorphic Brittle
Adult *X. laevis*[b]	2× Weekly6 pellets/frog	1:1 ratio4 mm BioTrout PelletsNasco Adult Brittle

[a]8–12 month old *X. laevis* and 5–8 month old *X. tropicalis*
[b]For J-strain adults feed 5× week if small, 3× week if large

clean it out as this detritus is not detrimental to animal health and the tadpoles feed off the settled food on the bottom of the tank. Furthermore, cleaning the detritus poses a risk of accidentally disposing of animals when they are relatively small. Therefore, we typically do not clean the excess detritus until the tadpoles reach late metamorphosis/early froglet stage. At this point, floating detritus can be removed using a net.

3.3.2 Feeding During and Post Metamorphosis

As the animals undergo metamorphosis and mature into froglets their feeding schedule and food size is adjusted. Eventually, slightly different feeding regimens and diets are used for maintenance of adult *X. laevis* and *X. tropicalis.*

1. *X. laevis* animals younger than 8 months should continue to be fed twice a day. At 8 months of age, feeding is changed to once a day, 5 days/week. After 12 months, outbred wild type *X.*

laevis are fed only twice a week. Inbred J strain *X. laevis* animals should continue to be fed 5 days/week; depending on their size feeding can eventually be reduced to 3 days/week.

2. *X. tropicalis* animals younger than 5 months should be fed twice a day. Once they reach 5 months of age, feeding should be changed to once a day. After 8 months, adult *X. tropicalis* should be fed 3 days/week.
3. Observe the animals few minutes after feeding. Add more food if all the food in the tank is gone and they are still actively feeding.
4. Typically, animals will ignore food that has been sitting in the tank water for extended amount of time. At 1–3 h post feeding visually inspect the tanks and if necessary net out any excess food settled on the bottom. Make sure to use a different, clean net for each tank.
5. Overfeeding can result in poor water quality, accumulation of waste, and additional system maintenance.

3.4 Transportation and Quarantine Procedures

Shipping animals is relatively straightforward but requires some planning and coordination, by both the shipper and the receiver, to limit the stress on the animals associated with the transport as well as prevent any potential problems. Any paperwork and health-tests that may be required by law or by the receiving institution should be done ahead of time. If possible, the animals should be shipped overnight and it is recommended that they are also shipped early in the week in case some problem occurs that causes their delivery to be delayed to a later date. Animals should be unpacked as soon as they arrive and any packing material such as shredded sponge or peat moss should be rinsed off before they are put in the system. Adult animals should be allowed to rest 1 to 3 weeks before being used for breeding. Finally, shipping cryopreserved sperm, testes, and tadpoles is also a possibility if the receiving facility cannot accommodate intake of adult animals.

3.4.1 Shipping Tadpoles

1. Prior to shipping tadpoles, it is important to check the temperature at the origin and the destination to assure that the tadpoles will not experience temperature conditions that may be detrimental to their survival. If necessary, ice or heating pack can be included in the shipping container. "*Handle with care*" and "*Live tadpoles*" stickers should be used to label the container and assure it is handled gently.
2. Use a 50 mL conical tube for shipping tadpoles younger than stage 48. Fill the tube with 35 mL–40 mL of 0.1× MMR plus 10 μg/mL of gentamicin. Put no more than 50 tadpoles per tube. Parafilm the whole tube to enhance its structural integrity and pack into a Ziploc bag as a secondary container. Ship

in either a Styrofoam box with bubble wrap or an insulated mailer envelope.

3. Tadpoles older than stage 48 can be shipped in 1 L–4 L volume plastic containers. We recommend the use of Cubitainers. Fill the Cubitainer 80% with frog system water or 0.1× MMR and add the tadpoles. Keep the density of tadpoles to 10–20 per 1 L; do not overcrowd. Parafilm the cap and ship secured in a Styrofoam box.

3.4.2 Shipping Adult Animals

1. As with tadpoles, prior to shipping adults check the temperature at the origin and destination to assure the conditions are viable for animal survival. *X. laevis* can tolerate temperatures as low as 5 °C and *X. tropicalis* can be shipped if the temperature is at least 10 °C; the animals can tolerate up to 30 °C. If the temperatures are outside the animal tolerance range, but it is absolutely necessary to ship the animals, then arrangements should be made with companies that specialize in shipping under climate controlled conditions. Label the shipping container with "*Handle with care*" and "*Live frogs*" stickers to assure they are handled gently.
2. To ship *X. laevis* use a 43 cm × 43 cm × 23 cm (17″ × 17″ × 9″) cardboard box with a compatible Styrofoam box and lid. With the Styrofoam box inside the cardboard box, and the cardboard box closed, drill 12 air holes through the top of both boxes with a $^{5}/_{16}$″ drill bit. Mark the lid of the Styrofoam box and the inside of the cardboard box to aid in orientation. Remove the Styrofoam box, fill half way with dried peat moss or shredded sponge, and let the moss/sponge soak in RO water for at least 1 h. Drain most of the water out leaving the moss/sponge damp but not soaking wet.
3. Density is critical to ensure the frogs arrive healthy to their destination. Place up to 10 adult *X. laevis* female wild type frogs (~130 g each) in the Styrofoam box and tape its edge. Up to 12 female J strain animals or 15 male frogs can be shipped per box since they are much smaller in size. Put the taped Styrofoam box back in the cardboard box, using the markings on the inside of the cardboard and the lid of the Styrofoam to align the air holes. Add any necessary shipping documents, tape the top of the cardboard box, label, and ship.
4. To ship *X. tropicalis* we use a 12 cup (3 L) plastic food container with a snap top. Drill air holes in the lid with a ¼″ drill bit. Fill the container half way with dried peat moss or shredded sponge and let it soak in RO water for at least 1 h. Drain the water so that the moss/sponge is damp but not soaking wet.
5. Place up to 5 adult frogs in the plastic container and seal it. Up to four containers can fit in a 43 cm × 43 cm × 23 cm

(17″ × 17″ × 9″) Styrofoam box. When shipping *X. tropicalis* always use a heat pack/hand warmer. Let the heat pack warm up, wrap it in damp paper towels, place it in a plastic bag with holes in it for ventilation, and tape it to the inside lid of the Styrofoam box. Make sure to also drill air holes in the lid of the Styrofoam box and the accompanying cardboard box.

3.4.3 Shipping Sperm and Testes

1. Shipping cryopreserved sperm on dry ice is a simple way of distributing distinct animal lines. Sperm can be used immediately or stored for extended period of time either at −80 °C or in liquid nitrogen. In general, fertilization from cryopreserved sperm is efficient and animals generated develop normally; however, it has also been demonstrated that cryopreserving sperm does cause some plasma membrane damage and DNA fragmentation in the spermatozoa [7, 8].
2. As an alternative to shipping cryopreserved sperm, fresh *Xenopus laevis* testes can be shipped. Fill a 50 mL conical tube with 35 mL to 40 mL of ice cold 1× MMR with 10 μg/mL of gentamicin. Add freshly isolated testes. Parafilm the entire tube to help maintain integrity. Put the tube in a Ziploc bag as a secondary container and wrap in bubble wrap. As a shipping container use either a Styrofoam box or an insulated mailer envelope. Put a freezer pack on the bottom of the container and cover with either paper towels or some bubble wrap to prevent direct contact between the freezer pack and the conical tube. Make sure the conical tube is secured and will not toss around during shipping by either taping it to the side of the shipping container or filling the container with bubble wrap.

3.5 Health

Daily visual checks are necessary for the maintenance of a healthy animal colony. Attention should be paid to both physical and behavioral indicators of potential health problems looking for discoloration or redness of the skin, apparent wounds, abnormal swimming, changes in body shape and size, bloating, and lethargic response to external stimulation. Any frogs showing these signs should be isolated from their tank. In contrast, frogs with visible wounds, bloating, or skin discoloration might recover after being placed for several days in a bath of 1× MMR with 10 μg/mL gentamicin, other animals should be euthanized and necropsied. Close attention should be paid with animals that go through sudden stressors such as temperature spikes, as it can lead to susceptibility to sickness or mortality. In bloated animals, pressure can sometimes be released by creating a small laterally located puncture in the posterior third of the ventral abdomen with a 16 gauge needle to release the accumulated fluid or gas. This should only be attempted by a trained individual or a veterinarian as care must be taken that the puncture goes through both skin and the abdominal muscle but is shallow enough to not damage any of the viscera. Animals that do not show improvement should be euthanized and necropsied.

Suboptimal quarantine protocols, in particular when bringing new animals into the colony, may lead to introduction of disease causing agents. Many frog pathogens are dormant with the disease outbreaks caused by stress, such as that associated with shipping or less than ideal husbandry conditions. Several known pathogens of *Xenopus* have been previously characterized in the literature and include *Ranavirus*, bacteria (*Mycobacterium* spp., *Chlamydophila* spp.), Chytrid fungus, and *Cryptosporidium* and even in colonies not showing any signs of sickness, regular tests (once or twice per year using qPCR) are recommended [9–13]. Frogs showing signs of disease as well as other frogs from the same tank should be immediately isolated from the system. If a disease is spread throughout the colony immediate aggressive treatment or euthanasia are recommended. If a colony is euthanized, the aquatic housing system should be sterilized before next use.

3.5.1 Euthanasia

When necessary, anesthetic overdose can be used as an efficient and humane way of euthanasia in *Xenopus*. Prepare a bath of 0.15% benzocaine in frog water, or 0.5% MS222 in water buffered to pH 7.0 with $NaHCO_3$. After 30 min, check that the frog is unresponsive to foot pinching and lacks a swallowing reflex when its throat is rubbed. Put the euthanized animal in a bag and freeze at −20 °C overnight. Discard the carcass following local and institutional regulations.

3.6 Record Keeping

Good record keeping is essential for the maintenance of a healthy frog colony. Records can be kept on paper or in a digital format, but it is essential that water parameters are recorded daily. In addition, a mortality log should be kept to document the tank from which the dead animal came, any particular abnormalities of the carcass, and whether it was a recently mated female. These records will allow establishment of an operational baseline that can be used to identify whether water parameters or frequency of animal death look abnormal, thus catching potential developing problems early. Figure 2 shows an example of such a log.

4 Notes

1. Some labs use home-made recirculating systems for housing their animals. In particular, in areas where municipal water is not highly chlorinated, tap water purified through a carbon filter can be used instead of reconstituted water. In such a configuration general water hardness provided by dissolved calcium and magnesium ions needs to be monitored as well, since

Date: ____________ AM Feeding: ____________________

Temp: ____________ PM Feeding: ____________________

Multi-rack system name

Temp:____________ pH:_______________ Cond:____________ %TGP:____________

Water Quality	Maintenance	
$NH_{3/4}$: ____________	Mech in:_________________psi	Dosing Res:
ALK: ____________	Mech out:_________________psi	Cond________L
NO_2: ____________	Carbon out:_______________psi	pH__________L
NO_3: ____________	Water level: _____________	UV light: ON/OFF
	Effluent counter:__________	
	Drum Filter counter: ______	

Notes:____

____

TANKS CLEANED: ____

____

Individual system name

Tank Temp:______________ pH:_________ Cond:_______ Water Level: ______

Water Quality	Maintenance	
$NH_{3/4}$: ____________	Carbon out:_______________psi	Dosing Res:
ALK: ____________	Sheet Filter Check ________	Cond________L
NO_2: ____________		pH__________L
NO_3: ____________		UV light: ON/OFF

Notes:____

TANKS CLEANED:____

Fig. 2 Record keeping. Two examples of format for record keeping of either multi-rack system or individual system. Various values/parameters are recorded daily

moderate to very hard water is required to produce high quality eggs and normal embryonic development [14].

2. Detergents should never be used to clean any equipment used for housing and handling adult frogs and embryos.

Acknowledgments

The National *Xenopus* Resource is supported by a grant from the National Institutes of Health (P40 OD010997).

References

1. Gurdon JB, Hopwood N (2000) The introduction of *Xenopus laevis* into developmental biology: of empire, pregnancy testing and ribosomal genes. Int J Dev Biol 44:43–50
2. Tandon P, Conlon F, Furlow JD, Horb ME (2017) Expanding the genetic toolkit in Xenopus: approaches and opportunities for human disease modeling. Dev Biol 426:325–335. https://doi.org/10.1016/j.ydbio.2016.04.009
3. Wlizla M, Falco R, Peshkin L et al (2017) Luteinizing hormone is an effective replacement for hCG to induce ovulation in Xenopus. Dev Biol 426:442–448. https://doi.org/10.1016/j.ydbio.2016.05.028
4. Pearl EJ, Grainger RM, Guille M, Horb ME (2012) Development of Xenopus resource centers: the National Xenopus Resource and the European Xenopus Resource Center. Genesis 50:155–163. https://doi.org/10.1002/dvg.22013
5. Hem LJ, Rusten B, Ødegaard H (1994) Nitrification in a moving bed biofilm reactor. Water Res 28:1425–1433. https://doi.org/10.1016/0043-1354(94)90310-7
6. Showell C, Conlon FL (2009) Natural mating and tadpole husbandry in the western clawed frog Xenopus tropicalis. Cold Spring Harb Protoc 2009: doi: https://doi.org/10.1101/pdb.prot5292
7. Pearl E, Morrow S, Noble A et al (2017) An optimized method for cryogenic storage of Xenopus sperm to maximise the effectiveness of research using genetically altered frogs. Theriogenology 92:149–155. https://doi.org/10.1016/j.theriogenology.2017.01.007
8. Morrow S, Gosálvez J, López-Fernández C et al (2016) Effects of freezing and activation on membrane quality and DNA damage in Xenopus tropicalis and Xenopus laevis spermatozoa. Reprod Fertil Dev 29:1556–1566. https://doi.org/10.1071/RD16190
9. De Jesús Andino F, Chen G, Li Z et al (2012) Susceptibility of Xenopus laevis tadpoles to infection by the ranavirus Frog-Virus 3 correlates with a reduced and delayed innate immune response in comparison with adult frogs. Virology 432:435–443. https://doi.org/10.1016/j.virol.2012.07.001
10. Godfrey D, Williamson H, Silverman J, Small PLC (2007) Newly identified Mycobacterium species in a Xenopus laevis colony. Comp Med 57:97–104
11. Reed KD, Ruth GR, Meyer JA, Shukla SK (2000) Chlamydia pneumoniae infection in a breeding colony of African clawed frogs (Xenopus tropicalis). Emerg Infect Dis 6:196–199. https://doi.org/10.3201/eid0602.000216
12. Tinsley RC, Coxhead PG, Stott LC et al (2015) Chytrid fungus infections in laboratory and introduced Xenopus laevis populations: assessing the risks for U.K. native amphibians. Biol Conserv 184:380–388. https://doi.org/10.1016/j.biocon.2015.01.034
13. Green SL, Bouley DM, Josling CA, Fayer R (2003) Cryptosporidiosis associated with emaciation and proliferative gastritis in a laboratory-reared South African clawed frog (Xenopus laevis). Comp Med 53:81–84
14. Godfrey EW, Sanders GE (2004) Effect of water hardness on oocyte quality and embryo development in the African clawed frog (Xenopus laevis). Comp Med 54:170–175

Chapter 2

Generation and Care of *Xenopus laevis* and *Xenopus tropicalis* Embryos

Marcin Wlizla, Sean McNamara, and Marko E. Horb

Abstract

Robust and efficient protocols for fertilization and early embryo care of *Xenopus laevis* and *Xenopus tropicalis* are essential for experimental success, as well as maintenance and propagation of precious animal stocks. The rapid growth of the National *Xenopus* Resource has required effective implementation and optimization of these protocols. Here, we discuss the procedures used at the National *Xenopus* Resource, which we found helpful for generation and early upkeep of *Xenopus* embryos and tadpoles.

Key words *Xenopus laevis*, *Xenopus tropicalis*, In vitro fertilization, hCG, oLH

1 Introduction

The strengths of *Xenopus laevis* and *Xenopus tropicalis* as experimental model systems include the ability to produce a large number of eggs year-round using injection of gonadotropic hormones to induce ovulation and the entirely external synchronous, stereotypical development following fertilization [1–4]. The fertilization protocols are straightforward and robust, and can be performed via in vitro fertilization (IVF), when a large number of synchronously developing siblings are required, or through natural mating if the parental male being used is precious and needs to be preserved for further experiments. During the first 2 weeks following fertilization, careful attention to culture density and cleanliness as well as frequency and amount of food provided is necessary to assure normal development and to promote animal survival.

The frog colony at the National *Xenopus* Resource (NXR), a national repository and distribution center for a large number of different *Xenopus* stocks, has grown in size from zero animals in 2010 to a current combined total of over 8000 *X. laevis* and *X. tropicalis*. This rapid growth has required us to optimize protocols necessary for efficient and rapid embryo production. Since many of the animals we stock have been derived from external sources, we have

Kris Vleminckx (ed.), *Xenopus: Methods and Protocols*, Methods in Molecular Biology, vol. 1865,
https://doi.org/10.1007/978-1-4939-8784-9_2,

incorporated procedures for embryo surface sterilization into our standard operating procedures to limit the potential of introducing pathogens into our colony. In this chapter, we present the techniques we use for efficient embryo generation in *X. laevis* and *X. tropicalis*, embryo care during the first 2 weeks following fertilization, and the embryo surface sterilization protocol.

2 Materials

2.1 Animals

Wild type outbred *X. laevis* and *X. tropicalis* can be purchased from a variety of commercial vendors, including Nasco, *Xenopus* One, and *Xenopus* Express. These companies are an excellent resource for obtaining large quantities of outbred *Xenopus* and they work closely with each researcher to provide animals at all stages. More specialized strains and lines of *Xenopus*, such as the inbred *X. laevis* J strain and the Nigerian *X. tropicalis* strain, are available for purchase through regional stock centers, including the National *Xenopus* Resource (NXR, RRID:SCR_013731, http://www.mbl.edu/xenopus), the European *Xenopus* Resource Centre (EXRC, RRID:SCR_007164, https://xenopusresource.org), the *Xenopus laevis* Research Resource for Immunobiology (XLRRI, https://www.urmc.rochester.edu/microbiology-immunology/xenopus-laevis.aspx) and the *Xenopus tropicalis* National BioResource Project of Japan (http://home.hiroshima-u.ac.jp/amphibia/xenobiores_en/iweb_en/Top.html). In addition, these stock centers distribute transgenic and mutant lines for both *X. laevis* and *X. tropicalis*. Detailed information about the various suppliers, including contact information, can be found online at Xenbase (RRID:SCR_003280, http://www.xenbase.org/other/obtain.do).

2.2 Food

1. Sera Micron Growth Food (sera North America, Montgomeryville, PA, USA).

2.3 Hormones

1. Pregnant Mare Serum Gonadotropin (PMSG) (e.g., BioVendor, Asheville, NC, USA): make a stock of 100 U/mL in 1× PBS and store at −20 °C.
2. Ovine Luteinizing Hormone (oLH) (e.g., National Hormone and Peptide Program, Los Angeles, CA, USA): make stock of 0.4 mg/mL in 1× PBS and store at −20 °C.
3. Human Chorionic Gonadotropin (hCG) (available from several sources including: National Hormone and Peptide Program, Los Angeles, CA, USA; Sigma-Aldrich, St. Louis, MO, USA; Chorulon brand, Merck Animal Health, Madison, NJ, USA): make a stock of 1000 U/mL in 1× PBS.

2.4 Anesthesia and Euthanasia Reagents

1. Ethyl 4-aminobenzoate, 98%; aka. Benzocaine (CAS# 94-09-7): make 10% stock in 95% Ethanol and store at room temperature.
2. Tricaine Methanesulfonate; aka. Tricaine-S, aka. MS 222 (CAS# 886-86-2).

2.5 Embryo and Tadpole Culture Solutions

1. Marc's Modified Ringer's Solution (MMR).
   Make 10× stock: 1 M NaCl, 20 mM KCl, 10 mM $MgSO_4{\cdot}7H_2O$, 20 mM $CaCl_2{\cdot}2H_2O$, 50 mM HEPES free acid; adjust the pH to 7.4–7.8 with NaOH; autoclave or filter sterilize (*see* **Note 1**).
   Dilute the 10× MMR stock with autoclaved ultrapure water to make culture solutions of required concentration. No additional pH adjustment is necessary (*see* **Note 2**).
2. Modified Barth's Saline (MBS).
   Make 10× stock: 800 mM NaCl, 10 mM KCl, 10 mM $MgSO_4{\cdot}7H_2O$, 50 mM HEPES free acid (*see* **Note 3**), 25 mM $NaHCO_3$; adjust the pH to 7.8 with NaOH; autoclave or filter sterilize.
   Make 0.1 M stock of $CaCl_2$.
   To make culture solutions dilute 10× MBS stock with autoclaved ultrapure water and add $CaCl_2$ stock. The final concentration of $CaCl_2$ in 1× MBS should be 0.7 mM.
3. Gentamicin Reagent Solution, 10 mg/mL in distilled water (e.g., ThermoFisher Scientific, Waltham, MA, USA).
   Where required use the 10 mg/mL solution as a 1000× stock (*see* **Note 4**).
4. L-Cysteine free base (CAS# 52-90-4) dejelly solution (*see* **Notes 5** and **6**).
   Make a 2% for *X. laevis* or 2.5% for *X. tropicalis* solution of L-Cysteine in 0.1× MMR, adjust the pH to 7.8–8.0 using NaOH.
5. Thimerosal (CAS# 54-64-8).
   Make minimum of 100 mL solution. Add 0.01 g thimerosal to 100 mL 0.1× MMR. Shake to dissolve. Store at 4 °C up to 7 days.
6. Phosphate Buffered Saline (PBS) tablets, 100 g.
   Make 1× PBS by dissolving the tablets in ultrapure water. Autoclave to sterilize. Use to resuspend hormones.

2.6 Egg Laying Solutions

1. 1× MMR works well as egg laying solution for *X. laevis*. Two alternative egg laying solutions are listed below.
2. 10× egg laying solution stock: 1.20 M NaCl, 24 mM KCl, 24 mM $CaCl_2{\cdot}2H_2O$, 12 mM $MgCl_2{\cdot}6H_2O$, 18 mM HEPES free acid; adjust the pH to 7.6 with NaOH; autoclave or filter sterilize.

Use autoclaved ultrapure water to dilute when making 1× egg laying solution.

3. 8× RAL's Ca_{2+} free High Salt Egg Laying Buffer for *X. laevis*: 880 mM NaCl, 120 mM Tris base, 16 mM KCl, 4 mM Na_2HPO_4, 16 mM $NaHCO_3$; adjust pH to 7.6 with glacial acetic acid; add 8 mM $MgSO_4 \cdot 7H_2O$.

   Use autoclaved ultrapure water to dilute when making 1× egg laying solution. Use fresh made 1× solution (*see* **Note 7**).

3 Methods

3.1 Generating Embryos

Both *X. laevis* and *X. tropicalis* are capable of generating well over a 1000 embryos per mating [2]. In both, injection of mammalian gonadotropins is used to stimulate ovulation and the male mating behavior, with the main difference between the two species being the amount of the hormone used. The experimental goals as well as the potential rarity of the parental male line used will influence whether the embryos are generated via a natural mating or through IVF. Finally, when generating embryos from parental individuals that have been kept in quarantine and might have an uncertain health record, cleavage stages are a good developmental period for producing a clean population of animals by performing a surface sterilization protocol.

3.2 Sexing Adults

Several sexually dimorphic phenotypic traits aid in distinguishing males and females. Same traits can be used for sexing both *X. laevis* and *X. tropicalis* (*see* **Note 8**).

1. Body size and shape is sufficient in most cases to distinguish male from female. In sexually mature individuals of the same age females are larger than males. Females have a pear-shaped body with large adult individuals showing a pair of distinct lateral ridges parallel to the anteroposterior body axis in the dorsal posterior region of the body.
2. The females have prominent cloacal labia. During mating these become reddish and engorged.
3. Sexually mature males have keratinized nuptial pads on the ventral surface of the forearm which they use to clasp onto the females back during amplexus. These pads tend to be darker in color than the surrounding epidermis. In individuals in which the color difference is not very prominent, running a finger along the ventral surface of the forearm can be used as an additional identification method; on male frogs this region of the skin will have a rough texture.

3.3 Gonadotropic Hormone Injection to Promote Ovulation in Females

A number of mammalian gonadotropic hormones have been identified as effective in inducing ovulation in females and mating behavior in males. These are generally injected subcutaneously into the dorsal lymph sacs of the animal, using a 27G–30G needle. Typically, females will be given two injections with the initial priming injection followed by a later boosting injection that induces ovulation (*see* **Note 9**). Boosting males seems to be of particular importance during natural mating as it seems to increase the likelihood of amplexus. Historically, human chorionic gonadotropin (hCG) has been used for boosting injections, but recent work has shown that both human and ovine luteinizing hormones can be used as well [2].

1. In large adult wild type *X. laevis* females, first prime them with 50 U of PMSG or 50 U of hCG between 24 h to 7 days before the boosting injection (priming is not essential for induction of ovulation). To induce ovulation boost them with 2 μg of oLH per 1 g of body mass (~200 μg of oLH) or with 500 U of hCG; some larger frogs may require greater amounts of hCG.
2. Inbred J strain *X. laevis* females are smaller than wild type and require lesser amounts of hormone. Prime them with 30 U of PMSG or 30 U of hCG and boost with 140 μg of oLH or 350 U of hCG.
3. *X. laevis* females kept at 18 °C should begin ovulating approximately 8–9 h following the boosting shot. Ovulation can be delayed by keeping the females at a lower temperature (*see* **Note 10**).
4. For *X. tropicalis,* prime the females with 20 U of PMSG or 10 U of hCG between 12 h to 2 days before the boosting injection. Boost the females with 5 μg of oLH per 1 g of body mass (~80 μg of oLH) or with 200 U of hCG.
5. *X. tropicalis* females kept at 25 °C will begin ovulating 3–5 h after the boosting injection.
6. Between ovulations females should be allowed to rest to let their oocytes mature. *X. laevis* require 3 months of rest and *X. tropicalis* require 2 months.
7. Frogs kept on a regular ovulation and rest schedule tend to produce high quality eggs.
8. Recently some labs have reported successful use of superovulation in *X. laevis* females to increase the number of eggs extracted from a single female. To do this, following ovulation let the female rest 1 week before giving it a priming hormone injection and then a boosting hormone injection 2 days later. After the second ovulation have the female rest 3 months.
9. After ovulation, the females should be transferred to a clean container with system water and allowed to recover for 24 h before being returned back into the system.

3.4 In Vitro Fertilization (IVF)

IVF allows generation of large numbers of simultaneously fertilized embryos, which is particularly useful for a number of experimental designs. The main drawback is that it requires the parental male to be euthanized which may be problematic if the male comes from a stock that is limited in number and will be used again (*see* **Note 11**).

1. The male needs to be euthanized prior to extraction of the testes. Put it in a bath of 0.15% benzocaine in frog water, or 0.5% of MS222 in water buffered to pH 7.0 with $NaHCO_3$. After 15–30 min check that the male is unresponsive to foot pinching and does not have a swallowing reflex when its throat is rubbed. Use scissors to perform cervical section and sever the spine immediately posterior to the skull as a secondary euthanasia method on the insensate male. Using dissection scissors cut through the skin and the abdominal muscle to open the lower abdominal cavity on the ventral side. The testes are found on the dorsal side of the cavity, on either side of the aorta. It may be necessary to move the fat bodies out of the way to make the testes visible. Cut the testes out and trim away any attached fat and viscera. Store the testes in 1× MMR with 10 μg/mL of gentamicin added to prevent the sperm from activating and limit growth of contaminants.
2. If stored at 4 °C, *X. laevis* testes will maintain their viability for at least a week.
3. *X. tropicalis* testes will remain viable for at least 2 days if stored at 25 °C–27 °C.
4. IMPORTANT: it is extremely important that the *X. tropicalis* testes are never kept on ice or at 4 °C. Even brief exposures to low temperatures cause *X. tropicalis* sperm to lose activity.
5. To collect eggs, the females need to be either squeezed to physically entice them to ovulate directly into an empty collection dish, or they can be placed into an egg laying bath of high salt and the eggs collected over time. Eggs ovulated directly into frog water lose viability shortly after having been laid and should not be used.
6. Eggs can be acquired from both *X. laevis* and *X. tropicalis* by squeezing. To do this grab the female with the dominant hand (the description given assumes the right hand as dominant, reverse sides if using left hand). The back of the female should face the palm of the hand with the head directed toward the wrist and the index finger between the legs. The index finger should be on the left side of the cloaca, with the left leg squeezed between the index and the middle fingers and pushed as close to the body as possible. The little, ring, and middle fingers wrap around the left side of the animal with the tips of the fingers resting along the ventral midline of the animal. The

thumb goes around the right side. Use the left hand to restrain the right leg, pulling it away from the cloaca along the right side of the animal. It may be necessary to let the animal relax before being able to cause a maximal leg extension. With legs braced against the body wall the frog is essentially immobilized in a joint lock, unable to get away. While holding the female, orient her vertically, with the cloaca over a dish to collect the oocytes, this will help prevent the eggs from running along the body of the female instead of directly into the collection dish. In *X. tropicalis* the female flexing against this hold is typically sufficient for egg expulsion through the cloaca. If that is not sufficient or if handling *X. laevis* it may be necessary to apply pressure to squeeze the eggs out. Use the right hand ring and middle finger tips to apply pressure to the belly with the thumb firmly around the left side. The thumb of the left hand should apply the pressure to the back of the animal. Shift the pressure in an anterior toward posterior direction as if squeezing paste out of a tube. Keep both index fingers as far from the cloaca as possible to prevent the eggs from collecting on fingers instead.

7. Egg laying baths (1× MMR or equivalent) work well with *X. laevis*. Frogs are kept in a bath at 18 °C–22 °C. The eggs are collected every 30–60 min and fertilized. Egg laying baths keep the eggs viable for at least 1 h following ovulation. Recipes for two alternative egg laying solutions are given in Subheading 2.6.

8. We have not heard of anyone using egg laying baths to collect eggs from *X. tropicalis*, and although the females can be comfortably kept in 1× MMR for at least 2 days we are not sure how long the eggs laid into 1× MMR would remain viable (*see* **Note 12**).

9. Fertilization of *X. tropicalis* eggs is relatively straightforward. Collect eggs into an empty petri dish. If any frog water dripped into the dish, remove as much of it as possible. Using a plastic transfer pipette try to spread the eggs into a monolayer. Place a half to a whole testis in a 1.5 mL microcentrifuge tube with 0.5 mL of 1× MMR. Use a plastic microcentrifuge tube pestle to crush the testis tissue in the tube and release the sperm. Using a transfer pipette apply the testis solution to the eggs and mix it in. Let it sit for 3 min, flood the dish with 0.1× MMR, and use a transfer pipette to mix again.

10. To fertilize *X. laevis* eggs collect them into an empty petri dish. Remove as much frog water or egg laying solution as possible. Put an eighth to a quarter of a testis into a 1.5 mL microcentrifuge tube with 1 mL of 1× MMR and use a plastic microcentrifuge pestle to crush it and release sperm. Using a transfer pipette add the sperm solution to the eggs and mix. Let it sit

for 5 min, then flood the dish with 0.1× MMR, and mix again with a transfer pipette. Alternatively, the testes piece can be placed in a glass dish and macerated using forceps; add a small amount of 0.1× MMR (5–10 mL) to the dish and then use a pipette and transfer the larger chunks into a 1.5 mL microcentrifuge tube; use a plastic microcentrifuge pestle to crush the larger pieces and release more sperm. Transfer the solution to the glass dish with the original macerated testes and add this solution to the laid eggs; let it sit for 5 min then flood with additional 0.1× MMR (*see* **Note 13**).

11. For smaller cultures, it may be fine to let the embryos develop in the jelly coat. However, removing the jelly helps maintain the cleanliness of the cultures and is absolutely necessary for microinjection, microsurgery, or small molecule treatments. To dejelly the embryos, make a 2% (*X. laevis*) or 2.5% (*X. tropicalis*) solution of L-Cysteine in 0.1× MMR, with the pH adjusted to 7.8–8.0 using NaOH. As early as 15 min after the embryos have been flooded with 0.1× MMR, decant all the 0.1× MMR from the dish and flood the dish with the L-Cysteine dejelly solution. Let it sit at room temperature occasionally gently moving the embryos around with a transfer pipette or by gently swirling the dish. Five to ten minutes after adding the dejelly solution they should be able to clump together, indicating dissolution of the jelly coat. Pipet or pour out as much of the cysteine as possible then wash with 0.1× MMR three to five times. These steps can be performed in the same petri dish or alternatively following the dissolution of the jelly coat the culture can be poured into a beaker and then gently washed with the 0.1× MMR. Following the final wash transfer the embryos to a clean petri dish with 0.1× MMR and 10 μg/mL gentamicin.

12. *X. tropicalis* embryos need a little bit extra care after removing the jelly coat. For the first 2 days following the fertilization or at least until they have hatched out of the vitelline envelopes and are sporadically swimming they should be kept in a dish coated with 2% noble agar (or agarose) in 0.05× MMR, or coated with 0.1% sterile BSA, or kept in 0.05× MMR with added BSA for a final concentration of 0.1%. *X. laevis* do not require this and can be kept in uncoated dishes without any BSA.

13. Adding dejelly solution less than 15 min after flooding with 0.1× MMR will result in abnormal development. Waiting 15 min is usually sufficient but the washes need to be done gently. Waiting 20 min may be better to assure maximal normal development.

14. Dejelly solution should be made fresh, as it will lose potency over time.
15. Embryos can tolerate a range of temperatures during early development with lower temperatures resulting in slower development. *X. laevis* can be kept at 15° C–23° C and *X. tropicalis* can be kept at 23° C–29° C. Middle of the range assures most consistently normal development especially during the more sensitive stages of gastrulation and neurulation and we recommend culturing *X. laevis* at 18°C and *X. tropicalis* at 23°C.

3.5 Natural Mating

When dealing with males that are from a limited and precious stock it may be prudent to generate embryos via natural mating. The main advantage is that the male is not sacrificed in the process and can be used repeatedly. The drawback is that the embryos produced this way will not be fertilized simultaneously and some sorting for embryos of a particular stage may be necessary depending on the experimental purpose they are being used for.

1. To induce the male mating behavior, inject the male with priming and boosting shots following that same schedule as for the female and outlined in Subheading 3.1, **step 2**. For the boosting injection use half the amount of hormone when injecting the male as that specified for the female.
2. After the boosting shot, put male and female together in a tank with system water. We generally use mouse breeding cages as these are easy to handle and make collecting embryos simple. Put the mouse cage in the dark (either covered on a bench top or in an incubator with no lights).
3. After approximately 4 h for *X. tropicalis* or 9 h for *X. laevis* it should be fine to begin checking on whether the frogs are in amplexus and if there are any eggs or embryos in the tank. Checking too early or too often can spook the animals preventing the male from going into amplexus or causing it to release.
4. Depending on the experimental design embryos can be collected at regular intervals as frequent as 30 min or the animals may be allowed to mate undisturbed for an extended period of time and all the embryos collected together at the end of the procedure.
5. If using these embryos for microinjection it may be useful to work in a pair with one person collecting, dejellying, sorting, and staging the embryos while the other one is solely dedicated to doing the injections.
6. After the mating is complete, the frogs should be transferred to a clean container with system water and allowed to recover for 24 h before being returned back to the system. Males and

females should be kept separate after mating is completed to avoid prolonged amplexus.

3.6 Surface Sterilization

When breeding quarantined animals in order to bring the new generation into the clean facility the following embryo surface sterilization protocol can be followed to limit the likelihood of introducing any pathogens into the colony. This protocol was originally devised for isolation of *Xenopus* neurons and muscle cells but has been adopted for use with embryo culture (Marilyn Fisher and Robert Grainger personal communication) [5]. The same procedure can be used for both *X. laevis* and *X. tropicalis.*

1. Before starting the procedure make the following solutions.
   - (a) 70% Ethanol in 0.1× MMR.
   - (b) 2.5% L-Cysteine free base.
   - (c) 0.1 mg/mL thimerosal in 0.1× MMR.
   - (d) 0.1% BSA in 0.1× MMR.
   - (e) Coat five 10 cm petri dishes with 0.1% BSA per fertilization.
   - (f) 1 L 0.1× MMR per fertilization.
2. The embryos can be generated either through natural mating or via IVF, however if any "clean" animals are required, for example males to provide the testes, they should be handled and processed first before handling any of the quarantined animals.
3. Begin the surface sterilization protocol before the onset of gastrulation to assure no pathogens present on the surface of the embryo are internalized, and in an isolated area to limit spread of any contaminants.
4. Change gloves and pipettes at each step.
5. Collect the embryos in a petri dish and drain any remaining culture solution. If they are clumped in a mass try to spread it out as much as possible without damaging the embryos.
6. Add 70% Ethanol solution and incubate for 10 s.
7. Remove 70% Ethanol and rinse three times with 0.1× MMR.
8. Dejelly to completion with 2.5% L-Cysteine, approximately 5–10 min at room temperature.
9. Remove the cysteine and wash embryos 8–10 times with 0.1× MMR. After the final wash place the dejellied embryos into a clean 0.1× MMR filled BSA coated petri dish.
10. Sort out any unfertilized eggs and dead embryos.
11. Transfer the clean embryos with as little fluid as possible to a BSA coated petri dish filled with thimerosal solution.

12. Incubate for 15 min with gentle agitation.
13. While the embryos are incubating fill the remaining three BSA coated dishes with 0.1× MMR with BSA and coat 3 transfer pipettes with BSA.
14. After the 15 min incubation is complete use a transfer pipette to move the embryos with as little fluid as possible into first of the three dishes. Change pipette and gloves and repeat two more times to move the embryos to the second and then the third dish.
15. Once the embryos are in the third dish they are considered sterile and can be grown in an incubator in the clean facility.
16. Sterilize all the surfaces used with 10% bleach and then 70% ethanol.
17. Following the protocol do not handle any clean animals or interact with any clean systems without showering and having a complete change of clothes first.

3.7 Embryo and Early Tadpole Care

Embryos and early tadpoles require attention and care. Following the guidelines listed here will help assure normal and healthy development.

1. Dejellied embryos should be kept in 0.1× MMR with 10 μg/mL gentamicin until they begin to feed after stage 45. At that point, they should be switched to gentamicin free 0.1× MMR.
2. Keep *X. tropicalis* embryos in 0.05× MMR with gentamicin, instead of 0.1× MMR, for the first day following fertilization.
3. Crowded cultures lose synchronous development over time and have a higher chance of showing abnormal development and death. We recommend keeping not more than 100 embryos younger than stage 35/36 and not more than 50 embryos stage 35/36 and older per 10 cm petri dish for *X. laevis* and *X. tropicalis*. As the embryos grow it may be necessary to transfer them to even larger plates, beakers, or flasks.
4. Cultures need to be checked every day. Dead or abnormal embryos should be removed; failure to remove the dead embryos may adversely affect the growth of other embryos. For the first 2 days of development if the cultures are particularly dirty it may be necessary to transfer all the healthy embryos to a clean dish with fresh 0.1× MMR. After that, once the embryos hatch, simply replacing the 0.1× MMR may be enough to keep the culture growing.
5. For the first 2 days following fertilization, or essentially until they begin swimming, *X. tropicalis* embryos need to be kept in dishes coated with 2% Noble agar (or agarose) in 0.1× MMR, or 0.1% BSA, or in 0.1× MMR with BSA added to a final concentration of 0.1%.

6. Begin feeding tadpoles after stage 45 by mixing Sera Micron into the culture medium [6]. At this point it may be useful to transfer the embryos into a larger container, like a beaker or a 15 cm petri dish. Once the embryos are in the beaker either add fresh 0.1× MMR to the beaker every day or do a 50% volume exchange. Use a pipette to clean out food and excretion solids accumulated on the bottom.
7. At 2 weeks post fertilization transfer the tadpoles to a nursery tank and follow the feeding directions outlined in the accompanying chapter 1.

4 Notes

1. Original formulation of 10× MMR also includes 10 mM EDTA. We omit it from our stock with no noticeable effect.
2. The pH of MMR will likely change following dilution of the 10× stock with ultrapure water, for example in our lab 0.1× MMR will usually have pH 6.95–7.00 when made from 10× MMR pH 7.4. Readjusting the pH of the diluted MMR is not necessary and we typically do not even trouble with checking it following dilution.
3. When making 1× MBS, HEPES sodium salt can be used instead of HEPES free acid.
4. 10 μg/mL of gentamicin in an embryo culture solution is sufficient to limit bacterial growth. Also, gentamicin is highly thermostable and can be added to solution prior to autoclaving if necessary [7].
5. When making dejelly solution we recommend using the L-Cysteine free base since it works efficiently in both *X. laevis* and *X. tropicalis.* Dejelly solution made with L-Cysteine hydrochloride salt works fine with *X. laevis* but is inefficient at degrading *X. tropicalis* jelly coat. A 3% solution takes as much as 12 min to remove the *X. tropicalis* jelly sufficiently for the embryos to be ready for microinjection.
6. We recommend making dejelly solution fresh. Although, it will work for several days when stored in a sealed bottle, it loses potency over time and will require longer incubations to completely remove the jelly coat. Eventually, a precipitate will form at which time the solution should no longer be used.
7. 8× stock of RAL's Ca_{2+} free High Salt Egg Laying Buffer can be stored at 4 °C for up to 8 weeks.
8. When sexing immature individuals in both *X. laevis* and *X. tropicalis*, and sometimes even in sexually mature *X. tropicalis*, the cloacal labia and nuptial pads may be somewhat incon-

spicuous. Priming and boosting such animals with gonadotropins may induce prominence of these features.

9. The priming gonadotropin injection aids in consistent production of a high number of mature eggs; however, females given only a boosting hormone injection are likely to lay large numbers of eggs as well.
10. Many labs do not feed the females during the period between the priming hormone injection and ovulation. The intent is to prevent excretion of solid waste concurrent with ovulation which may reduce oocyte durability and embryonic culture quality. Furthermore, frogs that are fed prior to ovulation often regurgitate the food.
11. Before IVF some labs prime, or prime and boost, the males with gonadotropic hormones following the same injection schedule but using half the amount of hormone as for the females. Evidence as to whether this in any way affects the quality of the testes, sperm viability, or fertilization efficiency is anecdotal.
12. Do not use 1× RAL's Ca_{2+} free High Salt Egg Laying Buffer with *X. tropicalis*, it will cause twitching and seizures within 30 min.
13. When making the testis solution and flooding the eggs and sperm mix during IVF, some labs substitute 1× MMR with 1× MBS and 0.1× MMR with 0.1× MBS. We exclusively use MMR as in our experience it works great for both *X. laevis* and *X. tropicalis* and unlike 0.1× MBS, 0.1× MMR does not need to be made fresh or have its pH frequently rechecked. The pH of 0.1× MBS tends to fall over time due to weak buffer strength.

Acknowledgments

The National *Xenopus* Resource is supported by a grant from the National Institutes of Health (P40 OD010997).

References

1. Gurdon JB, Hopwood N (2000) The introduction of Xenopus laevis into developmental biology: of empire, pregnancy testing and ribosomal genes. Int J Dev Biol 44:43–50
2. Wlizla M, Falco R, Peshkin L et al (2017) Luteinizing hormone is an effective replacement for hCG to induce ovulation in Xenopus. Dev Biol 426:442–448. https://doi.org/10.1016/j.ydbio.2016.05.028
3. Nieuwkoop PD, Faber J (1994) Normal table of Xenopus Laevis (Daudin). Routlege, New York
4. Amaya E, Offield MF, Grainger RM (1998) Frog genetics: Xenopus tropicalis jumps into the future. Trends Genet 14(7):253–255
5. Peng HB, Baker LP, Chen Q (1991) Tissue culture of Xenopus neurons and muscle cells as a model for studying synaptic induction. Methods Cell Biol 36:511–526

6. Showell C, Conlon FL (2009) Natural mating and tadpole husbandry in the western clawed frog Xenopus tropicalis. Cold Spring Harb Protoc 2009: doi: https://doi.org/10.1101/pdb.prot5292
7. Traub WH, Leonhard B (1995) Heat stability of the antimicrobial activity of sixty-two antibacterial agents. J Antimicrob Chemother 35:149–154

Chapter 3

Methods for CRISPR/Cas9 *Xenopus tropicalis* Tissue-Specific Multiplex Genome Engineering

Thomas Naert and Kris Vleminckx

Abstract

In this chapter, we convey a state-of-the art update to the 2014 Nakayama protocol for CRISPR/Cas9 genome engineering in *Xenopus tropicalis (X. tropicalis)*. We discuss in depth, gRNA design software and rules, gRNA synthesis, and procedures for tissue- and tissue-specific CRISPR/Cas9 genome editing by targeted microinjection in *X. tropicalis* embryos. We demonstrate the methodology by which any standard equipped *Xenopus* researcher with microinjection experience can generate F0 CRISPR/Cas9 mediated mosaic mutants (crispants) within one to two work-week(s). The described methodology allows CRISPR/Cas9 efficiencies to be high enough to read out phenotypic consequences, and thus perform gene function analysis, in the F0 crispant. Additionally, we provide the framework for performing multiplex tissue-specific CRISPR/Cas9 experiments generating crispants mosaic mutant in up to four genes simultaneously, which can be of importance for *Laevis* researchers aiming to target by CRISPR/Cas9 both the S and L homeolog of a gene simultaneously. Finally, we discuss off-target concerns, how to minimize these and ways to rapidly bypass reviewer off-target critique by exploiting the advantages of *X. tropicalis*.

Key words *Xenopus tropicalis*, CRISPR/Cas9, Multiplex, Tissue-specific, Disease model

1 Introduction

The fully diploid *Xenopus tropicalis* (*X. tropicalis*) allows for rapid loss-of-function genetic studies in the highly experimentally accessible *Xenopus* embryo [1]. Concurrently, the rapid advent of a novel generation of tools for fast, targeted genome editing (TALEN, CRISPR/Cas9) has matured *X. tropicalis* from its historically place as a developmental biologists tool and has put it firmly on the map as an alternative to established animal species for modeling human disease.

Since the adaptation of the prokaryotic CRISPR/Cas9 system as an eukaryotic-capable genome editing technology, it has been widely used to establish targeted genome knockouts across a plethora of experimental animal models. Consequentially, the amount of research work mentioning or employing CRISPR/Cas9 has

Kris Vleminckx (ed.), *Xenopus: Methods and Protocols*, Methods in Molecular Biology, vol. 1865,
https://doi.org/10.1007/978-1-4939-8784-9_3,

Table 1
Number of papers mentioning "CRISPR/Cas9" over time, data from before 2013 is omitted [3]

Number of papers	Year	% of total
27	2013	2.356
201	2014	16.743
484	2015	38.622
1052	2016	83.549

more than doubled each year since the first mention in 2013 (Table 1). Concurrently, an increasing trend for reviewers to put into question data obtained by morpholino experiments, and subsequent requests for CRISPR/Cas9 validation experiments, has left developmental biologist with a pressing need to adapt this system within their specific research units [2].

Over the last few years, early adaptors have employed the CRISPR/Cas9 system in *Xenopus* to answer research questions as diverse as craniofacial patterning [4], retinal degeneration [5], cancer [6], neural crest development [7], pronephric development [8], thyroid hormone receptor function [9], gene-dependent transcriptomics analysis [10], and congenital heart disease [11] among others [12, 13]. This combined body of work represents the still largely untapped possibilities to integrate CRISPR/Cas9 in solving both basic biology and developmental biology questions, up to accurate human disease modeling and even transcriptomics. While some of this previously cited work has been performed in *X. laevis*, we firmly believe that *X. tropicalis* due to its genuinely diploid genome will be increasingly used for genome editing experiments.

In this chapter, we aim to convey the detailed methodology to rapidly establish single or multiplex F0 mosaic mutant *X. tropicalis* embryos (crispants), thus reporting a state-of-the-art update to the pioneering methodologies published in 2014 by Nakayama et al. [14]. We will steer clear of simple introduction to the CRISPR/Cas9 system and its historical origins, about which there are many great reviews available [15]. Additionally, we will not go into detail on the many interesting additions to the classical CRISPR/Cas9 systems, such as gene mediated knock-in, base-editing systems, or CRISPR-Cpf1 system as these currently remain specialists application, which might be more challenging to rapidly implement [16–19].

Instead, we provide for (multiplex) gene knockout approaches: gRNA design software comparison and further design constraints, gRNA synthesis methods, and (targeted) delivery methodologies to allow the *Xenopus* researcher to rapidly, economically, and with minimal effort, acquire CRISPR/Cas9 mediated (tissue-specific) knockouts within his own experimental setups.

Conceptually, we would like to relay that due to the extremely rapid embryonic cell division within the *Xenopus* embryo, employing CRISPR/Cas9 in *X. tropicalis* embryos will lead to a mosaic pattern of insertion-deletion (INDEL) mutations within the developing animal [20]. In practice, employing CRISPR/Cas9 will thus lead to an F0 mosaic mutant crispant exhibiting either wild type or specific INDEL mutations on one or two of the targeted alleles. The ratio of cells exhibiting INDELs over wild type is dependent on genome editing efficiency of the employed gRNA. Presence of distinct variants of INDEL mutations will thus occur across the cells composing the F0 mosaic mutant animal, and presence of specific variants is highly variable across the organism. With the described methodology here, we have shown the possibility to generate crispants exhibiting clear phenotypic consequences such as cancer development [6]. However, even in the absence of a positive selection process, it is possible to obtain F0 phenotypes as this is mainly dependent on the efficiency of the employed gRNA and this can routinely surpass 80% with the described methodology (this is clearly delineated by the *tyrosinase* gene editing described in Subheading 3.3 and in Fig. 3) [14, 21–24].

We will also discuss the off-target concerns, how to minimize these and finally ways to rapidly bypass reviewer off-target critique by exploiting the advantages of *X. tropicalis*. Please note that this chapter will not include genotyping of resulting F0 crispants as the different methodologies for this are discussed in Chapter 5 of this book.

2 Materials

2.1 Generation of Microinjection Ready gRNA

2.1.1 Generation of gRNA DNA Template by PCR Method

1. 5′ oligo with the sequence: 'GAAT("oligo")ATAGC'. *See* Subheading 3.1 for explanation of target site choice and "oligo" sequence. Ordered as a desalted oligo.
2. 3′ primer with the sequence: 'AAAAGCACCGACTCGGTGCCACTTTTTCAAGTTGATAACGGACTAGCCTTATTTTAACTTGCTATTTCTAGCTCTAAAAC'. Ordered as a PAGE-purified oligo.
3. High fidelity proof-reading polymerase (e.g., Velocity—Bioline).
4. dNTP mix (10 mM each) (e.g., ThermoFisher Scientific).
5. RNase-free water.
6. PureLink® PCR Purification Kit (ThermoFisher Scientific).
7. A NanoDrop instrument (ThermoFisher Scientific).
8. BenchTop 1 kb DNA Ladder (e.g., Promega).

2.1.2 In-Vitro Transcription of gRNA

1. T7 MegaShortScript (ThermoFisher Scientific) or HiScribe™ T7 High Yield RNA Synthesis Kit (NEB).
2. TURBO™ DNase (ThermoFisher Scientific).
3. Phenol:chloroform:isoamylalcohol 25:24:1.
4. Chloroform EMSURE® (e.g., Merck).
5. Ammonium Acetate (5 M).
6. 100% EtOH (RNase-free).
7. Qubit® RNA BR Assay Kit (ThermoFisher Scientific).

2.1.3 gRNA Quality Control by Denaturing Gel Electrophoresis

All glassware employed should be baked at 180 °C for 2 h to remove any RNase activity.

1. Diethyl pyrocarbonate (DEPC) (e.g., Sigma-Aldrich).
2. DEPC-treated (RNase-free) water. Add 0.1% DEPC (v/v) to purified water and let it incubate at room temperature under the fume hood. Make sure to thoroughly mix the DEPC every 2 h by shaking the bottle. This shaking needs to be done at least three times and should be followed by an overnight incubation. The DEPC-treated water is then autoclaved to degrade the DEPC.
3. 10× MOPS buffer: 20 mM EDTA, 200 mM 3-(N-morpholino) propanesulfonic acid (MOPS), 50 mM sodium acetate, pH 7.0. Make this 10× stock with purified water. After setting correct pH by NaOH, add 0.1% DEPC (v/v) and perform the same steps as described above for correct DEPC treatment. The 10× MOPS buffer must then be autoclaved to break down the DEPC. The solution will turn straw-yellow and should be kept in the dark to prevent degradation. By this preparation method you inactivate any RNase present in your solid powder reagents used to prepare this buffer.
4. 37% formaldehyde (e.g., Merck).
5. RNA loading buffer. Dilute GelRed 10,000× (Biotium) to 1× working concentration in NorthernMax® Formaldehyde Load Dye (ThermoFisher Scientific).
6. RiboRuler High Range RNA Ladder (ThermoFisher Scientific).
7. UltraPure™ Agarose (Invitrogen) kept under RNase-free conditions.
8. RNase AWAY™ Surface Decontaminant (ThermoFisher Scientific).
9. Gel Doc™ XR+ Gel Documentation System (Bio-Rad).

2.1.4 gRNA Quantification

1. Image Lab™ Software 5.2.1 (Bio-Rad).
2. A Qubit™ fluorometer (e.g., Qubit™ 3.0 Fluorometer; ThermoFisher Scientific).
3. Qubit® RNA BR Assay Kit (ThermoFisher Scientific).

2.2 Delivery of gRNA/Cas9 Ribonucleoproteins by Microinjection to Xenopus tropicalis Embryos

1. General microinjection equipment consisting of a stereomicroscope (e.g., Stemi 2000, Zeiss) supplied with an ocular micrometer for the calibration of the injection volume, a cold-light source (e.g., KL750, Schott), a 3D micromanipulator (e.g., Narishige, M40B), and a microinjector (e.g., PLI-100, Harvard Bioscience).
2. Microinjection needles produced from borosilicate glass capillaries (with 1.0 and 0.58 mm as outer and inner diameters, respectively; Sutter instruments Co.) using a micropipette puller (e.g., Model P-97, Sutter instruments Co.).
3. Injection dish: a 60 mm petri-dish in which a mesh (e.g., 700 μm mesh; spectrum laboratory) is fixed with a few drops of chloroform.
4. Adult *Xenopus tropicalis* males and females.
5. Human Chorionic Gonadotropin (e.g., Pregnyl®, Organon).
6. 10× Marc's Modified Ringers (MMR): 1 M NaCl, 18 mM KCl, 20 mM $CaCl_2$, 10 mM $MgCl_2$, 50 mM Hepes, pH 7.6.
7. Ficoll 400 (e.g., Amersham Pharmacia).
8. 2% cysteine solution in 0.1× MMR, adjusted to pH 8.0 with NaOH.
9. Recombinant injection-ready Cas9 protein (e.g., Toolgen Cas9 WT protein; TGEN_CP) or home-made NLS-Cas9-NLS can be prepared as described in supplementary materials of Naert et al. [6].

3 Methods

3.1 Designing a Highly On-Target Efficient gRNA with a Minimal Off-Target Profile

Since the applications of the CRISPR/Cas9 system for genome editing were first demonstrated, a myriad of guide RNA (gRNA) design software tools have been published. Over the past years we successfully employed different tools such as CCTOP, CRISPRDesign, and CRISPRScan [25–27]. Earlier design software algorithms, in essence, just brute forced the input DNA sequence for 20 bp sites preceded by an NGG protospacer adjacent motive (PAM) sequence. This was followed by filtering genome-wide ("search and filter" algorithms) for potential off-targets of the output gRNA list while taking into account all regions of the genome that match the gRNA up to N number of mismatches at the target site (e.g., Cas-OFFinder, CRISPOR, etc.) [28, 29]. While these algorithms outputted gRNA on-target sites and associated off-target sites, there was no relation to bona-fide in vivo activity of the gRNA at its specific on- and off-target site. Newer, machine-learning-based algorithms employ large-scale genome on- and off-target CRISPR/Cas9 editing datasets to define rulesets of what a highly active gRNA should look like (base composition, thermodynamic features, loading to Cas9 in vivo, etc.), while some also learn the rules of which predicted "search and filter" off-target sites are in fact expected to yield substantial and unwanted in vivo activity [26, 27, 30].

Currently, we are consistently using, and we would thus recommend, the CRISPRScan design tool as this tool is prognostic by providing an in silico prediction of CRISPR/Cas9 modification efficiencies at the target sites and the employed machine-learning algorithm is validated in *X. tropicalis*. Recently, the CRISPRScan software suite has integrated the Cutting Frequency Determination (CFD) off-target calculator, thus incorporating the strength of a well-validated off-target determination tool to its own strength of determining gRNA with strong expected in vivo activities (*see* **Note 1**) [30]. One note of warning might be that the CFD score calculator has very recently received an update, is now named Elevation and is reported to outperform the original CFD algorithm. Elevation is, however, not (yet?) incorporated in CRISPRScan but can be used for manually double-checking for off-targets of your CRISPRScan generated gRNAs [31]. As such, the relevant output of the CRISPRScan algorithm for the *Xenopus* researcher are high-scoring gRNAs ("CRISPRScan score") with a low Cutting Frequency Determination (CFD) value.

However, blindly employing CRISPRScan can lead to unintended consequences, post-editing, on protein level, and as such we recommend following additional design constraints for gRNA (Fig. 1b–e). These constraints minimize the possibility to incidentally generate, by CRISPR/Cas9 editing of your gene, protein mutants that might retain functionality. In essence, CRISPR/Cas9 induced knockout of protein product for a gene is dependent on non-homologous end-joining (NHEJ) or microhomology mediated end joining (MMEJ or Alt-NHEJ) frameshift genome editing within a coding exon. This leads to transcription of an mRNA transcript that will, at the CRISPR/Cas9 edited site, go out-of-frame and as such the translation of this transcript will almost inevitably rapidly run into a stop codon within the alternative read-frame now presented (Fig. 1b). This then activates the non-sense mediated decay (NMD) pathway which will detect such a transcript with a premature stop codon (PTC) and will initiate decay of this aberrant mRNA, resulting in complete absence of protein product for the CRISPR/Cas9 targeted gene [32]. In order to achieve this, one should additionally conform to the following three rules.

Firstly, avoid targeting the first exon of a gene as this can give rise to potential alternative, in-frame, start codon usage and resulting protein product with N-terminal truncation (Fig. 1c).

Secondly, do not target the last exon or a sequence within 50 bp upstream of the last exon-intron boundary in order to avoid generating a C-terminally truncated protein due to the absence of NMD (Fig. 1d).

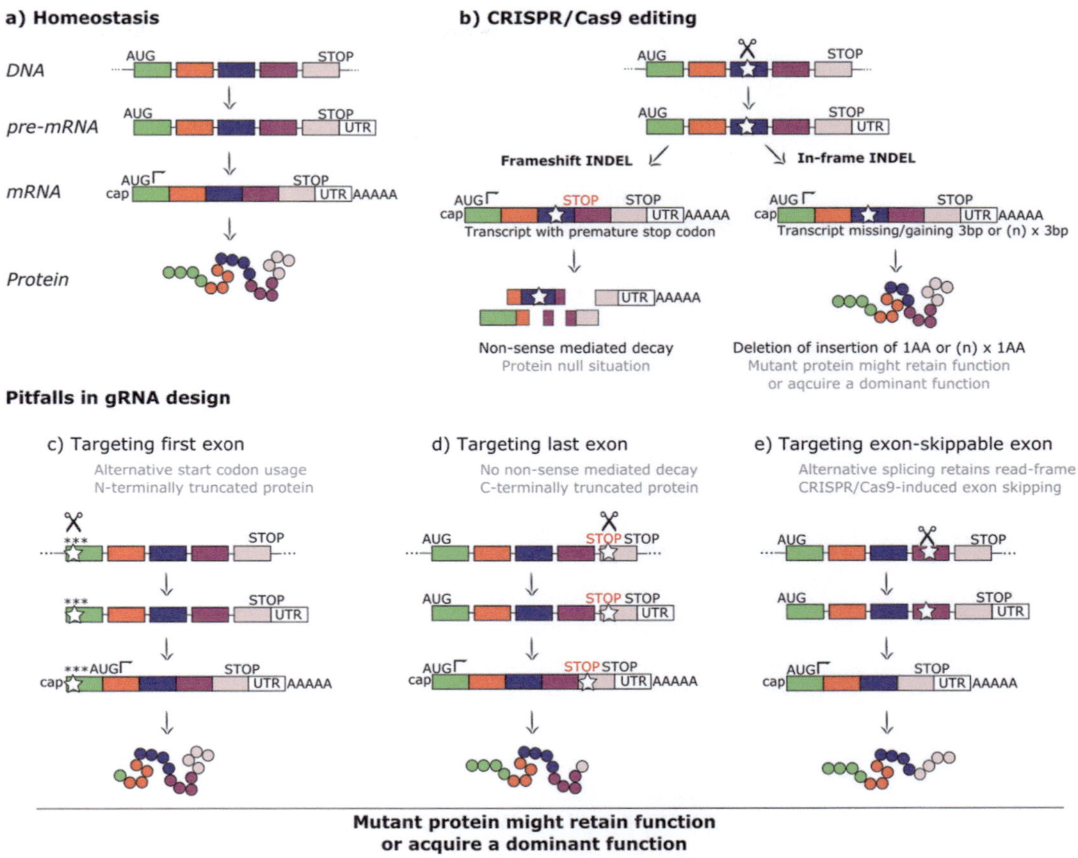

Fig. 1 Effects of CRISPR/Cas9 genome editing on the different levels of the central dogma of biology and pitfalls in gRNA design. (**a**) In homeostatic conditions, under the central dogma of biology, DNA is transcribed to pre-mRNA. Subsequent splicing and processing of pre-mRNA generates mature mRNA which will then be translated to a protein. (**b**) By editing a coding exon with the CRISPR/Cas9 technology, insertion-deletions (INDELs) will occur at the CRISPR/Cas9 target site (white star). When this INDEL leads to a frameshift mutation, the novel read frame will rapidly run into a premature stop codon. An mRNA transcript with premature stop codon will be recognized by the non-sense mediated decay (NMD) machinery and will be degraded. This results in a protein null situation. However, when the INDEL is an in-frame mutation the correct read frame will be conserved. As such the resulting mRNA transcript will miss/gain 3 bp or (***n***) × 3 bp culminating in a deletion or insertion of 1AA or (***n***) × 1AA within the resulting protein. Whether this protein is functional/foldable is dependent on its specific context (***see*** **Note 3**). (**c**–**e**) Pitfalls in gRNA design that can lead to generation of mutant protein that might retain function or acquire a dominant function. (**c**) Targeting the first exon of a gene should be avoided. It is possible that usage of an alternative start codon, in close enough proximity to regulatory elements, will lead to the formation of an N-terminally truncated protein. (**d**) Targeting the last exon should be avoided as a premature stop codon within the last exon will not be recognized by the NMD machinery and a C-terminally truncated protein will be generated. (**e**) Targeting an exon that is potentially exon-skippable should be avoided. It has been shown that CRISPR/Cas9 induced exon skipping can occur at these exons and as such mutant protein can be translated lacking the amino acids coded by the specific exon

Thirdly, CRISPR/Cas9 induced exon skipping can occur. This may result in the expression of a (partially) functional protein when alternative splicing of the pre-mRNA outsplices the CRISPR/Cas9 targeted exon without disturbing the read frame [33]. Therefore, if possible, it is best to avoid targeting an exon consisting of $(n) \times 3$ bp (Fig. 1e).

By obeying this rule-set you should be able to minimize your chances of aberrant, but possibly still functional protein expression. Evidently, in specific cases it may be required to target the last exon in order to generate a truncated protein—possibly acting as a dominant-active or dominant-negative—in order to mimic a specific clinical situation. This is the case, for example, when targeting the tumor suppressor gene *apc* [34, 35].

In either way, CRISPR/Cas9 mediated genome editing will lead, after NHEJ or MMEJ, to small INDEL mutations by which in theory 1/3 should be in-frame (IF) and 2/3 should be frameshifting (FS) (Fig. 1b). As such, one can appreciate that about 55% of targeted cells could retain at least one targeted allele expressing viable protein product, assuming that the occurring in-frame INDELs do not perturb protein function (*see* **Notes 2** and **3**). Indeed, it was rationalized, and we have experimentally shown that targeting a catalytical site should lead to an increased frequency of protein function null mutants. This, as small in-frame INDELs in regions of catalytic importance have a much higher chance to abrogate protein function. We showed this in practice, as targeting the E2F1 binding pocket of *rbl1* within our retinoblastoma model clearly increased tumor incidence at a lower basal level of genome editing [6]. These increased chances on protein null function, even in the absence of frameshift editing, when targeting essential protein domains have also been described within the CRISPR-Cas9 screening community [36].

As such the final rule-set for designing gRNA is in brief as follows.

Optional, find (if available) the coding sequence for an important (catalytical, structurally scaffolding, etc.) domain within your gene of interest and then design with the CRISPRScan tool a high scoring gRNA at the coding sequence for this domain (*see* **Notes 4** and **5**).

In general, employ the CRISPRScan software "by gene" function to output, the pre-calculated, gRNAs targeting the coding exons of your input gene.

Subsequently, validate that the location of your gRNA is not in any of the design exclusion exons (first exon, second-last exon 50 bp upstream to exon-intron boundary, last exon or potentially skippable exon) (Fig. 1c–e).

Finally, it is worth noting that even when using robust algorithms for off-target prediction (such as the CFD within the CRISPRScan program) one might still anticipate reviewer issues concerning off-targets contributing to your CRISPR/Cas9 induced phenotype following the "absence of evidence is not evidence of absence" logic statement. In either way, due to the advantages of *X. tropicalis* allowing rapid establishment of parallel knockout experiments we recommend simply targeting the same gene, with two distinct gRNAs in two parallel setups. As such it can be argued that, if the exact same phenotype is observed, it becomes extremely unlikely that this phenotypic output is the consequence of off-targets as the chance that two distinct gRNAs will show an overlap in off-target profile is almost non-existing [6]. Therefore, we recommend to immediately design two to three gRNAs targeting the same gene and to test in vivo to determine the ones with the best efficiencies (*see* Subheading 3.3).

1. *Optional*: Identify for the model organism of interest, if possible, the genetic sequence coding an important functional domain for your target. This can usually most easily be achieved using UniProt (http://www.uniprot.org/). By performing Clustal Omega pairwise alignment between the human protein (with annotated catalytical sites) and the *X. tropicalis* protein you can identify the coding sequence for the specific domains of your protein in the *X. tropicalis* genome (*see* **Note 4**).
2. Browse to the URL of the CRISPRScan design tool (http://www.crisprscan.org/; URL and website layout subject to change).
3. Use the CRISPRScan "by gene" option, to output all the gRNAs binding within the coding sequence for your gene of interest. *Remember* to manually curate these to exclude gRNAs in the design exclusion criteria (see Subheading 3.1 and Fig. 1c–e). *Optional*: When targeting a specific domain (see point 1 of this section), choose the "Submit sequence" tool at top left and input your target sequence as raw or FASTA data.
4. Select the appropriate genomic off-target calculator to the right, in our case *X. tropicalis.*
5. Export the output by "options"->"export"->"tab-delimited". The obtained sheet can be opened in the Microsoft Excel Suite (.tsv file) and should be then sorted (from high to low) according to CRISPRScan scores, which is a score assigned to the gRNA based on the predictive algorithm of CRISPRScan (higher = better). Take into account the CFD value which should be as low as possible and is a score representing the chance of off-targets. For more information on how these scores relate to in vivo effectiveness in our hands, *see* **Note 5**.
6. Make sure that for each gRNA design you intend to use, you archive not only the "oligo" output as shown in the CRISPRScan website prior to exporting as .tsv file, but also to save the matching "seq" column from within the .tsv file. Namely, due to the

nature of the CRISPRScan design software, the "oligo" output will not always correctly align to the *X. tropicalis* scaffold by BLAT, as the algorithm sometimes incorporates imperfect matches between the gRNA and the genome in order to increase stability of the gRNA itself in vivo. As such you will need the "seq" column data in order to remember where your designed gRNA is expected to induce editing on the *X. tropicalis* genome.

3.2 Generation of Microinjection–Ready gRNA

Below, we report our validated methodologies to obtain microinjection-ready gRNA. However, it must be stated that commercial kits exist to more rapidly obtain gRNA, such as the NEB enGEN® gRNA synthesis kit. While we have no direct hands-on experience with this, others (*see* Chapter 18) have employed this kit successfully. Nevertheless, the price-tag per gRNA will inevitably be higher when employing commercial solutions over the pipeline reported below, and as such one will have to balance these needs and costs within his own experimental efforts.

Generation of gRNAs can be rapidly performed by a cloning-free methodology shown schematically in Fig. 2a. This methodology is based on PCR-based amplification of two partly complementary DNA oligos followed by a T7 in vitro transcription reaction. Keep in mind that the first two nucleotides transcribed by the T7 RNA polymerase should be GG (*see* **Note 5**). Next to this, we demonstrate methods for quality control of gRNA and the best practice for quantification. During all steps of this protocol we recommend working as RNase-free as possible, this includes using only RNase-free reagents, consumables, glassware, etc., while working in an environment thoroughly cleaned by RNase Away and keeping any employed kits RNase-free.

3.2.1 Generation of gRNA DNA Templates by PCR Method

In order to obtain a linear DNA template for in vitro transcription of microinjection-ready gRNA, we employ a PCR-based strategy to anneal a variable forward 5′ primer (modified according to the desired gRNA target site) and a common 3′ primer. Please be aware that the CRISPRScan output "oligo" (*see* Subheading 3.1) contains all the necessary elements (T7 promotor, 20 nt gRNA seed sequence, overlap with the constant reverse 3′ primer) for both the annealing as well as the in vitro transcription reaction. However, in order to optimize the annealing reaction we add another stretch of overlapping sequence (ATAGC) 3′ on the 5′ primer. Additionally, we add an additional GAAT sequence 5′ on the 5′ primer for optimal T7 in vitro transcription [37]. Additionally, be aware that when using other gRNA design software that one will have to manually check if each necessary element is present when ordering oligos, with an emphasis on the 5′-GG necessary to initiate T7 in vitro transcription downstream of the T7 promotor (*see* **Note 6**).

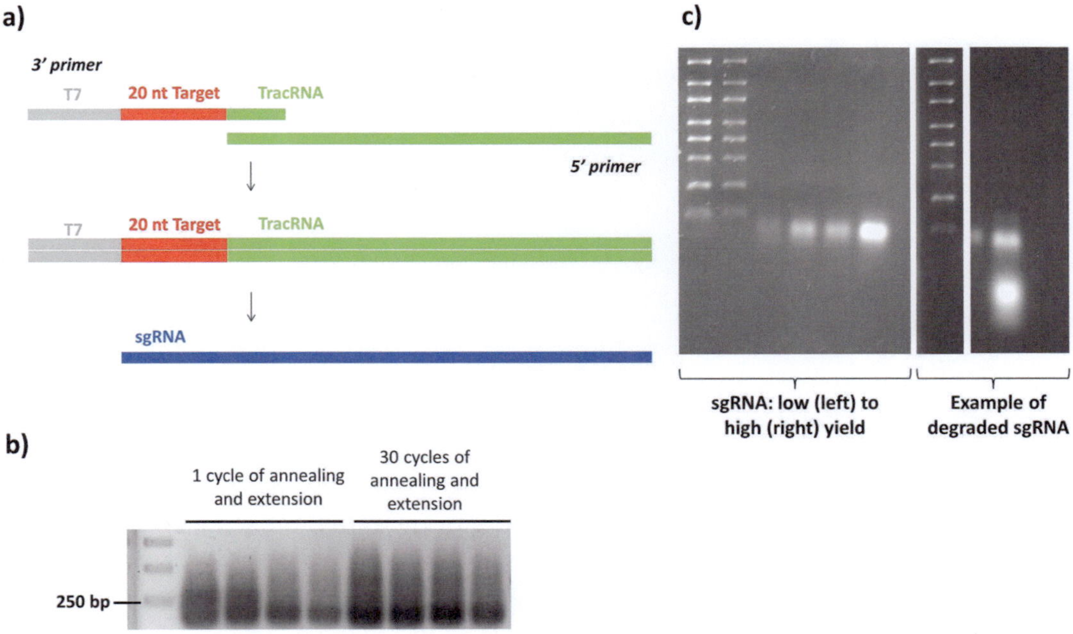

Fig. 2 Generation of gRNA DNA templates by PCR method. (**a**) A variable 3′ primer specific to the gRNA target site and a common 5′ primer are annealed and amplified in a PCR reaction generating large amounts of DNA template. This template will then be used to generate gRNA by in vitro transcription from the T7 promotor (gray). (**b**) Representative gel showing clear amplification of the correct DNA template fragment (~121 bp) when comparing 1 cycle of annealing and extension to 30 cycles of annealing and extension. (**c**) Representative denaturing RNA gel (left) showing two dilutions of the quantitative ladder with 4 gRNAs exhibiting low to high gRNA yield after in vitro transcription. Representative denaturing RNA gel (right; cut and pasted but ladder and sample are from the same gel) exhibiting gRNA that has undergone degradation due to inappropriate sample handling

1. Obtain following oligos.

   Variable forward gRNA targeting sequence containing 5' oligo

   5′-GAAT("oligo")ATAGC-3′ (*see* Subheading 3.1 on how to obtain this "oligo" sequence).

   Constant reverse 3' oligo

   5′-AAAAGCACCGACTCGGTGCCACTTTTT CAAGTTGATAACGGACTAGCCTTATTTT AACTTGCTATTT.

   CTAGCTCTAAAAC-3′

2. Set up the following DNA template generation reaction in a thermocycler (*see* **Note 7** on polymerase use). Please mind that 5′ and 3′ primers are added to this reaction at 100 μM concentration.

DNA template assembly reaction

5× Hi-Fi Reaction buffer	10 μL
10 mM dNTP mix	5 μL
5′ oligo (**100 μM**)	2 μL
3′ oligo (**100 μM**)	2 μL
Velocity Hi FI Polymerase	1 μL
Ultrapure water	30 μL

Run PCR reaction in a thermocycler with following program: 98 °C for 2′; 30 cycles of [98 °C for 30″; 60 °C for 30″; 72 °C for 15″]; 72 °C for 4′. Correct amplification of the DNA template (see below) can be demonstrated by loading a (negative) control sample that only underwent 1 cycle of annealing and extension (program: 98 °C for 2′; 1 cycle of [98 °C for 30″; 60 °C for 30″; 72 °C for 15″]; 72 °C for 4′.)

3. Use 2 μL (1/25 of the total volume) to perform a quality control by gel electrophoresis with an appropriate DNA ladder. The obtained band should be 121 base pairs in size and, if including the negative control, this band should be more intense, thus demonstrating correct amplification. A representative gel is shown in Fig. 2b.
4. Clean up the DNA template by PureLink® PCR Purification Kit or other equivalent column purification method. Mind to elute the template in RNase-free water (DEPC/Bidi) and not the kit suggested buffer. Performing traditional phenol/chloroform extraction and sodium acetate EtOH precipitation in order to obtain pure template DNA is also possible.
5. Quantify the DNA yield by spectrophotometry (we use NanoDrop). The resulting yield can range from 200 ng/μL to over 1 μg/μL and seems to be highly dependent on the exact molecular amounts (pipetting errors) of 5′ and 3′ primers added.

3.2.2 In Vitro Transcription of gRNA

In vitro transcription of gRNA from a DNA template is done by employing commercial kits.

1. Generate gRNA by employing a commercial T7 in vitro transcription kit such as T7 MegaShortScript or the HiScribe™T7 High Yield RNA Synthesis Kit (*see* **Note 8**). Add 1 μg of template DNA per reaction, as per manufacturer's instructions, and incubate at 37 °C for 4 h up to overnight. We recommend using a thermocycler with a heated lid to prevent evaporation during long-term incubations.
2. Ensure full removal of the DNA template by adding 1 μL TURBO™ DNase and incubate in a thermocycler at 37 °C for

15 min. Failure to do so will lead to DNA related embryo-toxicity upon injection of the gRNAs.

3. Purify the gRNAs by phenol/chloroform extraction and NH_4OAc/EtOH precipitation. The pellet should be clearly visible with the naked eye. Resuspend the pellet, after washing once with 70% EtOH, in 20 μL of DEPC/Bidi. Alternatively, more expensive commercial solutions such as the MEGAclear transcription clean-up kit (Ambion) can be used; however, we believe that traditional extraction and precipitation yields better quality gRNA for injection but this has never been thoroughly evaluated in a pair-wise comparison.
4. Aliquot the gRNAs in 1.5 μL portions in clean RNase-free Eppendorf tubes and place within an −70 °C ultra-freezer. For gRNA quality control and each gene editing experiment one aliquot can be thawed and used. Thawed gRNAs are never refrozen or used beyond the day of thawing to avoid degeneration of the transcript.

3.2.3 gRNA Quality Control by Denaturing Gel Electrophoresis

The generated gRNA needs to be quality controlled by running denaturing MOPS gel electrophoresis. This is necessary to verify that no degradation of the gRNA took place during purification and for visual estimation of in vitro transcription yield by comparison with a quantitative RNA marker. Due to the formation of secondary structures by gRNA under native conditions, it is necessary to run this gel under denaturing conditions. A representative image for injection ready gRNA under different concentrations and an example for degraded gRNA is shown in Fig. 2c.

1. Clean your entire electrophoresis system (including combs and casting system) with RNase AWAY and subsequently fill it with 1× MOPS running buffer.
2. Mix 0.7 g agarose gel UltraPure™ Agarose with 50 mL of DEPC-treated water. Note to take care to use baked (RNase-free) glassware for the preparation of this gel.
3. Boil the mixture by microwaving until agarose is dissolved and the mixture is completely translucent. Let it cool down to 60 °C.
4. Add 5.9 mL 10× MOPS and 1.8 mL 37% formaldehyde, pour the gel in the casting system, and let it solidify. Make sure to further mix and run this gel in the fume-hood.
5. Mix 1 μL of each gRNA with 4 μL of RNA loading buffer. Mix 2 μL of the RiboRuler High Range RNA Ladder with 8 μL of RNA loading buffer. Incubate in a thermoshaker at 65 °C for 5 min. Put on ice *immediately* after incubation for at least 2 min.
6. Load the gel with the samples and the ladders. Run until separation of the ladder is apparent on the employed electrophoresis system.

7. Visualize the gel by an appropriate method. As the GelRed is an in-sample stain, no post-stain is required and visualization can thus be performed immediately after running. Make sure to save this gel image digitally for gRNA quantification purposes (*see* Subheading 3.2.4).

3.2.4 gRNA Quantification

Accurate quantification of gRNAs after in vitro transcription is not at all straightforward. We have tested several methods such as NanoDrop (ThermoFisher Scientific), Qubit BR RNA assay (ThermoFisher Scientific), DropSense (Trinean), and Fragment Analyzer (Advanced Analytical). None of these seemed to accurately reflect (diverging by magnitudes of at least two) the amounts of gRNA that could be estimated by regression curve comparison using the two dilutions of the quantitative ladder on a denaturing gel electrophoresis (unpublished work). Furthermore, we have observed that the Qubit system underestimates the gRNA yield, but it does accurately measure the proportional differences in yield between different gRNA synthesis reactions (*see* **Note 9**). In conclusion, for absolute quantification we thus recommend regression curve-based quantification of the gRNA based on the comparison with known standards on a denaturing gel. For multiplex CRISPR/Cas9 experiments (co-injections of more than one gRNA), however, an additional concern is to provide equimolar amounts of each gRNA in the injection mixture. As such, we employ the Qubit system to accurately ensure a 1:1 molar ratio of each gRNA in the final injection mixture. Nonetheless, we are aware that the exact quantitative output of the qubit system is not correct.

1. Quantify your gRNA yield by comparing the intensity of the two dilutions of the RiboRuler High Range RNA Ladder bands of known concentration (*see* the manufacturer's instruction) to the intensity of the gRNA bands by digital analysis of the denaturing gel image obtained in Subheading 3.2.3. We use regression curve based "absolute quantification" under the "quantity tools" in Image Lab.
2. Quantify your gRNA yield by Qubit RNA BR Assay Kit according to the manufacturer's instructions.

3.3 Delivery of gRNA/Cas9 Ribonucleoproteins by Microinjection to *Xenopus tropicalis* Embryos

In the earlier days of CRISPR/Cas9 a lot of people choose to employ in vitro CRISPR/Cas9 cutting assays to determine those gRNAs which were effective at cutting the expected target sites [38]. We have previously performed these kinds of experiments and can say that, with some optimization, these do indeed work. However, considering the straightforwardness by which CRISPR/Cas9 can be introduced within the developing *X. tropicalis* embryo we recommend testing gRNA for efficiency immediately in an in vivo context. This as in vitro cutting efficiency does in essence tell you nothing about in vivo cutting efficiencies. As such, we

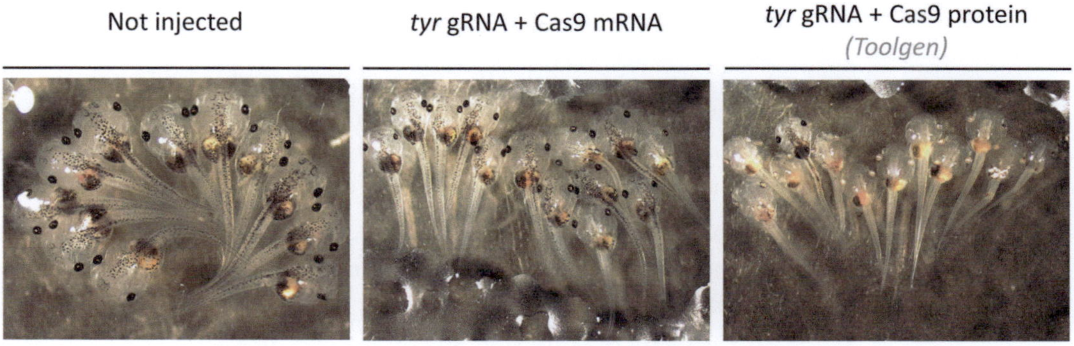

Fig. 3 Comparison of ***tyrosinase (tyr)*** CRISPR/Cas9 editing efficiencies between employing either Cas9 mRNA or Cas9 protein. (Left) Uninjected wild-type embryos exhibiting a normal pigmentation pattern. (Middle) Whole embryo injections of ***tyr*** gRNA together with Cas9 mRNA (at two cell-stage) results in a slight reduction of the pigmentation. (Right) Whole embryo injections (at two-cell stage) of ***tyr*** gRNA pre-complexed to Cas9 protein leads to a dramatic reduction in the pigmentation pattern and a clear phenotypic read-out of the consequence of knocking out ***tyr*** within the developing F0 crispant

recommend to skip the in vitro cutting assays and just to compare the efficiency of a couple (two to three) different gRNAs targeting the same gene in vivo by performing parallel microinjections of each respective gRNA/Cas9 and performing (semi-)quantitative genotyping methodologies (described in Chapter 5) or quantitative genotyping (described in Chapter 6) in order to identify the one(s) with the most promising in vivo efficiencies.

In the following, we will discuss delivery of the CRISPR/Cas9 system to the developing *Xenopus tropicalis* embryo. *Firstly*, we would like to mention that although efficient genome editing has been shown by co-injections of Cas9 mRNA together with gRNA, we and others have observed a dramatic increase in CRISPR/Cas9 mediated genome editing efficiency when employing recombinant Cas9 protein [39, 40]. This dramatic difference in efficiencies is demonstrated in Fig. 3, where it can be appreciated that, at injection of an equivalent amount of gRNA targeting the *tyrosinase* gene, Cas9 protein generates a much more severe albinism phenotype when compared to the Cas9 mRNA. Recombinant Cas9 protein can be bought commercially (we have good experience with Toolgen) or be made in-house as described in the supplementary of Naert et al. [6]. The gRNA is delivered together with Cas9 protein as a pre-complexed gRNA/Cas9 ribonucleoprotein (RNP). This RNP is obtained by short incubation at 37 °C and is essential because the gRNA/Cas9 RNP assembly is suboptimal at *X. tropicalis* embryo handling and rearing temperatures (around room temperature). *Secondly*, we would like to point out the possibilities for tissue-specific editing within *X. tropicalis*. The establishment of the *X. tropicalis* fate map allows some degree of tissue-specific genome editing [35, 41]. This can be quintessential in bypassing expected, or unexpected, embryonic lethality as a consequence of genome editing of essential genes. This can be much more efficient than titrating out gRNA in order to reduce efficiencies below lethal threshold as this will also heavily reduce your chance for F0 phenotype. *Thirdly*, we will demonstrate how to set up injection mixtures for targeting more than one gene in a single injection by multiplexing the CRISPR/Cas9 technology.

1. Obtain fertilized *X. tropicalis* embryos. Either by natural matings or by in vitro fertilization techniques (*see* Chapters 1 and 2).
2. Generate the injection mix by combining gRNA and Cas9 protein to obtain a final 1/5 to 1/20 dilution of gRNA in 800–1000 ng/μL (final concentration) of Cas9 protein. As a good starting point we suggest a 1/10 dilution of gRNA mixed with 800 ng/μL of Cas9 protein (*see* **Note 10**). We as thus routinely mix 0.2 μL of gRNA with 1.8 μL of Cas9.

   For multiplex CRISPR/Cas9 engineering employ the ratio between the Qubit quantification for each employed gRNA to calculate the desired dilution factor for the other gRNAs in the mix, thus ensuring a 1:1 molar ratio of the gRNAs (*see* **Note 11** for an example calculation).
3. Incubate the injection mixture at 37 °C for 2–5 min immediately before loading the mixture in the microinjection needle. This allows preformation of the gRNA/Cas9 ribonucleoprotein complex and maximization of mutagenesis as described by Burger et al. [42]. *For multiplex CRISPR/Cas9 engineering* it is important to establish each gRNA/Cas9 RNP separately, incubate at 37 °C, and only then mix 1.5 μL of each RNP complex together in a clean Eppendorf and mix by pipetting before immediately loading the multiplex engineering mixture into the microinjection needle (*see* **Note 11**).
4. Place embryos in 6% ficoll/0.1× MMR solution during the microinjection procedure.
5. Deliver by microinjection 1 nL of the RNP injection mix in one or two blastomere of the two-cell stage *X. tropicalis* embryo. Alternatively, deliver 0.5 nL of the injection mix in one or more of the cells in an four- or eight-cell *X. tropicalis* embryo. By employing the fate map it is at this stage possible to perform tissue-specific genome editing (*see* **Note 12**).
6. Both injected and non-injected embryos are kept overnight in 6% ficoll/0.1× MMR at room temperature.
7. The next day, carefully wash (3 times) surviving embryos to 0.1× MMR and remove any dead embryos. *X. tropicalis* should be raised further as described in Chapter 2.

At this point embryos will have to be genotyped in order to continue the experiment and to validate whether, and how efficient, the CRISPR/Cas9 editing was in vivo. This is further described in Chapter 5. For readers willing to employ this pipeline to generate *X. tropicalis* cancer models by targeting of tumor suppressors, we further point them to Chapter 11. Subsequently, the analysis of the resulting crispants depends on the specific biological question that is being addressed and could be, but is not limited to: histopathology (*see* Chapter 11), in situ hybridization, -omics, mass spectrometry, live reporter imaging, etc.

4 Notes

1. CFD (Cutting Frequency Determination). Doench et al. measured the cutting efficiency of potential off-targets and integrated them into the CFD score. Potential off-targets with up to four mismatches are scored with Doench et al matrix [30].
2. The chance of at least one allele containing an in-frame mutation, under the premise that both alleles are edited, is the sum of the chance that both alleles are in-frame (IF) and the chance that one allele is in-frame and one allele is frameshift (FS).

   $(p(IF*IF) = 0.33*0.33) + (p(IF*FS) = 0.33*0.66) + (p(FS*IF) = (0.33*0.66)) = 0.55.$
3. CRISPR/Cas9 induced in-frame mutation (insertion or deletion of $(n) \times 3$ bp) can lead to production of protein missing or gaining $(n) \times 1$ amino-acid. Whether or not this protein will be functional is a priori very hard to determine. One can make an educated guess on known protein 3D structure from structural biology work (if available) but in either way it remains hard to determine, based on DNA sequence information, the specific effect of the protein perturbation of the protein function.
4. It has to be noted that targeting a structurally important domain will result in mosaic crispants by which per definition some of the cells will express an in-frame variant of the protein product from your targeted gene (losing or gaining on or more amino acids). It is possible that these variants might acquire positive or negative dominant functions (gain-of-function) and as such some cells composing your mosaic crispant might not represent a full loss-of-function experiment. While still valuable in order to increase penetrance of resulting F0 phenotypes (due to loss of several evolutionary conserved amino acids in in-frame edited cells within the crispant probably leading to protein null in the majority of the situations), it might be better if intending to progress to (F1, F2, etc.) germ-line breeding to target other coding sequences for your gene. As you will, after breeding, select by genotyping anyway for animals exhibiting bi-allelic frameshift mutations.

5. CRISPRScan will output gRNAs with a color code, by which intense light-green is the preferential class for designing. If you cannot identify high-scoring targets it is, however, still feasible to continue with some lower scoring ones. As this is an in silico prediction, this indicates that the higher the score the higher the expected chance of success, defined as a highly efficient gRNA at that site. In line with this a gRNA with a lower score might have a reduced chance of success but can, however, be still highly active in vivo.
6. The CRISPRScan algorithm only outputs oligos with 5′-GG allowing for in vitro transcription. However, if employing a different gRNA design algorithm make sure to either: (1) select targets starting with 5′-GG, or (2) add GG in front of target sequence (we have experimentally observed that this overhang addition does not severely interfere with gRNA target site recognition).
7. Velocity HiFi polymerase and the corresponding reaction buffer can be replaced by another proof-reading and low error polymerase. The DNA template synthesis reaction works equally well by exchanging the Velocity in this reaction by Phusion polymerase.
8. We have tested the HiScribe™T7 High Yield RNA Synthesis Kit (NEB) and the MEGAshortscript T7 (Ambion) kits in parallel (same DNA input) in order to quantify difference in gRNA yield and have found these to be in the same order (data not shown). We expect in vitro T7 transcription kits from other manufacturers to most likely yield acceptable outcomes; however, we cannot vouch for these as they have not been tested in our hands.
9. We suspect this underestimation to be the consequence of the Qubit fluorescent dyes not binding to the gRNA target molecules due to the specific inherent secondary structures of the gRNA molecule. As such, the measurement will severely underestimate the gRNA yield. Nevertheless as the inherent secondary structure formation of the gRNAs should be, more or less, identical between different gRNAs, we believe this to be a powerful method for accurately determining 1:1 molar ratios for gRNAs.
10. If following our pipeline described in Subheading 3.2 for generating injection ready gRNA, we usually obtain gRNA that, when loaded in a 1/200 dilution in the Qubit BR RNA assay (*see* Subheading 3.2.4), will give rise to *raw* QF value measurements of between 200 and 600. If the gRNA that you have generated, has a comparable yield, we suggest to dilute your resulting gRNA 1/10 in Cas9 protein for initial in vivo experiments. We refrain from mentioning absolute concentrations of

gRNA to add to the injection mix, as absolute quantification still suffers from major limitations (*see* Subheading 3.2.4). We believe it to be better, to report here the 1/10 dilution of your gRNA in Cas9 protein when it has correctly passed the different gRNA quality control steps. These dosages can then be further refined in additional experimental setups, but we have, in our hands, observed that this (CRISPRScan designed) gRNA to Cas9 ratio leads to a good percentage of gRNAs (>50%) exhibiting a good efficiency profile (>70% mutant reads on next-generation sequencing).

11. We want to co-inject gRNA x and y and thus want to obtain an injection mix containing both gRNA at a 1:1 molar ratio. For instance, assume that gRNA x is quantified at 170 raw QF (1/200 dilution) on Qubit BR RNA assay and gRNA y at 240 raw QF (1/200 dilution). We dilute both gRNA separately with Cas9 protein, making sure that each gRNA is present in the same concentration:

    0.2 μL of gRNA x with 1.8 μL of Cas9 protein (1000 ng/μL stock) = 17 ng/μL final gRNA concentration

    0.14 μL of gRNA y with 1.8 μL of Cas9 protein (1000 ng/μL stock) = + − 17 ng/μL final gRNA concentration

    (Note that this 17 ng/μL is not the correct absolute quantification of the gRNA added to the injection mix, see Subheading 3.2.4)

    Both these mixtures should be pre-incubated at 37 °C for 2–5 min and then 1.5 μL of each of these pre-complexed RNPs should be mixed together in a new Eppendorf tube equilibrated at room temperature. This mixture should *stay at room temperature* and can subsequently be loaded in the microinjection needle. Because of the pre-complexing prior to the mixing of the different RNPs there is no possibility of one of the two gRNAs showing a thermodynamically more favorable loading within the RNP. This can lead to out-crowding of one gRNA to the other and subsequent reduction in multiplex CRISPR/Cas9 engineering efficiencies. This methodology can be employed to mix more than two gRNAs together. We have experimentally shown this methodology to be sound for the mixing of up to four gRNAs (data not shown).

12. The *Xenopus* fate map can be consulted on Xenbase: (http://www.xenbase.org/anatomy/static/xenbasefate.jsp)

    One should identify the tissues/organs one wishes to engineer by CRISPR/Cas9 and then employ the reverse fate map to determine the blastomere to inject in order to ensure major contribution (red) to the tissue lineage you wish to affect.

Acknowledgments

The authors would like to thank Dr. Tom Van Nieuwenhuysen and Dr. Hong Thi Tran for the collaborative effort in the initial establishment of the CRISPR/Cas9 system within the research unit. Furthermore, we would like to acknowledge the work of Robin Colpaert in comparison of gRNA quantification methodologies and their specific limitations. We would like to acknowledge Dr. Tom van Nieuwenhuysen and Sarah Geurs for the pairwise comparison of Cas9 protein versus mRNA shown in Fig. 3. Finally we would like to thank Marjolein Carron and Dieter Tulkens for critical proof-reading of this manuscript. Research in the authors' laboratory is supported by the Research Foundation—Flanders (FWO-Vlaanderen) (grants G0A1515N and G029413N), by the Belgian Science Policy (Interuniversity Attraction Poles—IAP7/07) and by the Concerted Research Actions from Ghent University (BOF15/GOA/011). Further support was obtained by the Hercules Foundation, Flanders (grant AUGE/11/14) and the Desmoid Tumor Research Foundation.

References

1. Abu-Daya A, Khokha MK, Zimmerman LB (2012) The hitchhiker's guide to Xenopus genetics. Genesis 50:164–175. https://doi.org/10.1002/dvg.22007
2. Kok FO, Shin M, Ni C-W et al (2015) Reverse genetic screening reveals poor correlation between Morpholino-induced and mutant phenotypes in Zebrafish. Dev Cell 32:97–108. https://doi.org/10.1016/j.devcel.2014.11.018
3. Alexandru Dan Corlan (2004) Medline trend: automated yearly statistics of PubMed results for any query. In: Online—own website. http://dan.corlan.net/medline-trend.html
4. Lansdon LA, Darbro BW, Petrin AL, et al (2017) Identification of *Isthmin 1* as a novel Clefting and craniofacial patterning gene in humans. Genetics. doi: https://doi.org/10.1534/genetics.117.300535
5. Feehan JM, Chiu CN, Stanar P et al (2017) Modeling dominant and recessive forms of retinitis Pigmentosa by editing three rhodopsin-encoding genes in Xenopus Laevis using Crispr/Cas9. Sci Rep 7:6920. https://doi.org/10.1038/s41598-017-07153-4
6. Naert T, Colpaert R, Van Nieuwenhuysen T, et al (2016) CRISPR/Cas9 mediated knockout of rb1 and rbl1 leads to rapid and penetrant retinoblastoma development in *Xenopus tropicalis*. Sci Rep 6. doi: https://doi.org/10.1038/srep35264
7. Liu Z, Cheng TTK, Shi Z et al (2016) Efficient genome editing of genes involved in neural crest development using the CRISPR/Cas9 system in Xenopus embryos. Cell Biosci 6:22. https://doi.org/10.1186/s13578-016-0088-4
8. DeLay BD, Corkins ME, Hanania HL, et al (2017) Tissue-specific gene inactivation in *Xenopus laevis*: knockout of *lhx1* in the kidney with CRISPR/Cas9. Genetics. doi: https://doi.org/10.1534/genetics.117.300468
9. Sakane Y, Iida M, Hasebe T et al (2018) Functional analysis of thyroid hormone receptor beta in *Xenopus tropicalis* founders using CRISPR-Cas. Biol Open 7. https://doi.org/10.1242/bio.030338
10. McQueen C, Pownall ME (2017) An analysis of MyoD-dependent transcription using CRISPR/Cas9 gene targeting in *Xenopus tropicalis* embryos. Mech Dev 146:1–9. https://doi.org/10.1016/j.mod.2017.05.002
11. MacColl Garfinkel A, Khokha MK (2017) An interspecies heart-to-heart: using Xenopus to uncover the genetic basis of congenital heart disease. Curr Pathobiol Rep 5:187–196. https://doi.org/10.1007/s40139-017-0142-x

12. Ledford KL, Martinez-De Luna RI, Theisen MA et al (2017) Distinct cis-acting regions control six6 expression during eye field and optic cup stages of eye formation. Dev Biol 426:418–428. https://doi.org/10.1016/j.ydbio.2017.04.003
13. Jaffe KM, Grimes DT, Schottenfeld-Roames J et al (2016) c21orf59/kurly controls both cilia motility and polarization. Cell Rep 14:1841–1849. https://doi.org/10.1016/j.celrep.2016.01.069
14. Nakayama T, Blitz IL, Fish MB et al (2014) Cas9-based genome editing in *Xenopus tropicalis*. Methods Enzymol 546:355–375. https://doi.org/10.1016/B978-0-12-801185-0.00017-9
15. Jiang F, Doudna JA (2017) CRISPR–Cas9 structures and mechanisms. Annu Rev Biophys 46:505–529. https://doi.org/10.1146/annurev-biophys
16. Park D-S, Yoon M, Kweon J et al (2017) Targeted base editing via RNA-guided Cytidine Deaminases in Xenopus laevis embryos. Mol Cells 40:823–827. https://doi.org/10.14348/molcells.2017.0262
17. Aslan Y, Tadjuidje E, Zorn AM, Cha S-W (2017) High-efficiency non-mosaic CRISPR-mediated knock-in and indel mutation in F0 Xenopus. Development 144(15):2852–2858
18. Moreno-Mateos MA, Fernandez JP, Rouet R et al (2017) CRISPR-Cpf1 mediates efficient homology-directed repair and temperature-controlled genome editing. Nat Commun 8:2024. https://doi.org/10.1038/s41467-017-01836-2
19. Nakade S, Tsubota T, Sakane Y et al (2014) Microhomology-mediated end-joining-dependent integration of donor DNA in cells and animals using TALENs and CRISPR/Cas9. Nat Commun 5:5560. https://doi.org/10.1038/ncomms6560
20. Naert T, Van Nieuwenhuysen T, Vleminckx K (2017) TALENs and CRISPR/Cas9 fuel genetically engineered clinically relevant *Xenopus tropicalis* tumor models. Genesis 55. doi: https://doi.org/10.1002/dvg.23005
21. Lei Y, Guo X, Liu Y et al (2012) Efficient targeted gene disruption in Xenopus embryos using engineered transcription activator-like effector nucleases (TALENs). Proc Natl Acad Sci U S A 109:17484–17489. https://doi.org/10.1073/pnas.1215421109
22. Guo X, Zhang T, Hu Z et al (2014) Efficient RNA/Cas9-mediated genome editing in *Xenopus tropicalis*. Development 141(3):707–714
23. Nakayama T, Fish MB, Fisher M et al (2013) Simple and efficient CRISPR/Cas9-mediated targeted mutagenesis in *Xenopus tropicalis*. Genesis 51:835–843. https://doi.org/10.1002/dvg.22720
24. Blitz IL, Biesinger J, Xie X, Cho KWY (2013) Biallelic genome modification in F(0) *Xenopus tropicalis* embryos using the CRISPR/Cas system. Genesis 51:827–834. https://doi.org/10.1002/dvg.22719
25. Stemmer M, Thumberger T, del Sol Keyer M et al (2015) CCTop: an intuitive, flexible and reliable CRISPR/Cas9 target prediction tool. PLoS One 10:e0124633. https://doi.org/10.1371/journal.pone.0124633
26. Moreno-Mateos MA, Vejnar CE, Beaudoin J-D et al (2015) CRISPRscan: designing highly efficient sgRNAs for CRISPR-Cas9 targeting in vivo. Nat Methods 12:982–988. https://doi.org/10.1038/nmeth.3543
27. Doench JG, Hartenian E, Graham DB et al (2014) Rational design of highly active sgRNAs for CRISPR-Cas9–mediated gene inactivation. Nat Biotechnol 32(12):1262–1267. https://doi.org/10.1038/nbt.3026
28. Haeussler M, Schönig K, Eckert H et al (2016) Evaluation of off-target and on-target scoring algorithms and integration into the guide RNA selection tool CRISPOR. Genome Biol 17:148. https://doi.org/10.1186/s13059-016-1012-2
29. Bae S, Park J, Kim J-S (2014) Cas-OFFinder: a fast and versatile algorithm that searches for potential off-target sites of Cas9 RNA-guided endonucleases. Bioinformatics 30:1473–1475. https://doi.org/10.1093/bioinformatics/btu048
30. Sullender M, Hegde M, Vaimberg EW et al (2016) Optimized sgRNA design to maximize activity and minimize off-target effects of CRISPR-Cas9. Nat Biotechnol 34(2):184–191. https://doi.org/10.1038/nbt.3437
31. Listgarten J, Weinstein M, Kleinstiver BP et al (2018) Prediction of off-target activities for the end-to-end design of CRISPR guide RNAs. Nat Biomed Eng 2:38–47. https://doi.org/10.1038/s41551-017-0178-6
32. Baker KE, Parker R (2004) Nonsense-mediated mRNA decay: terminating erroneous gene expression. Curr Opin Cell Biol 16:293–299. https://doi.org/10.1016/j.ceb.2004.03.003
33. Mou H, Smith JL, Peng L et al (2017) CRISPR/Cas9-mediated genome editing induces exon skipping by alternative splicing or exon deletion. Genome Biol 18:108. https://doi.org/10.1186/s13059-017-1237-8

34. Kwong LN, Dove WF (2009) APC and its modifiers in colon cancer. Adv Exp Med Biol 656:85–106
35. Van Nieuwenhuysen T, Naert T, Tran HT et al (2015) TALEN-mediated apc mutation in *Xenopus tropicalis* phenocopies familial adenomatous polyposis. Oncoscience 2(5):555–566. https://doi.org/10.18632/oncoscience.166
36. Shi J, Wang E, Milazzo JP et al (2015) Discovery of cancer drug targets by CRISPR-Cas9 screening of protein domains. Nat Biotechnol 33:661–667. https://doi.org/10.1038/nbt.3235
37. Baklanov MM, Golikova LN, Malygin EG (1996) Effect on DNA transcription of nucleotide sequences upstream to T7 promoter. Nucleic Acids Res 24(18):3659–3660
38. Anders C, Jinek M (2014) In vitro enzymology of Cas9. Methods Enzymol 546:1–20. https://doi.org/10.1016/B978-0-12-801185-0.00001-5
39. Shigeta M, Sakane Y, Iida M et al (2016) Rapid and efficient analysis of gene function using CRISPR-Cas9 in *Xenopus tropicalis* founders. Genes Cells 21:755–771. https://doi.org/10.1111/gtc.12379
40. Bhattacharya D, Marfo CA, Li D et al (2015) CRISPR/Cas9: an inexpensive, efficient loss of function tool to screen human disease genes in Xenopus. Dev Biol 408:196–204. https://doi.org/10.1016/j.ydbio.2015.11.003
41. Moody SA (1987) Fates of the blastomeres of the 32-cell-stage Xenopus embryo. Dev Biol 122:300–319
42. Burger A, Lindsay H, Felker A et al (2016) Maximizing mutagenesis with solubilized CRISPR-Cas9 ribonucleoprotein complexes. Development 143:2025–2037. https://doi.org/10.1242/dev.134809

Chapter 4

Targeted Genome Engineering in *Xenopus* Using the Transcription Activator-Like Effector Nuclease (TALEN) Technology

Tom Van Nieuwenhuysen and Kris Vleminckx

Abstract

Targeted genome engineering technologies are revolutionizing the field of functional genomics and have been extensively used in a variety of model organisms, including *X. tropicalis* and *X. laevis*. The original methods based on Zn-finger proteins coupled to endonuclease domains were initially replaced by the more efficient and straightforward transcription activator-like effector nucleases (TALENs), adapted from plant pathogenic *Xanthomonas* species. Although functional genomics are more recently dominated by the even faster and more convenient CRISPR/Cas9 technology, the use of TALENs may still be preferred in a number of cases. We have successfully implemented this technology in *Xenopus* and in this chapter we describe our working protocol for targeted genome editing in *X. tropicalis* using TALENs.

Key words TALEN design, Golden gate cloning, Web tools

1 Introduction

Genome engineering is evidently a very powerful tool in functional genomics research. The advent of programmable nucleases such as Transcription Activator-Like Effector Nucleases (TALEN) and Clustered Regularly Interspaced Short Palindromic Repeats (CRISPR)/Cas9 has omitted the need for extensive cloning of knock-out vectors and ES cell-based selection of correctly recombined clones, making fast and precise genome engineering available in nearly any model organism, including *X. laevis* and *X. tropicalis* [1–5]. The use of targeted genome engineering in *Xenopus* has a number of important specific advantages: One mating yields high numbers of embryos, which are large and develop

The published version of the book has the family name of author that was erroneously captured as given name in the XML. This has now been updated.

The updated online version of this book can be found at
https://doi.org/10.1007/978-1-4939-8784-9

Kris Vleminckx (ed.), *Xenopus: Methods and Protocols*, Methods in Molecular Biology, vol. 1865,
https://doi.org/10.1007/978-1-4939-8784-9_4,

externally, making them ideally suited for large-scale microinjections with these programmable nucleases. Due to the high efficiency of the programmable nucleases, phenotypes can often already be analyzed in F0 mosaic mutant animals, omitting the need for crossing or inbreeding to obtain mutant lines. Additionally, the programmable nucleases will only be active after a number of cell divisions, and will remain active during early development. As a consequence, embryos will be mosaic for different mutations at the target sites, and more importantly, some cells will be homozygous mutant, whereas others will still be wild type [6]. This mosaicism can thus allow development and survival even when essential genes are targeted [7]. Additionally, the use of fate maps, which indicate the destination tissues of the different blastomeres of the early embryo, allows injecting only a subset of target tissues, leaving the rest of the animal unaffected [8, 9]. These targeted injections may thus serve as a crude alternative to the CRE-LoxP system for tissue-specific genome engineering. We were for example able to enhance survival and separate different tumor types by targeted injection of *apc* TALENs [10]. Although the TALEN technology has successfully been applied both in *Xenopus laevis* and *tropicalis*, we prefer to use *X. tropicalis* for genome editing because of its diploid genome.

Although CRISPR/Cas9 is mostly dominating the field of targeted genome engineering due to its higher simplicity, throughput, and multiplexing capability, the TALEN technology may still be useful for a number of reasons. When a very specific target needs to be edited for example, it may be necessary to opt for the TALEN technology. CRISPR/Cas9 requires the target sequence to be followed by NGG-3′, whereas each TALEN binding site needs to be preceded by a T residue. Especially AT-rich sequences are thus more easily targeted using TALENs. Additionally, TALENs are reported to have less off-target effects and may thus be used for more specific genome engineering when important off-target effects are predicted using CRISPR/Cas9 [11, 12]. Finally, TALENs can be used to confirm results obtained during for example large-scale CRISPR/Cas9 screens.

The TALEN technology was adapted from plant pathogenic *Xanthomonas* species, which introduce TAL effectors into the host cells, which function as transcription factors and induce expression of genes favoring bacterial survival and growth [13, 14]. These effectors contain a central repeat domain consisting of nearly identical 33–34 amino acid tandem repeats, differing almost exclusively at amino acid positions 12 and 13, termed the repeat variable diresidue (RVD). The last repeat typically only consists out of 20 amino acids and is therefore termed a half repeat. The central repeat domain is responsible for specific host DNA binding and each of the 33–34 amino acid repeats interacts with one consecutive nucleotide in the target sequence [15, 16] (Fig. 1). The spe-

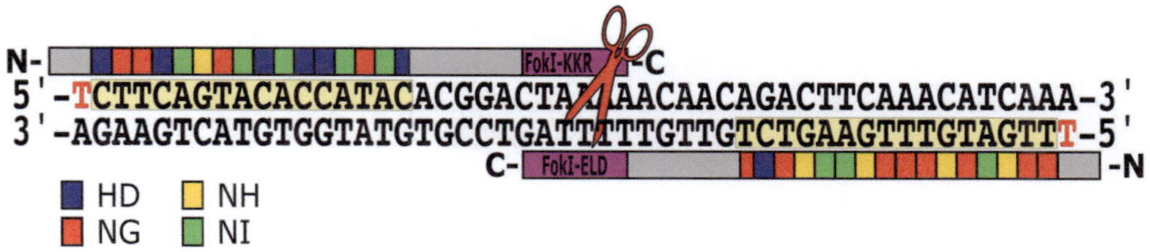

Fig. 1 Basic design of a TALEN construct. Each TALEN monomer consists of 15–20 TAL repeats, each recognizing a specific nucleotide as determined by the repeat variable diresidue (RVD; HD-C; NG-T; NI-A; NH-G), fused to one of the obligate heterodimeric FokI endonuclease domains. Combination of two TALEN monomers yields a functional endonuclease at the target site in the genome. Target sites are indicated in yellow and need to be preceded by a T nucleotide (red)

cific amino acids at the RVD positions determine the nucleotide recognized and thus specific DNA target recognition is conferred by the order of these specific RVD repeats. This one to one relationship between RVD and nucleotide implies that essentially four RVD repeat variants are required to target any DNA sequence with as only restriction that a T residue precedes the target site. The optimal four consensus RVDs were found to be NI for A, NG for T, HD for C, and NH for G [17, 18]. Coupling of a custom made central repeat domain to the obligate dimeric FokI endonuclease domain generates a TALEN monomer. The association of two of these monomers at the target DNA results in the formation of a functional endonuclease, inducing double stranded breaks (DSBs) specifically at the target site (Fig. 1). Although FokI is normally active as a homodimer, mutated obligate heterodimeric forms (FokI-ELD and FokI-KKR) have been developed to increase specificity [19]. TALEN induced double stranded breaks are by default repaired by the error-prone non-homologous end joining (NHEJ) pathway, resulting in small insertions and deletions and thus often frame-shifts and gene loss-of-function. Alternatively, DNA repair templates can be introduced together with the TALEN to obtain targeted knock-in of a desired DNA sequence. In this section, we will describe how to use the TALEN technology in *Xenopus tropicalis* for targeted gene knock out from design up until injection. For evaluation of cutting efficiency we refer to Chapter 5.

2 Material

2.1 TALEN Vector Cloning and mRNA Synthesis

1. Golden Gate TALEN and TAL effector kit 2.0 (addgene #1000000024).
2. Golden Gate cloning manual (https://media.addgene.org/cms/filer_public/98/5a/985a6117-7490-4001-8f6a-24b2cf7b005b/golden_gate_talen_assembly_v7.pdf) and all material described therein.
3. Heating block or thermo shaker for eppendorf tubes.

4. Nanodrop spectrophotometer for nucleic acid quantification (or equivalent).
5. Midi or Maxi plasmid preparation kit (e.g., Qiagen or equivalent).
6. Ambion mMessage machine® sp6 kit (Thermo Fisher Scientific).
7. pCS2-Flag-TALEN-ELD and pCS2-Flag-TALEN-KKR plasmids [2].
8. NotI restriction enzyme.
9. Rnase ZAP™ (Merck or equivalent).
10. Phenol:Chloroform:Isoamyl Alcohol (25:24:1, v/v).
11. Chloroform, Ethanol, and isopropanol.
12. 3 M Na Acetate.
13. 0.5 M EDTA.

2.2 TALEN mRNA Microinjection

1. 10× Marc's Modified Ringer's (MMR) stock solution: 1 M NaCl, 20 mM KCl, 10 mM $MgSO_4$, 20 mM $CaCl_2$, 50 mM HEPES, pH 7.6.
2. 0.1× MMR.
3. 6% Ficoll in 0.1× MMR.
4. 2% cysteine in 0.1× MMR, adjusted to pH 8.0 with NaOH (*see* **Note 1**).
5. Human chorionic gonadotropin (hCG).
6. Micro-injector with foot pedal and adjustable injection pressure and time.
7. Mechanical micro-manipulator.
8. Stereomicroscope with microscopic ruler.
9. Injection dish: a 60 mm petri-dish in which a nylon mesh (e.g., 700 μM mesh) is fixed with a few drops of chloroform.
10. Micropipette puller (e.g., Sutter Instruments).
11. Glass capillaries (with 1.0 and 0.58 mm as outer and inner diameter, respectively).
12. Microloader™ tips (e.g., Eppendorf).

3 Methods

3.1 TALEN Design

Several free online tools have been developed that allow convenient design of TALEN pairs (Table 1). These tools scan a user submitted sequence or gene identification number for suitable TALEN binding sites and their corresponding RVD sequences. Ranking of the results in these tools is often based on on-target and off-target binding prediction. On-target binding prediction is usually based on extrapolation of data from large-scale

Table 1
Comparison of commonly used tools for TALEN design. This table is not exhaustive

	Sequence input	Annotation target available (species)	Off-target prediction	Off-target (species)	Ranking
TALE-NT [26, 27]	FASTA	None	# Mismatches + RVD binding frequencies	Any sequenced genome (NCBI ID) or sequence (FASTA)	Cut site; Streubel guidelines [18]
ZiFit [28]	FASTA	None	None	NA	Cut site
Mojo hand [29]	FASTA or NCBI ID	Any	None	NA	Unranked list per exon, out-of-frame score
E-TALEN [30]	FASTA or ENSEMBL ID	Vertebrates	# Mismatches	Extensive preset list and FASTA	Graphical, ranking based on off-target and efficiency scores [18]
CHOPCHOP [31, 32]	FASTA, Refseq, ENSEMBL ID, NCBI ID	Any	# Mismatches	Extensive preset list	Graphical, list ranked by off-target score and target position
SAPTA [20]	FASTA	None	None	NA	List ranked by own scoring algorithm
PROGNOS [33]	TALEN target or RVD sequence	NA	# Mismatches + RVD binding frequencies	Most common model organisms	List of potential off-target sites and score

TALEN-binding studies [18, 20], whereas off-target binding prediction can be based merely on a number of mismatches between the RVD sequences and the off-target sequence, but is often also based on experimental data on binding frequencies (Table 1). Some of these tools also provide the possibility to target exons or intron-exon boundaries only. The number of RVD repeats and the spacer sequence can usually be user defined and by default we set 16 RVD repeats and 16 bp spacer sequence, based on Lei et al [2].

These constraints can be relaxed if necessary, preferably using the aforementioned guidelines [18]. The output is usually a ranked list of potential target sites and corresponding RVD sequences.

3.2 TALEN Vector Cloning and mRNA Synthesis

Many different approaches have been described for the generation of TALEN encoding plasmids [21–25], but we use the method developed by Cermak and colleagues [26], requiring minimal laboratory equipment and expertise. This method is a 5 day protocol using Golden Gate cloning, and we have only adapted the final destination vector to the pCS2+ based vector (pCS2-Flag-TALEN-ELD and pCS2-Flag-TALEN-KKR) generated by the Hui Zhao lab [2], which is suited for in vitro mRNA transcription from the *sp6* promotor. In Golden Gate cloning, typeIIS restriction endonucleases are employed, which cut target DNA a few bases away from the recognition site, generating customizable sticky ends. The Golden Gate TALEN and TAL effector kit 2.0 (addgene #1000000024) contains a library of 86 vectors, including 11 vectors per RVD (NI, NG, HD, and NH) and intermediary pFUS destination vectors (Fig. 2).

1. The cloning protocol is very extensively described in great detail in [26] and in the Golden Gate cloning manual (https://media.addgene.org/cms/filer_public/98/5a/985a6117-

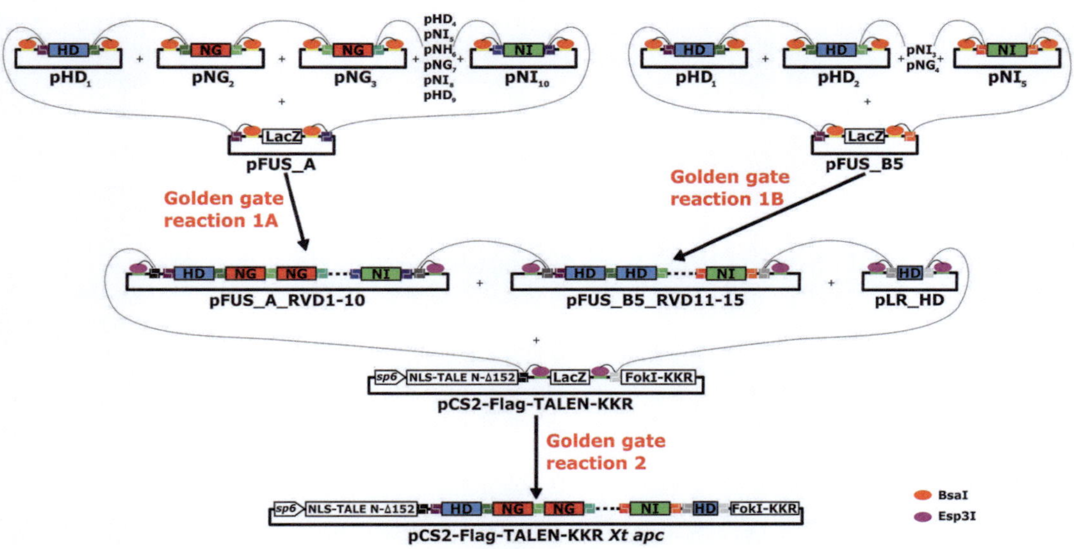

Fig. 2 Generation and use of a 16 RVD TALEN targeting the ***X. tropicalis apc*** gene [10], schematic overview. During Golden Gate assembly reaction 1, single RVD coding modules are cut out of their backbone vector using the typeIIS restriction enzyme BsaI, generating unique sticky ends. This allows directional assembly of multiple (up to 10) RVD modules in a single new backbone vector (pFUS_X). These module arrays are then combined in a second Esp3I mediated Golden Gate cloning reaction with the final half RVD repeat (pLR_RVD) and a final backbone vector, suited for in vitro mRNA transcription from the ***sp6*** promoter and with an SV40 poly-A signal (pCS2-Flag-TALEN-ELD/KKR) [2]. After in vitro transcription and injection, the mRNA is translated in vivo

7490-4001-8f6a-24b2cf7b005b/golden_gate_talen_assembly_v7.pdf) and we have successfully followed this protocol with the only adaptation of the final destination vector suited for in vitro transcription, which was kindly provided by Prof. Dr. Hui Zhao [2].

2. After obtaining the desired TALEN RVD sequence in the pCS2-Flag-TALEN-ELD and pCS2-Flag-TALEN-KKR vector backbones, prepare sufficient amounts of plasmid DNA using a plasmid Midi Kit or plasmid Maxi Kit.
3. Linearize 20 μg of each plasmid using 50 units of NotI enzyme in the recommended buffer and with acetylated BSA (0.1 μg/μL) for at least 3 h.
4. Check complete plasmid linearization by agarose gel electrophoresis. Incomplete linearization will interfere with mRNA transcription.

   After linearization, stop the restriction digest by adding 1/20th volume of 0.5 M EDTA, 1/10th volume of 3 M Na Acetate, and 2 volumes of 100% ethanol (*see* **Note 2**).
5. Mix and leave at −20 °C for at least 15 min and pellet DNA in a microcentrifuge at top speed (>14,000 × *g*).
6. Remove the supernatant by decanting and wash pellet with 70% ethanol. Remove residual drops by pipetting with a fine pipet tip. Keep the tube inverted and dry DNA pellet for 10–15 min. Resuspend pellet in 20 μL of nuclease-free water.
7. Use 1 μg of linearized plasmid DNA as template for in vitro mRNA transcription from the *sp6* promoter using the mMessage machine sp6 kit. After the transcription reaction, add TURBO DNase as described in the kit manual.
8. Purify the mRNA using phenol:chloroform extraction: add 115 μL nuclease-free water and 15 μL of Ammonium Acetate Stop Solution (provided in the mMessage machine sp6 kit) to the 20 μL reaction mixture. Mix thoroughly and add an equal volume of phenol:chloroform. Spin down in a microcentrifuge for 10 min at >14,000 × *g* and transfer aqueous phase to a new tube. Add an equal volume of chloroform, spin down and transfer aqueous phase to a new tube.
9. Precipitate mRNA by adding an equal volume of isopropanol and mix well. Chill the mixture at −20 °C for at least 15 min and spin down at 4 °C at >14,000 × *g* for 15 min. Decant the supernatant and wash and dry pellet as described in **step 7**. Resuspend pellet in 10 μL–25 μL of nuclease-free water.
10. Measure mRNA concentration using a Nanodrop or equivalent. Aliquot mRNA and store at −80 °C.

3.3 TALEN mRNA Micro-Injection

1. Needle preparation (*see* **Note 3**) conditions will be dependent on the specific needle puller that is used.
2. Prime male and female *Xenopus tropicalis* with 10 IU and 20 IU of hCG, respectively, by injection into the dorsal lymph pouch 2 days before micro-injection.
3. On the day of the injection, boost the male and female frogs in the morning with 100 IU and 150 IU, respectively, and form couples to allow mating. Egg-laying should begin around noon.
4. Collect fertilized eggs using a plastic pasteur pipette. Remove remaining aquarium water and incubate the embryos with a 2% cysteine solution to remove the jelly coat. The disappearance of the jelly coat can easily be monitored by swirling the embryos and visually inspecting the proximity of the embryos. Wash at least 5 times with 0.1× MMR and transfer to a petri dish.
5. Prepare injection mixtures with forward and reverse TALEN mRNA in nuclease-free water (*see* **Note 4**). We have had successful genome editing with total mRNA injections of 600 pg/embryo or down to 40 pg/embryo.
6. Transfer a few microliters of injection mixture into the injection needle by using microloader tips, connect to the micro-injector, and place in the micro-manipulator.
7. Calibrate the injection volume by measuring the drop size at the tip of the injection needle using a microscopic ruler and calculate the corresponding volume. Adjust injection time and pressure to obtain an injection volume of 1 nL for injections in 1- or 2-cell stage or 0.5 nL for injections in later stages. Keep the tip of the injection needle in 6% ficoll to avoid drying at the tip of the needle, which can lead to needle blocking.
8. Select embryos of the desired stage for injection (1-cell up to 32-cell stage) and transfer to a dish with injection grid containing 6% ficoll in 0.1× MMR and inject one or multiple cells with the TALEN mRNA. Allow embryos to recover in 6% ficoll in 0.1× MMR for at least 1 h (can be overnight) and then transfer to 0.1× MMR. Select out surviving embryos the morning after injection and refresh 0.1× MMR.

4 Notes

1. 2% cysteine solution in 0.1× MMR should be made fresh the day of injection.
2. Make sure to work RNase free from now on. Prepare buffers with nuclease-free water, use nuclease-free tubes and tips and spray surfaces and material with Rnase ZAP™ or equivalent.

3. The needle hallmarks should be the same as those used for standard RNA injections (e.g., having a long taper).
4. If desired a fluorescent tracker such as rhodamin dextran can be added to the mixture to identify injected region later in development.

Acknowledgments

Research in the authors' laboratory is supported by the Research Foundation–Flanders (FWO-Vlaanderen) (grants G0A1515N and G029413N), by the Belgian Science Policy (Interuniversity Attraction Poles—IAP7/07) and by the Concerted Research Actions from Ghent University (BOF15/GOA/011). Further support was obtained by the Hercules Foundation, Flanders (grant AUGE/11/14) and the Desmoid Tumor Research Foundation.

References

1. Guo X, Zhang T, Hu Z, Zhang Y, Shi Z, Wang Q, Cui Y, Wang F, Zhao H, Chen Y (2014) Efficient RNA/Cas9-mediated genome editing in Xenopus tropicalis. Development 141(3):707–714. https://doi.org/10.1242/dev.099853
2. Lei Y, Guo X, Liu Y, Cao Y, Deng Y, Chen X, Cheng CHK, Dawid IB, Chen Y, Zhao H (2012) Efficient targeted gene disruption in Xenopus embryos using engineered transcription activator-like effector nucleases (TALENs). Proc Natl Acad Sci U S A 109:17484–17489. https://doi.org/10.1073/pnas.1215421109
3. Naert T, Van Nieuwenhuysen T, Vleminckx K (2017) TALENs and CRISPR/Cas9 fuel genetically engineered clinically relevant *Xenopus tropicalis* tumor models. Genesis 55:e23005. https://doi.org/10.1002/dvg.23005
4. Suzuki K-IT, Isoyama Y, Kashiwagi K, Sakuma T, Ochiai H, Sakamoto N, Furuno N, Kashiwagi A, Yamamoto T (2013) High efficiency TALENs enable F0 functional analysis by targeted gene disruption in Xenopus laevis embryos. Biology Open 2:448–452. https://doi.org/10.1242/bio.20133855
5. Tandon P, Conlon F, Furlow JD, Horb ME (2016) Expanding the genetic toolkit in Xenopus: approaches and opportunities for human disease modeling. Dev Biol 426(2): 325–335. https://doi.org/10.1016/j.ydbio.2016.04.009
6. Ratzan W, Falco R, Salanga C, Salanga M, Horb ME (2016) Generation of a Xenopus laevis F1 albino J strain by genome editing and oocyte host-transfer. Dev Biol 426(2):188–193. https://doi.org/10.1016/j.ydbio.2016.03.006
7. Naert T, Colpaert R, Van Nieuwenhuysen T, Dimitrakopoulou D, Leoen J, Haustraete J, Boel A, Steyaert W, Lepez T, Deforce D, Willaert A, Creytens D, Vleminckx K (2016) CRISPR/Cas9 mediated knockout of rb1 and rbl1 leads to rapid and penetrant retinoblastoma development in Xenopus tropicalis. Sci Rep 6:35264
8. Moody SA (1987) Fates of the blastomeres of the 16-cell stage Xenopus embryo. Dev Biol 119:560–578. https://doi.org/10.1016/0012-1606(87)90059-5
9. Moody SA (1987) Fates of the blastomeres of the 32-cell-stage Xenopus embryo. Dev Biol 122:300–319. https://doi.org/10.1016/0012-1606(87)90296-X
10. Van Nieuwenhuysen T, Naert T, Tran HT, Van Imschoot G, Geurs S, Sanders E, Creytens D, Van Roy F, Vleminckx K (2015) TALEN-mediated apc mutation in Xenopus tropicalis phenocopies familial adenomatous polyposis. Oncoscience 2:555–566
11. Frock RL, Hu J, Meyers RM, Ho Y-J, Kii E, Alt FW (2014) Genome-wide detection of DNA double-stranded breaks induced by engineered nucleases. Nat Biotechnol 33:179–186. https://doi.org/10.1038/nbt.3101

12. Wang X, Wang Y, Wu X, Wang J, Wang Y, Qiu Z, Chang T, Huang H, Lin R-J, Yee J-K (2015) Unbiased detection of off-target cleavage by CRISPR-Cas9 and TALENs using integrase-defective lentiviral vectors. Nat Biotechnol 33:175–178. https://doi.org/10.1038/nbt.3127
13. Kay S, Bonas U (2009) How Xanthomonas type III effectors manipulate the host plant. Curr Opin Microbiol 12:37–43. https://doi.org/10.1016/j.mib.2008.12.006
14. Szurek B, Marois E, Bonas U, Van den Ackerveken G (2001) Eukaryotic features of the Xanthomonas type III effector AvrBs3: protein domains involved in transcriptional activation and the interaction with nuclear import receptors from pepper. Plant J 26:523–534. https://doi.org/10.1046/j.0960-7412.2001.01046.x
15. Boch J, Scholze H, Schornack S, Landgraf A, Hahn S, Kay S, Lahaye T, Nickstadt A, Bonas U (2009) Breaking the code of DNA binding specificity of TAL-type III effectors. Science (New York, NY) 326:1509–1512. https://doi.org/10.1126/science.1178811
16. Moscou MJ, Bogdanove AJ (2009) A simple cipher governs DNA recognition by TAL effectors. Science (New York, NY) 326:1501. https://doi.org/10.1126/science.1178817
17. Cong L, Zhou R, Kuo Y-C, Cunniff M, Zhang F (2012) Comprehensive interrogation of natural TALE DNA-binding modules and transcriptional repressor domains. Nat Commun 3:968. https://doi.org/10.1038/ncomms1962
18. Streubel J, Blücher C, Landgraf A, Boch J (2012) TAL effector RVD specificities and efficiencies. Nat Biotechnol 30:593–595. https://doi.org/10.1038/nbt.2304
19. Doyon Y, Vo TD, Mendel MC, Greenberg SG, Wang J, Xia DF, Miller JC, Urnov FD, Gregory PD, Holmes MC (2011) Enhancing zinc-finger-nuclease activity with improved obligate heterodimeric architectures. Nat Meth 8:74–79. https://doi.org/10.1038/nmeth.1539
20. Lin Y, Fine EJ, Zheng Z, Antico CJ, Voit RA, Porteus MH, Cradick TJ, Bao G (2014) SAPTA: a new design tool for improving TALE nuclease activity. Nucleic Acids Res 42:e47. https://doi.org/10.1093/nar/gkt1363
21. Briggs AW, Rios X, Chari R, Yang L, Zhang F, Mali P, Church GM (2012) Iterative capped assembly: rapid and scalable synthesis of repeat-module DNA such as TAL effectors from individual monomers. Nucleic Acids Res 40:e117–e117. https://doi.org/10.1093/nar/gks624
22. Reyon D, Khayter C, Regan MR, Joung JK, Sander JD (2012) Engineering designer transcription activator-like effector nucleases (TALENs) by REAL or REAL-Fast assembly. Current protocols in molecular biology/edited by Frederick M Ausubel et al Chapter 12:Unit 12.15. doi:https://doi.org/10.1002/0471142727.mb1215s100
23. Reyon D, Tsai SQ, Khayter C, Foden JA, Sander JD, Joung JK (2012) FLASH assembly of TALENs for high-throughput genome editing. Nat Biotechnol 30:460–465. https://doi.org/10.1038/nbt.2170
24. Schmid-Burgk JL, Schmidt T, Kaiser V, Höning K, Hornung V (2013) A ligation-independent cloning technique for high-throughput assembly of transcription activator–like effector genes. Nat Biotechnol 31:76–81. https://doi.org/10.1038/nbt.2460
25. Weber E, Gruetzner R, Werner S, Engler C, Marillonnet S (2011) Assembly of designer TAL effectors by golden gate cloning. PLoS One 6:e19722. https://doi.org/10.1371/journal.pone.0019722
26. Cermak T, Doyle EL, Christian M, Wang L, Zhang Y, Schmidt C, Baller JA, Somia NV, Bogdanove AJ, Voytas DF (2011) Efficient design and assembly of custom TALEN and other TAL effector-based constructs for DNA targeting. Nucleic Acids Res 39:e82. https://doi.org/10.1093/nar/gkr218
27. Doyle EL, Booher NJ, Standage DS, Voytas DF, Brendel VP, Vandyk JK, Bogdanove AJ (2012) TAL effector-nucleotide targeter (TALE-NT) 2.0: tools for TAL effector design and target prediction. Nucleic Acids Res 40:W117–W122. https://doi.org/10.1093/nar/gks608
28. Sander JD, Zaback P, Joung JK, Voytas DF, Dobbs D (2007) Zinc finger targeter (ZiFiT): an engineered zinc finger/target site design tool. Nucleic Acids Res 35:W599–W605. https://doi.org/10.1093/nar/gkm349
29. Neff KL, Argue DP, Ma AC, Lee HB, Clark KJ, Ekker SC (2013) Mojo hand, a TALEN design tool for genome editing applications. BMC Bioinformatics 14:1. https://doi.org/10.1186/1471-2105-14-1
30. Heigwer F, Kerr G, Walther N, Glaeser K, Pelz O, Breinig M, Boutros M (2013) E-TALEN: a web tool to design TALENs for genome engineering. Nucleic Acids Res 41:e190. https://doi.org/10.1093/nar/gkt789

31. Labun K, Montague TG, Gagnon JA, Thyme SB, Valen E (2016) CHOPCHOP v2: a web tool for the next generation of CRISPR genome engineering. Nucleic Acids Res 44:W272–W276. https://doi.org/10.1093/nar/gkw398
32. Montague TG, Cruz JM, Gagnon JA, Church GM, Valen E (2014) CHOPCHOP: a CRISPR/Cas9 and TALEN web tool for genome editing. Nucleic Acids Res 42:W401–W407. https://doi.org/10.1093/nar/gku410
33. Fine EJ, Cradick TJ, Zhao CL, Lin Y, Bao G (2014) An online bioinformatics tool predicts zinc finger and TALE nuclease off-target cleavage. Nucleic Acids Res 42:e42. https://doi.org/10.1093/nar/gkt1326

Chapter 5

Genotyping of CRISPR/Cas9 Genome Edited *Xenopus tropicalis*

Thomas Naert and Kris Vleminckx

Abstract

The targeted nuclease revolution (ZFN, TALEN, and CRISPR/Cas9) has led to a myriad of reports describing genotyping methodologies for genome edited founders (F0—crispants) and their offspring (F1). As such, choosing a specific genotyping methodology for your *Xenopus* CRISPR/Cas9 experiments can be challenging. In this chapter we will discuss, with emphasis on *Xenopus tropicalis* (*X. tropicalis*), different methods for assessing genome editing efficiencies within F0 CRISPR/Cas9 founders and for identification of their hetero-, compound hetero-, and homozygous mutant F1 offspring. For F0 crispants, we will provide the protocols and the respective (dis)advantages of genotyping with heteroduplex mobility assay (HMA), subclone Sanger sequencing, and sequence trace decomposition. Furthermore, we provide a previously unpublished pipe-line for rapid genotyping of F1 offspring—high resolution melting analysis (HRMA) and sequence trace decomposition—procured from breeding with F0 crispants. As such, we report here the current state-of-the-art cost- and time-effective approaches to perform genotyping of CRISPR/Cas9 experiments for the *Xenopus tropicalis* researcher.

Key words CRISPR/Cas9, *Xenopus*, *Tropicalis*, Genotyping, Genome editing, HMA, HRMA, TIDE

1 Introduction

Across the pioneering efforts of zinc-finger nucleases (ZFN), through the advent of the targeted activator-like effector nuclease (TALEN) and culminating in the recent CRISPR/Cas9 revolution, the identification of genome engineered animals by different genotyping strategies has been reported [1–3]. These include surveyor assay [4], T7 endonuclease I assay (T7E1) [5], sequence trace decomposition (e.g., TIDE) [6], fragment analysis [7], high resolution melting curve analysis (HRMA) [8], and heteroduplex mobility assay (HMA) [9], among others.

For genotyping of CRISPR/Cas9 genome edited F0 mosaic animals (crispants) we routinely employ HMA for qualitative (yes/no) assessment of genome editing, as this method has certain distinct advantages. Firstly, it does not require the purchase of

Kris Vleminckx (ed.), *Xenopus: Methods and Protocols*, Methods in Molecular Biology, vol. 1865,
https://doi.org/10.1007/978-1-4939-8784-9_5,

expensive enzymatic reagents nor purification of the PCR reaction. Secondly, it allows for detection of very small genome editing efficiencies (below 2%) while for surveyor assay, T7E1 and HRMA, respectively, detection limits of 5%, 20%, and 4.7% have been reported [8, 10, 11]. We believe that when targeted CRISPR/Cas9 injections—for tissue-restricted genome engineering—become more popular, sensitive techniques for F0 genotyping will gain in importance (*see* Chapter 3 for tissue-restricted CRISPR/Cas9 genome engineering) [12, 13]. After qualitative assessment of genome editing, we routinely perform quantitative assessment (%mutant read variants/total read) by targeted deep sequencing (*Illumina Miseq*) and BATCH-GE bio-informatics analysis (*see* **Note 1** and Chapter 6) [14]. While next-generation sequencing approaches are the optimal method of defining bona-fide quantitative genome editing efficiencies, we realize that not all labs may have routine access to such equipment. As such, for quantitative assessment of genome editing in the F0, after initial qualitative assessment, we will delineate a subcloning and Sanger sequencing protocol as well as a sequence trace decomposition methodology for rough quantification of genome editing efficiencies within the F0.

For rapid identification of hetero-, compound hetero-, and homozygous mutant F1 animals, procured from breeding with genome edited F0 crispants, we recommend a previously unpublished genotyping pipe-line based on a combination of HRMA analysis and subsequent sequence trace decomposition (TIDE or ICE analysis) of capillary electrophoresis Sanger sequencing data [15].

2 Materials

2.1 General

1. Lysis buffer: 50 mM Tris-HCl pH 8.8, 1 mM EDTA, 0.5% Tween-20, 200 μg/mL proteinase K (Bioline).
2. Phusion polymerase (ThermoFisher Scientific).
3. dNTP (10 mM each).
4. 10× Marc's Modified Ringers (MMR): 1 M NaCl, 18 mM KCl, 20 mM $CaCl_2$, 10 mM $MgCl_2$, 50 mM Hepes, pH 7.6 set with 10 M NaOH.
5. Binocular stereo microscope.
6. Agarose gel electrophoresis system and associated reagents.

2.2 Heteroduplex Mobility Assay

1. 5× Tris-borate-EDTA (TBE): 45 mM Tris-borate, 1 mM EDTA, pH 8.3 set with 10 M NaOH.
2. Polyacrylamide gel electrophoresis system (e.g., Mini-PROTEAN Tetra Cell).

3. 10% Ammonium persulfate (APS) solution.
4. Tetramethylethylenediamine (TEMED).
5. Acrylamide/Bis-Acrylamide 37.5:1, 40% Molecular biology.
6. DNA Gel Loading Dye (ThermoFisher Scientific).
7. DNA marker (e.g., BenchTop 1 kb DNA Ladder (Promega)).
8. GelRed 10,000× in water (Biotium).

2.3 Subcloning and Capillary Sequencing

1. NEB PCR cloning kit (NEB).
2. GoTaq® G2 Master Mixes (Promega).
3. NucleoSpin Plasmid (Machery-Nagel).
4. Standard liquid Lysogeny Broth (LB) medium; supplemented with carbenicillin.
5. Standard LB agar plates; supplemented with carbenicillin.

2.4 Obtaining DNA from F1 Tadpole

1. Scalpels.
2. Gentamycin (liquid) (Gibco).
3. Ethyl 3-aminobenzoate methanesulfonate (MS-222) (Sigma-Aldrich). 1% (w/v) stock solution (100×) aliquoted per 500 μL in 1.5 mL eppendorf tubes. Store at −20 °C and use immediately upon thawing. Dilute to 1× working solution for sedation of tadpoles.
4. Hard-Shell 96-Well Semi-Skirted PCR Plates, High-Profile (Bio-Rad).
5. 24-well plates.

2.5 High Resolution Melting Analysis

1. LCGreen Plus+ Melting Dye (Idaho Technology Inc.).
2. LightCycler 480 Instrument (Roche).
3. LightCycler 480 Multiwell plate 96, White (Roche).
4. LightCycler 480 release 1.5.0 SP3 software package (Roche).

3 Methods

3.1 Genotyping of F0 *X. tropicalis* Crispants

3.1.1 Obtaining F0 Genomic DNA by Embryo Lysis

Embryos can be lysed as early as 24 h post-injection, however due to practical constraints (yolk) optimal lysis is around stage 33–36 (*see* **Note 2**). In order to average out possible fluctuations in genome editing efficiency between different tadpoles within the same setup, we perform genotyping on pools of three embryos.

1. Transfer three embryos from your experimental setup (CRISPR/Cas9 microinjected) to 100 μL lysis buffer. Generate at least two of these sample preparations. The number of embryos to lyse will depend on the number injected and the

survival rate. Ideally, you generate three sample preparations containing three embryos each.

2. Transfer three wild type embryos (from the same clutch as the embryos lysed in point 1) to 100 μL lysis buffer. Generate two of these sample preparations (negative control).
3. Perform embryo lysis overnight at 55 °C on a thermoshaker with gentle agitation (around 500 rpm) in order to obtain DNA for downstream analysis.
4. The next day, boil all the samples for 5 min at 99 °C within a thermoshaker to inactivate proteinase K within the lysis buffer.
5. Spin down at 14,000 × *g* in a benchtop centrifuge for 1 min to pellet remaining cellular debris and generate a transparent supernatant for input in downstream experiments. If a layer of yolk forms on the top of this supernatant, *see* **Note 2**.

3.1.2 Qualitative Genotyping of F0 Crispants by Heteroduplex Mobility Assay (HMA)

Heteroduplex mobility assay (HMA) hinges on the phenomenon that the cells of the mosaic mutant F0 crispant will exhibit either wild type (WT) or small insertion-deletion (INDEL) variants, in one or both of the alleles targeted by the CRISPR/Cas9 system. After PCR amplifying the CRISPR/Cas9 genomic target site, the PCR products are boiled to denature and subsequently a slow cooling is performed. This allows different INDEL-containing DNA fragments to anneal, even if they are not completely identical, thus forming heteroduplex DNA in addition to homoduplex DNA [16]. The mixture is subsequently separated on a standard polyacrylamide gel electrophoresis (PAGE) system. By this methodology, multiple heteroduplex bands can be detected in CRISPR/Cas9 genome edited samples when compared to WT samples due their specific differential migration patterns (*see* Fig. 1). As such, HMA is a fast and short hands-on time method for qualitative assessment (**yes**/no) of CRISPR/Cas9 genome editing. Additionally, most of these reagents should be available in a standard molecular biology laboratory setting abolishing the need to buy more expensive commercial solutions such as T7E1 assay kits.

1. Design primers to amplify a region of 400–600 bp around the CRISPR/Cas9 target site (*see* **Note 3** for details on primer design). Make sure to keep your gRNA target site located in the relative center of your PCR amplicon.
2. Quality control designed primers by performing a standard proof-reading PCR on WT DNA and separate the resulting product by 1% agarose gel electrophoresis. Confirm presence of on-target fragment and absence of any off-target fragments. This step is quintessential as aspecific products can interfere with correct interpretation of your HMA gel and can render a false negative result.

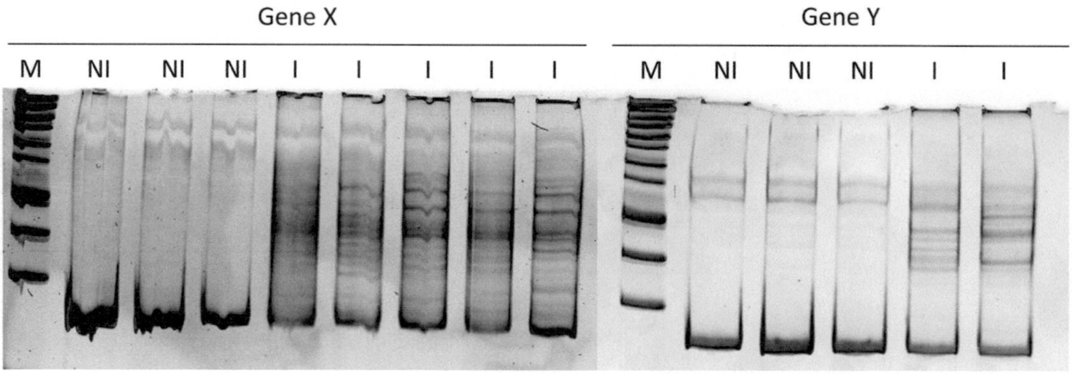

Fig. 1 Heteroduplex mobility assay (HMA) demonstrates in a qualitative manner genome editing by CRISPR/Cas9. When comparing the "not injected" (NI) and "CRISPR/Cas9 injected" (I) samples, extra heteroduplex bands can be observed in the injected samples. As such, successful qualitative (Yes/No) CRISPR/Cas9 genome editing for both gene X and gene Y can be concluded. Two gels have been digitally cut and pasted together for this representative image

3. Perform HMA PCR. We recommend taking at least two samples for reference negative control lanes (WT DNA—*see* **step 2** in Subheading 3.1.1) and at least three test samples (DNA from crispants—*see* **step 1** in Subheading 3.1.1) as this simplifies interpretation of the resulting data. Run following PCR program with a proof-reading polymerase (e.g., Phusion polymerase) in a thermocycler: 98 °C for 3′; 35 cycles of [98 °C for 10″; x °C for 30″; 72 °C for 15″]; 72 °C for 4′; 98 °C for 5′; controlled cooldown to 4 °C at a ramp rate of 1 °C/s. The annealing temperature × is evidently dependent on the designed primer pair.
4. Generate a 8% TBE-buffered polyacrylamide gel on a PAGE casting system in following proportions:

1× TBE buffer (prepared from 5× stock)	12 mL
40% Acrylamide/Bisacrylamide	3 mL
10% APS	140 μL
TEMED	14 μL

5. Wait for the gel to polymerize within the casting system and subsequently transfer it to the PAGE running system according to the instructions of the manufacturer. Fill up the tank with 1× TBE running buffer.
6. Mix the HMA PCR products with DNA loading dye and load these on the gel alongside a DNA marker.
7. Run the TBE-buffered polyacrylamide gel for 3–4 h at 50–60 V.

8. Post-stain the gel with GelRed at working stock diluted in purified water and visualize gel by appropriate methods (*see* **Note 4** for alternative gel staining procedures).
9. When CRISPR/Cas9 genome editing was successful one should clearly observe the presence of additional heteroduplex bands in the injected samples when compared to the non-injected samples. These heteroduplex bands are indicative of effective CRISPR/Cas9 genome editing. A representative gel image for a qualitative check of successful multiplex genome editing in gene X and gene Y is shown in Fig. 1.

3.1.3 Quantitative Genotyping of F0 Crispants

In absence of access to next-generation sequencing equipment (*see* Chapter 6), it might be necessary to obtain a rough quantitative estimation of CRISPR/Cas9 genome editing efficiencies within F0 crispants. Namely, this information will be quintessential to make a decision whether to grow up your F0 crispants and attempt germline transmission of genome edited alleles. Evidently, the higher the efficiency of genome editing within the F0 crispant, the higher the chance of germline transmission. Quantitative estimation of CRISPR/Cas9 genome editing efficiencies can be performed by either subcloning and Sanger sequencing or Sanger sequencing followed by sequence trace decomposition. Both methodologies are described below with their respective (dis)advantages. We recommend to employ the sequence decomposition methodology initially, and employ the subcloning protocol as a validation or a backup methodology if the prior methodology fails to render the required data.

Methodology 1: Subcloning and Sanger Sequencing

Subcloning with commercial cloning kits is a rather expensive and time-consuming method of assessing CRISPR/Cas9 genome editing efficiencies. Additionally, It has been reported that low efficiencies of genome editing (<25%) will not yield reliable genotyping results with this methodology [17]. As such, this is a low sensitivity genotyping method but it could be argued that, considering the fact that low genome editing efficiencies will give rise to lower rates of germline transmission, this is actually a non-issue. Described in brief, subcloning of the CRISPR/Cas9 genome target site, picked up by PCR amplification, within a vector backbone will give rise to vectors containing either wild type sequences or CRISPR/Cas9 genome edited INDEL-containing sequences in a ratio that is indicative of the efficiency of the CRISPR/Cas9 genome editing within your F0 crispants.

1. Using the primers designed and validated earlier (*see* **steps 1** and **2** in Subheading 3.1.2), perform a standard proof-reading PCR amplifying the CRISPR/Cas9 genome target site from the test samples (DNA from crispants—*see* **step 1** in Subheading 3.1.1).

2. Employ a commercial subcloning kit (e.g., NEB PCR Cloning kit) to clone the resulting PCR amplicons within a vector backbone per manufacturer's instructions. Obtain colonies on a selective plate the next day.

3. Perform a colony PCR using GoTaq G2 DNA Polymerase to identify colonies that correctly obtained inserts. Generate GoTaq G2 master-mix, per manufacturer's instructions, employing the forward and reverse sequencing primers included in the NEB® PCR Cloning Kit (or employ respective primers flanking the insert for your vector). For DNA input, pick up a colony with a sterile 10 μL pipette tip, streak on a selective antibiotic containing replica plate and directly add this tip to a PCR tube containing colony PCR master mix (Fig. 2a) [18]. Remove tips after short (1 min) incubation, close the PCR strip, and perform PCR reaction under following conditions: 95 °C for 5′; 35 cycles of [95 °C for 1′; 55 °C

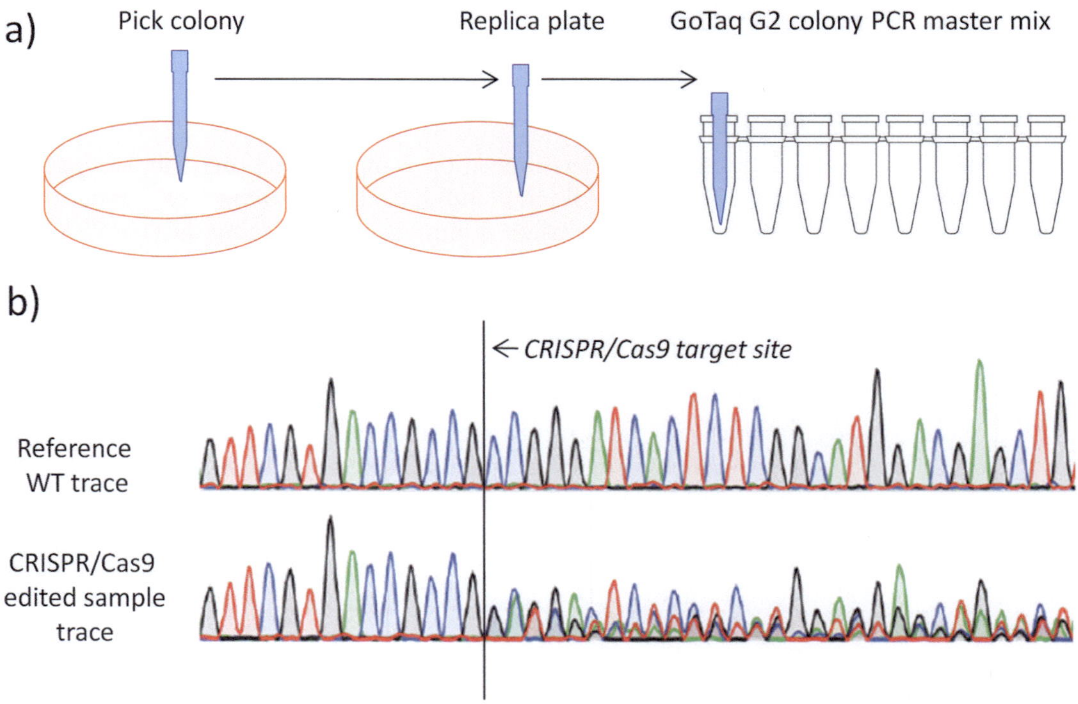

Fig. 2 Correct way of assembling a colony PCR reaction and representative Sanger sequence trace for sequence trace decomposition. (**a**) In order to correctly assemble a colony PCR reaction, employ a sterile 10 μL pipette tip to pick up a colony, subsequently use this tip to streak on a replica plate. Finally, transfer this pipette tip to the colony PCR reaction mixture and let it incubate for at least 1 min. (**b**) Representative Sanger sequencing trace files comparing a WT reference trace and a CRISPR/Cas9 edited sample trace. Note the exact matching of the sequence traces upstream to the CRISPR/Cas9 target site. Downstream of this target site, however, CRISPR/Cas9 edited sample trace shows mixed base calls. Sequence trace decomposition allows to deconvolute this into a quantitative estimation of CRISPR/Cas9 genome editing efficiencies with associated allelic variant calls

for 30″; 72 °C for 2′]; 72 °C for 10′; hold at 12 °C. Incubate the plate containing the reference streaks overnight at 37 °C.

4. Separate the resulting colony PCR products by 1% agarose gel electrophoresis. Identify colonies containing the correct insert by comparing the length of the PCR amplicon on the gel to the expected theoretical length. Note that larger deletions or insertions may shift these PCR amplicons slightly on the gel.
5. Next day, pick up a correct bacterial streak (containing an insert) from the replica plate and start an overnight liquid culture (4 mL of selective LB medium) with this clone. Set up at least six of these liquid colonies for different insert-containing clones.
6. Next day, prepare plasmid DNA from the liquid cultures (e.g., NucleoSpin Plasmid) and subsequently submit each insert-containing vector for Sanger sequencing with the correct sequencing primer (e.g., NEB forward primer).
7. Align the resulting Sanger sequencing reads with the *X. tropicalis* reference genome and calculate the ratio INDEL containing reads over wild type reads as the estimation of your CRISPR/Cas9 genome editing efficiencies.

Methodology 2: Direct Sanger Sequencing of PCR Amplicons and Sequence Trace Decomposition

Sequence trace decomposition is more novel than employing subcloning methodologies and has the advantage of being much cheaper (no expensive commercial subcloning kits) and much faster (no cloning and associated bacterial overnight incubations). In essence, the CRISPR/Cas9 genome target region is amplified and Sanger sequenced for both a control and a CRISPR/Cas9 edited sample. By designing the PCR so that the projected CRISPR/Cas9 target site is 200–300 bp from the sequencing primer, the resulting Sanger trace consists of a read in which the base pairs upstream to the target site are of high Phred quality and identical between the reference (WT) sample and the CRISPR/Cas9 edited sample. However, downstream of the CRISPR/Cas9 target site bases will be mixed in a CRISPR/Cas9 edited population (Fig. 2b). Computationally, by comparing the reference (WT) trace to this mixed trace, deconvolution of the CRISPR/Cas9 edited trace can be performed resulting in an relatively accurate estimation of CRISPR/Cas9 genome editing efficiencies with output of specific frequencies of different INDEL variants. TIDE analysis has been accepted as an alternative for targeted next-generation deep sequencing as an estimation of genome editing efficiencies [13].

1. Design primers to amplify a fragment between 600 and 1000 bp containing your CRISPR/Cas9 genome target site. The projected mutation site should be located ~200 bp downstream of the forward primer binding site. If encountering

problems downstream in this protocol (e.g., bad quality sequence traces disallowing correct decomposition), redesign primers or employ a nested PCR approach to generate a cleaner amplicon (*see* **Note 5**).

2. Perform a normal proof-reading 20 μL PCR reaction with these primers. For a reference WT sequence trace employ WT input DNA (*see* **step 2** in Subheading 3.1.1). For each test sample use DNA input from crispants (*see* **step 1** in Subheading 3.1.1).
3. Verify, by using a fraction of each PCR product (5 μL), that the correct fragment is amplified by performing an 1% agarose gel electrophoresis.
4. Send (a fraction of) the remaining PCR product for Sanger sequencing with the forward primer (~200 bp upstream of mutation site) per sequencing facility instructions.
5. Obtain genotyping results by running the TIDE analysis (https://tide-calculator.nki.nl/) according to instructions on the website. Make sure to use advanced settings to increase the indel size range to at least 25 bp in order not to underestimate the genome editing efficiencies. Alternatively, employ the recently published Inference of CRISPR Edits (ICE) (https://ice.synthego.com/#/) tool which is reported to be an upgraded version of the TIDE algorithm, allowing for batch-analysis with less necessity for user-defined, advanced settings, input [15].

3.2 Genotyping and Identification of F1 Hetero-, Compound Hetero- and Homozygote Mutant Animals Procured from Breeding with F0 Crispants

The following part of this chapter will be fully dedicated to the optimal methodology of identifying F1 offspring after inter- or outcrossing F0 mosaic crispants. We will describe the protocol for recovering hetero-, compound hetero-, or homozygous mutant F1 animals and for obtaining correct allelic calls for each recovered mutant in a cost-effective manner. Obtaining correct allelic calls is of importance in order to identify and remove F1 animals bearing in-frame mutant alleles from further analysis. This method is based on non-lethal harvesting of DNA from stage 43 to 46 tadpoles followed by direct genotyping by high resolution melting analysis (HRMA) methodology and subsequent identification of germline transmitted mutant alleles by sequence trace decomposition of Sanger sequencing reads. This pipe-line is schematically represented in Fig. 3.

This protocol can also be applied to genotype post-metamorphic F1 frogs, in which case you cut toes in a pattern that allows identification of the animal post-genotyping and follow the methodology below (excluding Subheading 3.2.2) to identify animals carrying your mutations of interest.

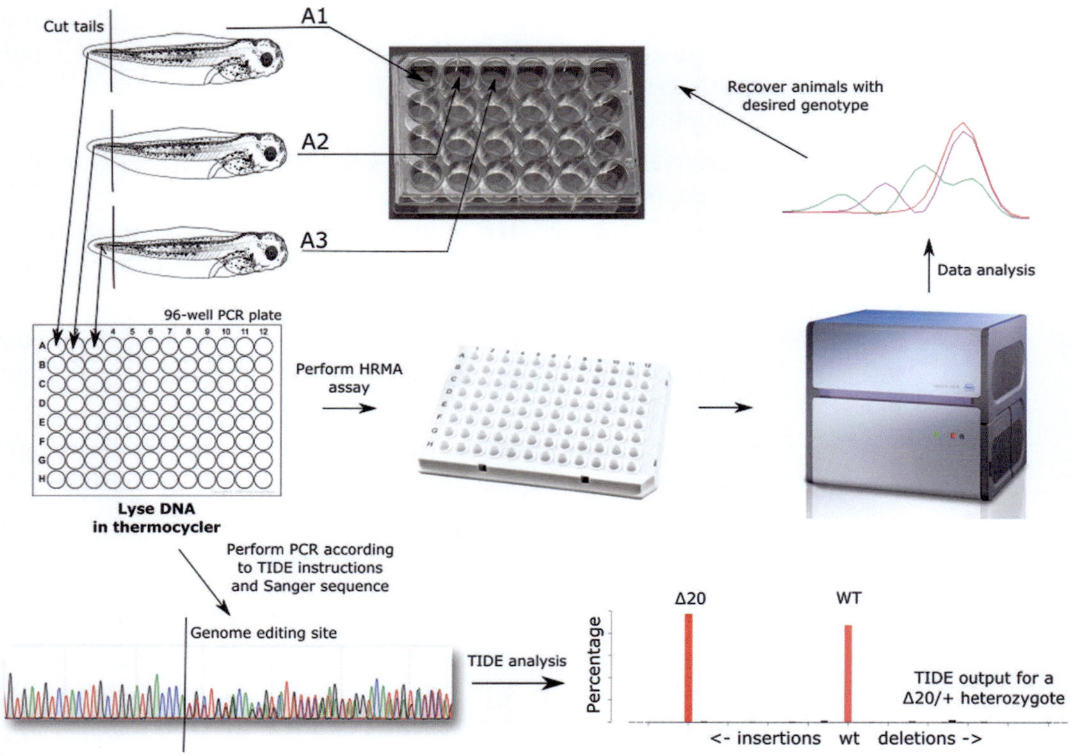

Fig. 3 Schematic representation workflow: genotyping and identification of F1 hetero-compound hetero- and homozygote mutant animals procured from breeding with F0 crispants. For each tadpole to be genotyped, a small piece of tail is isolated. This piece is subsequently lysed within a 96-well PCR plate and as such assigned a coordinate (A1, A2, etc.). The corresponding tadpole is kept in a 24-well plates under the same coordinates. HRMA analysis is performed in order to identify those tadpoles carrying mutant alleles and these can then be recovered from the 24-well plate. In order to achieve accurate allelic variant calls for each mutant tadpole, TIDE analysis can be performed. PCR amplification is performed on DNA from the 96-well lysis plate on those animals that the HRMA demonstrated mutant allele presence. Subsequent Sanger sequencing and comparison of the resulting sequence file to a wild type reference trace allows exact identification of allelic variants (***see*** Subheading 3.1.3)

3.2.1 Setting up the High Resolution Melting Analysis (HRMA)

1. Design two to three primer pairs spanning the CRISPR/Cas9 genome target site, with the cut site optimally situated in the middle of the amplicon, thus generating a 80–120 bp amplicon (optimal size is 100 bp).
2. Set up a test HRMA run within a LightCycler 480 apparatus to validate recovery of a single PCR fragment exhibiting a clean melting curve profile.

   Mix following HRMA reaction mix (10.5 μL per reaction) using a wild type DNA preparation (*see* **step 2** in Subheading 3.1.1).

Ultrapure water	6.2 μL
5× HF buffer	2 μL
dNTP (10 mM each)	0.2 μL
LCGreen+ melting dye	1 μL
Primer mix (10 μM each)	0.5 μL
Wild type DNA	0.5 μL
Phusion polymerase	0.1 μL

Pipette 10.5 μL of this master mix in the well of a LightCycler 480 96-well multiwell plate. Cover the plate with the provided foil, perform a short spin to collect fluid at bottom of wells, and run following reaction on the LightCycler 480 instrument:

Detection format: SYBR Green I/HRM Dye. Cycle parameters: 98 °C for 3′; 35 cycles of [98 °C for 10″; x °C for 30″; 72 °C for 15″] in quantification analysis mode; 72 °C for 5′; 98 °C for 5′; 37 °C for 1″ and finally go from 37 °C to 98 °C at a ramp rate of 0.04 °C/s (25 acquisitions per °C) in melting curve analysis mode.

3. Data is analyzed by employing the LightCycler 480 release 1.5.0 SP3 software package. To perform this analysis, load the run data (.ixo file) by "new experiment" -> "import." Then click on the "analysis" button and select the "melt curve analysis" option. Create a new analysis for your run samples. Subsequently, use the "settings" tab to set the temperature range to span the observed melting temperature of your WT sample by minus and plus five (5)°C. Press calculate to allow the software to calculate the melting curves. Only one clean peak should be observed for any primer pair you wish to continue using further for HRMA.

3.2.2 Obtaining F1 Genomic DNA by Tail Lysis

1. Anaesthetize stage 43–46 tadpoles by immersion in a solution of 0.01% MS-222.
2. Transfer sedated tadpole to a petri-plate containing 0.1× MMR, 0.01% MS-222, and 10 μg/mL gentamycin.
3. Under a binocular stereomicroscope cut a very small piece of tail with a scalpel as described before in tail regeneration assays [19]. Make sure to clean this scalpel between animals by wiping with a paper tissue to avoid cross contaminating your samples.
4. Spot on the clean lid of a petri dish a drop of 30 μL lysis solution with a 50 μL pipette. Transfer, under the stereomicroscope, with a 50 μL pipette (clean tip and set at 25 μL volume) the cut tail to the 30 μL drop taking care to transfer as little 0.1× MMR as possible to the lysis buffer. Rapidly (in order to

avoid attachment of tail piece to the plastic of the tip) pipette out the tail piece within the 30 μL drop of lysis solution.

5. Pipette, under the stereomicroscope, up and down to generously mix the inevitably transferred 0.1× MMR buffer with the lysis solution. Finally, in a fluent stroke pipette up the tail and 25 μL lysis buffer and subsequently transfer this to a 96-well hard PCR plate.
6. Transfer the tadpole to a 24-well plate filled with 0.1× MMR (supplemented with 10 μg/mL gentamycin) clearly marked with the coordinates (A1, A2, etc.) of where you pipetted its respective tail in the 96-well PCR plate. Repeat until satisfied with number of embryos or until 90 tails have been cut (*see* Fig. 3).
7. Cover the 96-well PCR plate with a suitable thermo-stable lid. Short spin the 96-well PCR plate to collect all fluid at bottom of wells and place the 96-well PCR plate in a thermocycler at following program: 55 °C for 1 h; 98 °C for 10′; hold at 12 °C forever. After this incubation short spin the 96-well PCR plate to collect all fluid at bottom of the wells.

3.2.3 Identification of F1 Hetero-, Compound Hetero-, and Homozygote Mutant Animals by High Resolution Melting Analysis (HRMA)

1. Set up a 10 μL HRMA reaction mix (*omitting the WT DNA*) as described above (*see* **step 2** in Subheading 3.2.1) with primers tested to generate one clear peak on WT DNA (*see* **step 3** of Subheading 3.2.1). Transfer 10 μL of HRMA reaction mix to each well of a 96-well LightCycler plate.
2. Transfer 0.5 μL from each well of the 96-well PCR plate, containing F1 genomic DNA (*see* Subheading 3.2.2), to the 96-well LightCycler plate (A1 to A1, A2 to A2, etc.) (*see* **Note 6**). Furthermore, load 3 wells of the HRMA plate with wild type negative control DNA (*see* **step 2** in Subheading 3.1.1). Subsequently, if possible, load 3 wells with positive control DNA you know contains INDELs within CRISPR/Cas9 genome target region (from the parent F0 crispants employed for the breeding that resulted in this F1 clutch) as a positive control.
3. Cover the plate with the provided foil, perform a short spin to collect fluid at bottom of wells, and run the HRMA assay in a LightCycler 480 at the program validated before (*see* **step 2** of Subheading 3.2.1).
4. Run melting curve analysis as performed before (*see* **step 3** in Subheading 3.2.1). The WT samples within the F1 will group together with your wild type negative control DNA. This peak Tm can be considered the WT DNA melting peak (red peaks in Fig. 4) (*see* **Note 7**). Mutant samples will show a differential melting pattern when compared to the WT sample, this should be apparent when comparing your positive control melting

curves with the WT DNA melting curve. Heterozygous mutant animals will display additional melting peaks (black arrows) due to melting of different heteroduplexes formed during the controlled cool down step, while still retaining a peak at the exact Tm of the wild type sample (orange arrow) (Fig. 4b). Additionally, identification of homo- and compound heterozygous F1 mutants is possible. Namely, these samples show absence of a peak at the location of the wild type melting peak and one or more peaks are observed on the left (deletions) or right (insertions) of the WT peak (Fig. 4c). These represent differential melting of the specific heteroduplexes formed in compound heterozygous mutant F1 animals or either the peak of the differential melting of a homoduplex in a fully homozygous mutant F1 animal. Interestingly, we have observed that different homo- or heterozygous mutant F1 animals can cluster together based on the similarity of their melting curve profile. We have seen (unpublished data) that these clusters, ultimately, are animals of the exact same genotype (same INDEL mutation(s)). As such, the cost of downstream sequencing and TIDE-analysis (*see* below) can theoretically be reduced by only analyzing one or two animals from each cluster and projecting the resulting genotype within the cluster.

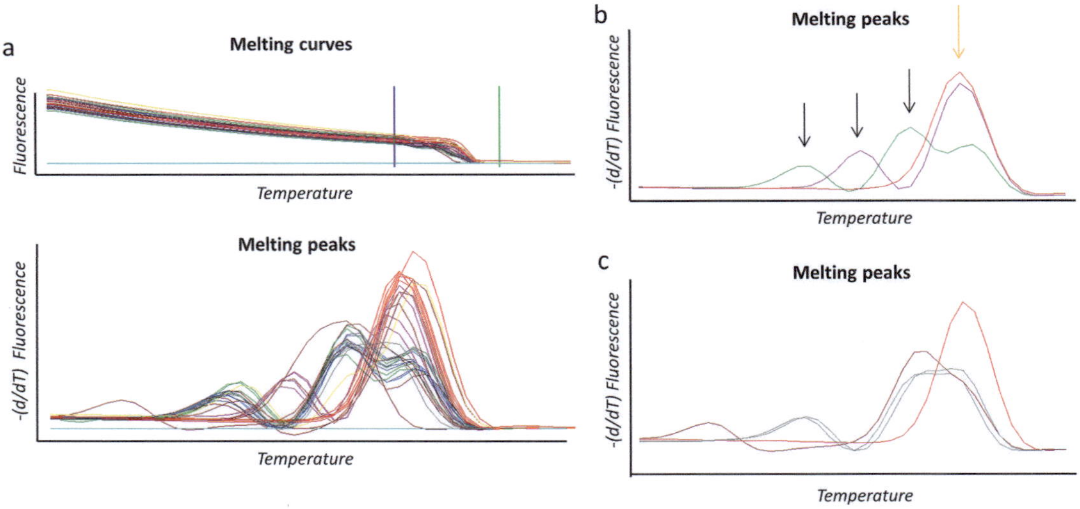

Fig. 4 Melt curve genotype analysis for F1 procured from intercrossing two F0 mosaic founders. (**a**) Melting curve analysis for the analyzed samples. (**b**) Melting curve comparison between wild type (red curve) and two heterozygous mutants. Note that for all the heterozygotes at least one melting temperature of a peak maxima overlaps with the melting temperature of a peak maxima of the wild type (at location of orange arrow). In contrast, additional peaks can be observed for the heterozygotes (at locations of black arrows) but not for the wild type. (**c**) Melting curve pattern comparison between wild type (red) and compound heterozygote mutants. Note that none of the melting temperature of the peak maxima of the compound heterozygotes overlap with the melting temperature of the peak maxima of the wild type. Namely, there is no longer a wild type homoduplex formed within compound heterozygous samples. All genotype calls reported in **b** and **c** were subsequently validated on sequence level by targeted next-generation sequencing and BATCH-GE analysis (data not shown)

5. After separation of wild type animals from mutant animals by HRMA you can employ Sanger sequencing and sequence trace decomposition (see methodology 2 in Subheading 3.1.3) to generate accurate allelic variant calls for each F1 homo- or heterozygous mutant animal or cluster of mutant animals. Sequence trace decomposition works especially well on this as the decomposition of the homo-, hetero-, or compound heterozygous sequence reads is rather straightforward compared to the complexity of an F0 mosaic crispant.

4 Notes

1. For readers interested in setting up a fully quantitative assessment of genome editing, the NGS methodology and subsequent BATCH-GE analysis is described in detail in Chapter 6.
2. When lysing early stage embryos, there will be a visually apparent layer of yolk present on top of the aqueous phase containing your genomic DNA of interest. Beware not to transfer this into downstream reactions (PCRs, etc.), as it will interfere with these. Fast pipetting can ensure clean transfer of DNA-containing aqueous phase while yolk attached on the outside of the pipette tip should be removed by wiping with a Tork paper.
3. When encountering difficulties to design in silico primers for amplification of your CRISPR/Cas9 target sequence, it is possible to go as high as an amplicon size of 900 bp. Beware, however, that when running your HMA, the homoduplexes and heteroduplexes will need longer to separate and thus respect the rule that: the larger your amplicon, the longer your HMA will have to run in order to give unambiguous data (can be up to 5 h for large amplicons). Designing primers amplifying fragments smaller than 300 bp are also not recommended considering that larger deletions might be excluded from being picked up by PCR and as such an underestimation of the genome editing efficiencies within your setup can occur.
4. Evidently, alternative post-stain methodologies can be employed to stain your HMA gel. Toxicological concerns within our department have led to the establishment of this protocol using GelRed. However post-staining with EtBr at normal working concentration is also an option. We would however like to point out that some other stains (e.g., Midori Green) did not render good results for DNA staining in polyacrylamide gels in our hands.
5. We have noted that sometimes it can be a bit of a hassle to get the sequence decomposition assay working due to bad quality

sequence traces as a consequence of sending in unpurified DNA amplicons for Sanger sequencing. This is, however, just a question of persistence in finding a good primer pair for amplifying the CRISPR/Cas9 genome target region, or designing a nested PCR approach in order to generate an on-target amplicon uncontaminated by off-target amplicons.

6. It is ergonomically advised, if available, to employ a 0.5 μL capable multi-channel pipette.
7. Beware not to trust the auto-clustering algorithm of the LightCycler Software in melting curve analysis mode, make sure to investigate each peak manually for shifts in order to avoid wrong genotyping calls.

Acknowledgments

The authors would like to thank Dr. Tom Van Nieuwenhuysen for the initial establishment of the HMA assay and the subcloning/Sanger sequencing methodologies within the research unit. Furthermore, the authors would like to thank Lana Hellebaut for critical proofreading of this chapter. Research in the authors' laboratory is supported by the Research Foundation – Flanders (FWO-Vlaanderen) (grants G0A1515N and G029413 N), by the Belgian Science Policy (Interuniversity Attraction Poles - IAP7/07) and by the Concerted Research Actions from Ghent University (BOF15/GOA/011). Further support was obtained by the Hercules Foundation, Flanders (grant AUGE/11/14) and the Desmoid Tumor Research Foundation.

References

1. Young JJ, Cherone JM, Doyon Y et al (2011) Efficient targeted gene disruption in the soma and germ line of the frog Xenopus tropicalis using engineered zinc-finger nucleases. Proc Natl Acad Sci U S A 108:7052–7057. https://doi.org/10.1073/pnas.1102030108
2. Lei Y, Guo X, Liu Y et al (2012) Efficient targeted gene disruption in Xenopus embryos using engineered transcription activator-like effector nucleases (TALENs). Proc Natl Acad Sci U S A 109:17484–17489. https://doi.org/10.1073/pnas.1215421109
3. Nakayama T, Fish MB, Fisher M et al (2013) Simple and efficient CRISPR/Cas9-mediated targeted mutagenesis in Xenopus tropicalis. Genesis 51:835–843. https://doi.org/10.1002/dvg.22720
4. Qiu P, Shandilya H, D'Alessio JM et al (2004) Mutation detection using surveyor nuclease. BioTechniques 36:702–707
5. Babon JJ, McKenzie M, Cotton RGH (2003) The use of Resolvases T4 endonuclease VII and T7 endonuclease I in mutation detection. Mol Biotechnol 23:73–82. https://doi.org/10.1385/MB:23:1:73
6. Brinkman EK, Chen T, Amendola M, van Steensel B (2014) Easy quantitative assessment of genome editing by sequence trace decomposition. Nucleic Acids Res 42:e168–e168. https://doi.org/10.1093/nar/gku936
7. Bhattacharya D, Marfo CA, Li D et al (2015) CRISPR/Cas9: an inexpensive, efficient loss of function tool to screen human disease genes in Xenopus. Dev Biol 408:196–204. https://doi.org/10.1016/j.ydbio.2015.11.003

8. Thomas HR, Percival SM, Yoder BK et al (2014) High-throughput genome editing and Phenotyping facilitated by high resolution melting curve analysis. PLoS One 9:e114632. https://doi.org/10.1371/journal.pone.0114632
9. Upchurch DA, Shankarappa R, Mullins JI (2000) Position and degree of mismatches and the mobility of DNA heteroduplexes. Nucleic Acids Res 28:E69
10. Vouillot L, Thélie A, Pollet N (2015) Comparison of T7E1 and surveyor mismatch cleavage assays to detect mutations triggered by engineered nucleases. G3 (Bethesda) 5:407–415. https://doi.org/10.1534/g3.114.015834
11. Zhu X, Xu Y, Yu S et al (2015) An efficient genotyping method for genome-modified animals and human cells generated with CRISPR/Cas9 system. Sci Rep 4:6420. https://doi.org/10.1038/srep06420
12. Naert T, Van Nieuwenhuysen T, Vleminckx K (2017) TALENs and CRISPR/Cas9 fuel genetically engineered clinically relevant Xenopus tropicalis tumor models. Genesis 55. https://doi.org/10.1002/dvg.23005
13. DeLay BD, Corkins ME, Hanania HL et al (2018) Tissue-specific gene inactivation in *Xenopus laevis*: knockout of *lhx1* in the kidney with CRISPR/Cas9. Genetics 208(2):673–686. https://doi.org/10.1534/genetics.117.300468
14. Boel A, Steyaert W, De Rocker N et al (2016) BATCH-GE: Batch analysis of next-generation sequencing data for genome editing assessment. Sci Rep 6:30330. https://doi.org/10.1038/srep30330
15. Hsiau T, Maures T, Waite K et al (2018) Inference of CRISPR edits from sanger trace data. bioRxiv. https://doi.org/10.1101/251082
16. Ota S, Hisano Y, Muraki M et al (2013) Efficient identification of TALEN-mediated genome modifications using heteroduplex mobility assays. Genes Cells 18:450–458. https://doi.org/10.1111/gtc.12050
17. Nakayama T, Blitz IL, Fish MB et al (2014) Cas9-based genome editing in Xenopus tropicalis. Methods Enzymol 546:355–375. https://doi.org/10.1016/B978-0-12-801185-0.00017-9
18. Bergkessel M, Guthrie C (2013) Colony PCR. Methods Enzymol 529:299–309
19. Lin G, Slack JMW (2008) Requirement for Wnt and FGF signaling in Xenopus tadpole tail regeneration. Dev Biol 316:323–335. https://doi.org/10.1016/j.ydbio.2008.01.032

Chapter 6

BATCH-GE: Analysis of NGS Data for Genome Editing Assessment

Wouter Steyaert, Annekatrien Boel, Paul Coucke, and Andy Willaert

Abstract

Due to its simple nature, the clustered regularly interspaced short palindromic repeats (CRISPR)—Cas9 technique is massively used nowadays to modify genomic loci in a wide range of model systems. The possibility to interrogate gene function on a genome-wide scale is revolutionizing fundamental life sciences and will lead to new clinical breakthroughs. Its strength is even more pronounced when it is used in tandem with next-generation sequencing (NGS). The high throughput and low cost cause NGS to be the method of choice for exploring CRISPR-Cas9 experimental results. To analyze the NGS reads from genome editing experiments only few bioinformatics tools are available. BATCH-GE is a flexible and easy-to-use tool, which is especially useful for dealing with large amounts of data. It detects and reports indel mutations and other precise genome editing events and calculates the corresponding mutagenesis efficiencies for a large number of samples in parallel.

Key words CRISPR-Cas9, NGS, Linux, Cutsite, Data-analysis, Mutagenesis efficiency, HDR

1 Introduction

In order to determine genome editing efficiency in CRISPR/Cas9 or TALEN experiments, several methods are available where the targeted region in the genome is PCR amplified and the amplicon is subjected to assays such as T7 endonuclease I (T7E1), heteroduplex mobility assay (HMA), and the Surveyor's assay, in order to evaluate the cutting efficiency. However, these assays do not reveal details on the actual range of mutations (insertions, deletions, substitutions) that are present in the different cells that have been targeted. To this extent, the PCR amplicon can be ligated in a recipient vector and transformed in bacteria. Several individual bacterial clones can then be analyzed via Sanger sequencing of the plasmid insert to give an idea of the modifications that were introduced. As an alternative, in order to get a more detailed picture of the genome modifications present in the organism, the PCR amplicon covering the targeted region can be subjected to next-generation

Kris Vleminckx (ed.), *Xenopus: Methods and Protocols*, Methods in Molecular Biology, vol. 1865,
https://doi.org/10.1007/978-1-4939-8784-9_6, © Springer Science+Business Media, LLC, part of Springer Nature 2018

sequencing (NGS). This can be done by spiking the amplicon in a routine diagnostic NGS run at a local genetic center. To analyze the data, we developed BATCH-GE, a tool that provides detailed analysis of genome editing experiments and supports batch analysis of multiple samples [1, 2]. The tool is implemented as a freely available Perl script and can be run on any Linux-based server or personal computer, ensuring easy accessibility. A detailed overview of the type, length, and frequency of the generated indel mutations is generated. In addition, BATCH-GE enables assessment of homology-directed repair (HDR)-mediated precise genome editing experiments by returning efficiencies for one or multiple intended base pair substitutions. Multiple genome editing experiments can be analyzed in a batch-wise manner with limited and simple input requirements. BATCH-GE does not only aid in the analysis of CRISPR/Cas9-based experiments, but can assess the outcome of any genome editing experiment.

As a first step, DNA derived from a genome editing experiment is amplified via singleplex PCR for the corresponding targeted region. Multiple PCR products covering different genomic regions in one specific or different genomes can be pooled together for DNA library preparation and NGS (Fig. 1). Following NGS, BATCH-GE directly uses raw NGS data (FASTQ file format) as input to generate a detailed report on genome editing efficiencies. We here provide a detailed protocol for downloading and installing the BATCH-GE software and for performing analysis of CRISPR/Cas9 or TALEN experiments in *Xenopus tropicalis*.

2 Installation Guidelines

2.1 Setting Up Your Operating System

The BATCH-GE software is developed for Linux systems. We advise Windows and Mac users to set up a virtual Linux environment. Oracle VM VirtualBox is a free and open-source virtualizer that can be used for this purpose. After downloading and installing VirtualBox you can create a new virtual machine with one of the many freely available Linux images as a boot disk.

Fig. 1 (continued) panel, E (Experiment.csv) and C (Cutsites.bed) icons), supplying experimental specifications to the tool. Individual sample analysis is initiated by data conversion into the SAM file format. Next, the reads in the SAM file are screened for their coverage of the region(s) of interest, which are user-defined regions supplied via the input files, encompassing the theoretical CRISPR/Cas9 cut site (middle panel, red sequence). Reads meeting this criterion (indicated by a tick) are selected and screened for indel mutations initiated within the indicated user-defined region of interest (middle panel, red dash-lined box). The detected indel variants, along with information about their position, type, length, and frequency, are listed in a text file (Variants.txt). Reads not containing indel mutations are subsequently screened for the presence of (multiple) intended precise base pair substitutions (indicated via the input files). The RepairReport file lists the encountered types of partial and full repair, with their concordant frequencies. In addition, calculated total indel and repair rates are written to the 'Efficiencies' file

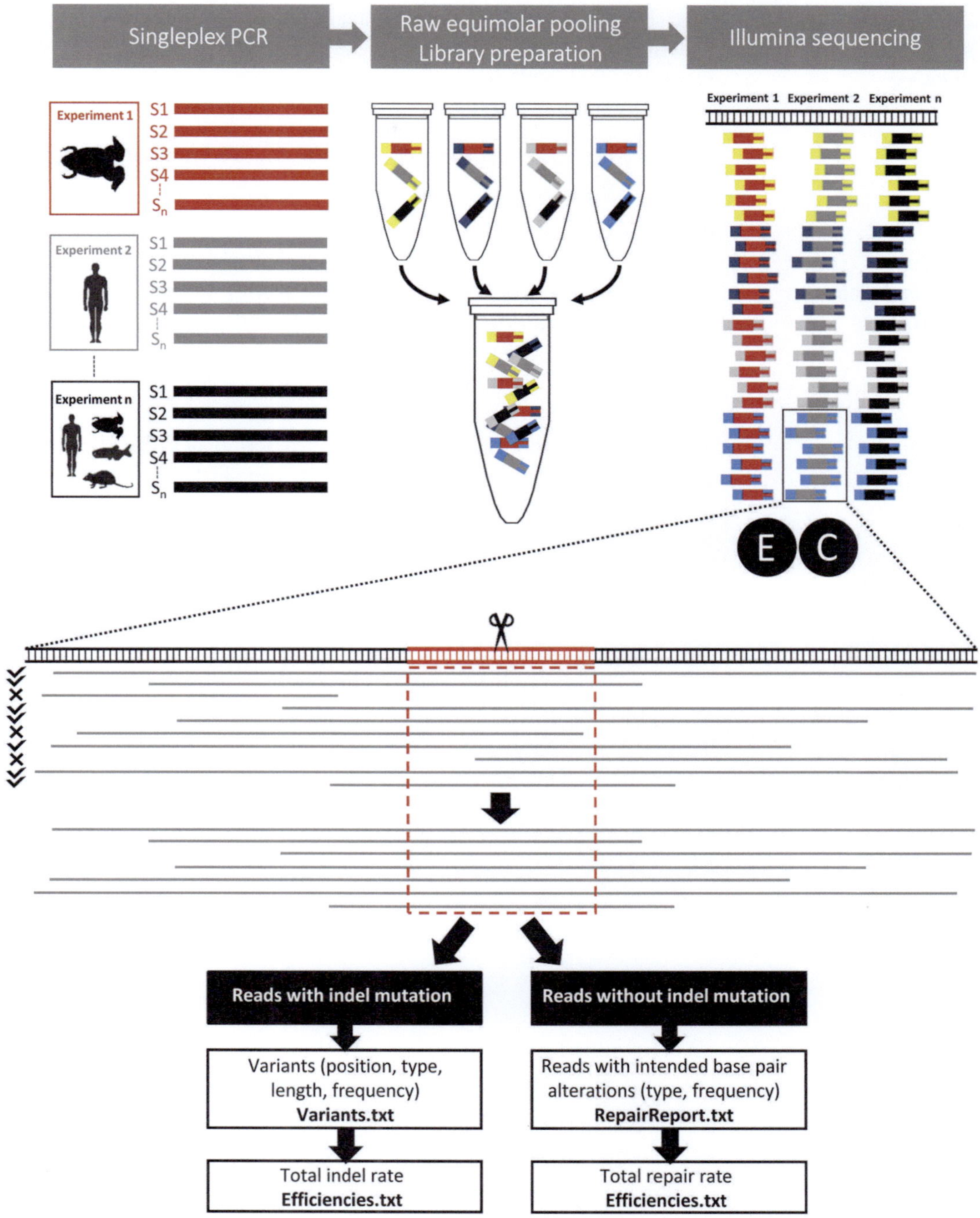

Fig. 1 Implementation of the BATCH-GE tool. NGS sample preparation consists of three steps, depicted in the upper panel. First, multiple singleplex PCR products (S1, S2, ..., Sn), are generated from different genomic sequences in one specific or in different genomes (upper panel, left), Next, equimolar pooling of these PCR products is followed by NGS library preparation, in which the input DNA is simultaneously fragmented and tagged (depicted by yellow, gray, light, and dark blue bars) (upper panel, middle) after which Illumina NGS sequencing is carried out (upper panel, right). Raw sequencing output is analyzed by the BATCH-GE tool, which evaluates the data sample-by-sample in an automated batchwise manner, guided by two input files (middle

2.2 Downloading the BATCH–GE Software

The BATCH-GE software is available as a git repository. To download the repository make sure that git is installed on your machine. To install git, open a terminal on your machine and enter:

sudo apt-get install git

BATCH-GE can be downloaded in a directory of choice by entering:

git clone https://github.com/WouterSteyaert/BATCH-GE.git

2.3 Installation of BATCH-GE

If your system has a web accessible folder in which you have write access, enter "perl BATCH-GE/Install.pl --WebAccFolder= <LOCATION WEB ACCESSABLE FOLDER>" and replace <LOCATION WEB ACCESSABLE FOLDER> by the full path to this directory on your machine. If your system does not have any web accessible folder simply run " perl BATCH-GE/Install.pl." The consequence will be that the URLs generated by BATCH-GE will not work. It has no consequences to the rest of the script.

Some Linux distributions are not accompanied with the appropriate libz. In that case the compilation of bwa, which is part of the above installation script, will fail. To install libz, enter sudo apt-get install libz-dev.

2.4 Preparation of the Reference Sequence

The software needs a reference sequence to map the reads onto. This reference sequence must be located in the subfolder 'genomes' of the BATCH-GE package. Follow the underneath instructions to download and prepare the Xenopus tropicalis genome (xenTro3) in order to run the examples 3 and 4 which are included in the package. Other genomes can be downloaded and prepared in exactly the same manner. The steps only need to be performed once for each reference sequence in the package.

1. Change directory to the subfolder "genomes" by entering "cd genomes."
2. Download the appropriate genome. For xenTro3 enter:

   wget ftp://hgdownload.cse.ucsc.edu/goldenPath/xenTro3/bigZips/xenTro3.fa.gz.

   This will download xenTro3 from the UCSC ftp site.
3. Change directory to BATCH-GE by entering "cd ..."
4. Prepare the genome for analysis by entering " perl PrepareGenome.pl --Genome=xenTro3."

   This script will create a sequence dictionary and an index necessary in the further analysis.

3 Practical Use

The BATCH-GE package contains four examples to illustrate its use and to make the user more familiar with the software. Examples 1 and 2 are Danio rerio examples. Instructions and output for those examples can be found in the README document in the BATCH-GE folder. Here, we will work out a Xenopus tropicalis example, i.e., example 3 in the package.

In order to run BATCH-GE, 2 files need to be prepared: an experiment file that contains all experiment-specific variables and a cutsite file. Both files can have any name and can be placed in any directory on your machine. The file type and extension, however, needs to be .csv (semicolon separated) for the experiment file and .bed (tab delimited) for the cutsite file. The full file paths to the respective files in example 3 are /PREFIX/BATCH-GE/_EXAMPLE_3/ExperimentFile.csv and /PREFIX/BATCH-GE/_EXAMPLE_3/XenopusCutsites.bed where PREFIX is the file path to the directory in which BATCH-GE is cloned.

3.1 Cutsite File

The cutsite file contains the genomic locations of the cutregions corresponding to the cutsites used in the experiment. Other cutregions for the same genome can also be stored in the file, as such the cutsite file can be used as a library of cutsites for a given genome. As is illustrated in Fig. 2 the first column contains the chromosome, the second and third column respectively represent the chromosomal start and end of the cutsite region. The fourth column finally contains the user specified name of the cutsite.

For most experiments a cutsite region of 60 bp is suitable, i.e., 30 bp up and downstream of the theoretical cutsite. An evaluation of the size of the cutsite region is made in [1]. Only reads spanning the full cutsite region are used in the efficiency calculation. This is done to avoid an artificial increase in mutagenesis efficiency due to artifacts in the sequencing reads.

3.2 Experiment File

As mentioned, the experiment file contains all experiment-specific parameters that are necessary for the script in order to analyze the data. It needs to have the following header:"**FastqDir;SampleNumbers;Genome;CutSite;OutputDir;CutSitesFile;RepairSequence;**" (*see* Fig. 3).

In the following section the different fields are explained:

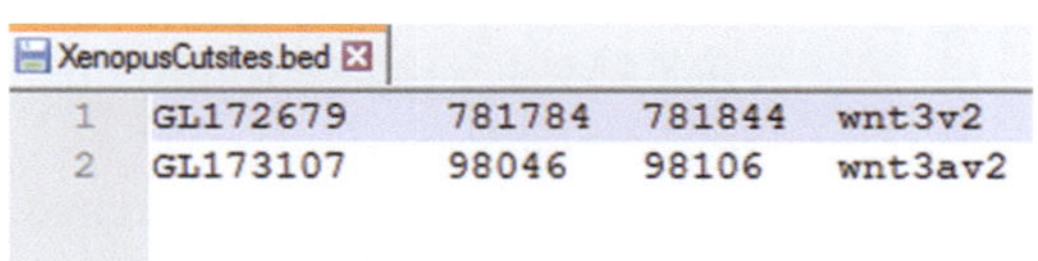
XenopusCutsites.bed

1	GL172679	781784	781844	wnt3v2
2	GL173107	98046	98106	wnt3av2

Fig. 2 The ***Xenopus tropicalis*** cutsite file as it is present in example 3 of BATCH-GE-package

	A	B	C	D	E	F	G
1	FastqDir	SampleNumbers	Genome	CutSite	OutputDir	CutSitesFile	RepairSequence
2	/PREFIX/BATCH-GE/_EXAMPLE_3/Fastq/	2	xenTro3	wnt3v2	/PREFIX/BATCH-GE/_EXAMPLE_3/Output/	/PREFIX/BATCH-GE/_EXAMPLE_3/XenopusCutsites.bed	

Fig. 3 Experiment file as it is present in example 3 of BATCH-GE-package

3.2.1 FastqDir

The full path to the directory where the fastq.gz files are located in. This folder is /PREFIX/BATCH-GE/_EXAMPLE_3/Fastq/ in the example.

3.2.2 SampleNumbers

The identifier(s) for the sample(s) to be analyzed in the specified FastqDir. The notation x,y ensures that both samples x and y will be analyzed. If numerical identifiers are used the notation x-y means that all samples from x to y will be analyzed. The sample number in a conventional Illumina fastq.gz file name is the number following the S. In the example under study, the fastq.gz filename is Pool-2-Run182_S2_wnt3v2_sorted_rg_1.fastq.gz for the first read file. The number 2 is following the S, so that's the sample number.

3.2.3 Genome

The name of the reference sequence onto which the reads should be aligned to. The installation guidelines contain instructions for downloading and preparing xenTro3. Other genomes can be used in exactly the same way. In fact, also user-specified contigs can be used as a reference as long as its name under which it is stored in the genome subdirectory of the BATCH-GE folder exactly corresponds to the name in the genome field in the experiment file.

3.2.4 CutSite

An identifier of choice for a particular cutsite. This identifier needs to be the same as the identifier used in the BED file (i.e., fourth column). In example 3 the name of the cutsite is wnt3v2. The chromosomal coordinates of this cutsite region are stored in the cutsite file.

3.2.5 OutputDir

The directory where BATCH-GE should write its output to. Running the example will lead to the creation of a subdirectory called 'Output' in the EXAMPLE_3_ directory.

3.2.6 CutSitesFile

To full path to the cutsite file (/PREFIX/BATCH-GE/_EXAMPLE_3/XenopusCutsites.bed).

3.2.7 RepairSequence [Optional]

The HDR template sequence. Placing square brackets around certain bases of the repair template indicates that these base pair alterations need to be introduced in the genome under study. Round brackets, on the other hand, indicate base pair alterations that do not necessarily need to be introduced, e.g., alterations needed for codon optimization of the template. In examples 1 and 2 of the BATCH-GE package HDR is used. The screen output can be found in the README document. Be aware that those are examples of Danio rerio experiments and therefore the Danio rerio genome (danRer7) needs to be downloaded and prepared.

3.3 Script Execution

When both files are prepared in the proper way, the script can be executed. This is done by entering perl /PREFIX/BATCH-GE/BATCH-GE.pl --ExperimentFile /PREFIX/BATCH-GE/_EXAMPLE_3/ExperimentFile.csv in the Linux command window. Replace PREFIX by the folder in which you have cloned BATCH-GE. The script is written for Perl5.

3.4 Script Output

The screen output of the script is displayed in Fig. 4. During execution the different steps in the analysis are written to the screen in order for the user to see the progress of the analysis. At the end of the analysis the script displays the mutagenesis efficiency.

In the output folder that is specified in the experiment file (OutputDir field), additional output can be found.

3.4.1 Variants.txt

This file contains the different indel variations that are present in the reads spanning the entire cutsite region. It lists the chromosome name and chromosomal location of the variant, type and length of the variant, the reference sequence surrounding the indel (10 bp upstream and 10 bp downstream of the indel) with [] marking the inserted sequence or with [deleted base pairs] marking the deleted sequence, and absolute and relative frequency of the variants.

3.4.2 RepairReport.txt

In case of HDR analysis, the reads that do not contain any indel, are screened for the presence of the intended base pair alterations. BATCH-GE can distinguish between full and partial repair, in case multiple base pair alterations are intended to be introduced in the region of interest. If partial repair is encountered, the specific sequence of the partial repair is listed.

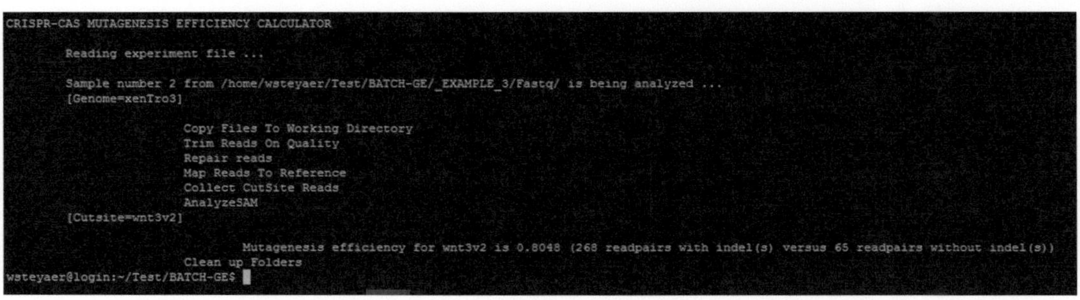

Fig. 4 Screen output of example 3 in the BATCH-GE package

3.4.3 *Efficiencies.txt*

General indel and repair rates are shown in the efficiencies file.

3.4.4 *URL.txt*

In this file URLs can be found which allow the user to visualize the reads in the UCSC Genome Browser. This will only work if the user has a web accessible directory on his machine and BATCH-GE is installed as such (see installation guidelines). If so, just paste the URL in a web browser and the reads under study become visible.

References

1. Boel A, Steyaert W, De Rocker N, Menten B, Callewaert B, De Paepe A, Coucke P, Willaert A (2016) BATCH-GE: Batch analysis of next-generation sequencing data for genome editing assessment. Sci Rep 6:30330. https://doi.org/10.1038/srep30330
2. Naert T, Colpaert R, Van Nieuwenhuysen T, Dimitrakopoulou D, Leoen J, Haustraete J, Boel A, Steyaert W, Lepez T, Deforce D, Willaert A, Creytens D, Vleminckx K (2016) CRISPR/Cas9 mediated knockout of rb1 and rbl1 leads to rapid and penetrant retinoblastoma development in Xenopus tropicalis. Sci Rep 6:35264. https://doi.org/10.1038/srep35264

Chapter 7

A Simple Knock-In System for *Xenopus* via Microhomology Mediated End Joining Repair

Ken-ich T. Suzuki, Yuto Sakane, Miyuki Suzuki, and Takashi Yamamoto

Abstract

Following completion of the genome sequences of *Xenopus tropicalis* and *X. laevis*, gene targeting techniques have become increasingly important for the further development of *Xenopus* research in the life sciences. Gene knockout using programmable nucleases, such as TALEN and CRISPR/Cas9, has reached a level whereby we can readily and routinely perform loss-of-function analysis of genes of interest in these species. However, there is still room for improvement in gene knock-in techniques owing to some technical problems. To overcome these problems, several knock-in techniques have been developed. Among them, we introduce in this chapter a simple knock-in system mediated by microhomology mediated end joining repair. This protocol allows us to produce knock-in animals for in vivo tagging, promoter/enhancer traps, and transgenesis in both of these *Xenopus* species.

Key words *Xenopus tropicalis*, *Xenopus laevis*, CRISPR/Cas9, Knock-in, TALEN, Transgenesis, Microhomology mediated end joining repair (MMEJ)

1 Introduction

The completion of the genome sequences of *Xenopus tropicalis* and *X. laevis* in the last decade [1, 2] has opened a new era in *Xenopus* research. Researchers can now readily retrieve genomic information on genes of interest (GOIs), such as their exons, introns, promoters, and putative regulatory sequences, from the online database Xenbase, which is based on substantial genome, epigenome, and transcriptome data [3]. Moreover, remarkable advances in genome editing have transformed experimental strategies for *Xenopus* research [4–7]. As a result, even beginners are now able to easily produce *Xenopus* with knockout of a GOI using a programmable nuclease, particularly the CRISPR/Cas9 system owing to its convenience and efficiency in both of the above-mentioned *Xenopus* species [8–13]. Currently, Cas9 ribonucleoprotein complex (RNP) is used to create animals carrying biallelic mutations in almost all somatic cells, hence generating virtual knockout animals

Kris Vleminckx (ed.), *Xenopus: Methods and Protocols*, Methods in Molecular Biology, vol. 1865,
https://doi.org/10.1007/978-1-4939-8784-9_7,

in the founder generation (Crispants), at least for *X. tropicalis*, which in contrast to *X. laevis* has a diploid genome [14].

Despite these developments, there is still room for improvement in the integration of exogenous DNA into a target chromosomal site, i.e., generating knock-ins, which is still somewhat limited compared with the practical utility of knockout. Prior to the genome editing era, the conventional knock-in strategy based on homologous recombination (HR) was the only option available, but it had limitations in mouse embryonic stem cells because intrinsic HR at a target site occurs with extremely low frequency in somatic cells. In addition, targeted knock-in animals could only be generated in the mouse, and later on in rats.

However, this difficulty has recently been overcome by the introduction of novel genome editing techniques. These techniques feature programmable nucleases such as TALENs and CRISPR/Cas9, which can efficiently introduce DNA double strand breaks (DSBs) at a target site. Importantly, the efficiency of DSB induction by TALENs and CRISPR/Cas9 is much higher than the incidence of intrinsic HR. Most DSB sites are repaired by three mechanisms in somatic cells: nonhomologous end joining (NHEJ), microhomology mediated end joining (MMEJ), and HR, although this latter mechanism occurs at a very low frequency [15]. In *Xenopus*, a knock-in technique mediated by NHEJ repair has been reported [16]. This technique is the simplest to apply of the various knock-in methods: The target site in the genome and a circular donor vector are simultaneously cut and linearized by the same nuclease in embryos, and then the exogenous DNA is integrated into the DSB site by NHEJ repair. However, this method is associated with the problem that the donor vector is randomly integrated in either the forward or the reverse direction.

MMEJ repair is known to repair a DSB site via 5–25 bp microhomologous sequences in both DNA ends resection [15]. Bae et al. previously reported that MMEJ repair often occurs at DSB sites induced by TALENs and CRISPR/Cas9 [17]. Prompted by these findings, we established a new knock-in technique based on MMEJ repair, referred to as the PITCh system (Precise Integration into Target Chromosome) [18]. The principle of this MMEJ mediated knock-in system involves microhomologous annealing between the DSB site and a donor vector: the target site in the genome and a circular donor vector are simultaneously cut and linearized by the programmable nuclease(s) in vivo. As a result, the exogenous DNA is correctly integrated into the DSB site via MMEJ repair. When knock-in is completed using this technique, the original target sequence is altered by MMEJ and the target site can no longer be cut off by the same Cas9 RNP or TALEN pair. Moreover, exogenous DNA can be integrated into the target site in the desired direction. In addition, the donor vector can be constructed very easily; researchers only need to perform inverse

PCR. This system has already been applied successfully in various animals, such as silkworm, zebrafish, mouse, and human cell lines, as well as in *Xenopus* [18–20].

In this chapter, we present this simple strategy to generate knock-in *Xenopus* based on MMEJ repair. This protocol can be applied for the rapid and efficient functional analysis of a GOI in *Xenopus* research, such as in vivo tagging, promoter/enhancer trap, and transgenesis [18, 21].

2 Materials

2.1 Reagents for Construction of Donor Vectors

1. Vectors and GOI cDNA. We often use pEGFP-1 (promoter-less for in vivo tagging and promoter/enhancer trap, Clontech) and pCS2+ (CMV promoter + cDNA of GOI for transgenesis) as donor vectors [18, 21].
2. High-fidelity PCR enzymes (e.g., PrimeStar max, TaKaRa).
3. A pair of oligonucleotides for inverse PCR (*see* Subheading 3.1, **step 1**).
4. T4 polynucleotide kinase and its buffer (e.g., TaKaRa).
5. 10 mM ATP.
6. *Dpn*I (e.g., TaKaRa).
7. DNA ligase or ligation kit (e.g., DNA Ligation Kit Mighty Mix, TaKaRa).
8. Plasmid purification kit (e.g., QIAprep Spin Miniprep Kit, Qiagen).
9. PCR purification kit (e.g., QIAquick PCR Purification Kit, Qiagen).

2.2 Reagents for Injection in Xenopus

1. 10× Marc's modified Ringer's (MMR): 1 M NaCl, 20 mM KCl, 10 mM $MgCl_2$, 20 mM $CaCl_2$, 50 mM HEPES, and it is adjusted to pH 7.4 using NaOH. Sterilize with a 0.2 μm filter or autoclave. This is a stock solution. Dilute to 0.1× or 0.3× MMR with sterilized water and add 1/1000 streptomycin/penicillin (stock: 50 mg/mL in sterilized water) for use.
2. Ficoll buffer: 5% (w/v) Ficoll 400 dissolved in 0.3× MMR containing streptomycin/penicillin. Sterilize with a 0.2 μm filter.
3. Equipment for injection: puller for microneedles, microinjector, stereomicroscope, micromanipulator, microneedle, mineral oil, and incubator [22].

2.3 Reagents for Genotyping and Sequencing Analysis

1. Genomic DNA purification kit (e.g., DNeasy Blood & Tissue Kit, Qiagen).
2. High-fidelity and robust performance PCR enzymes (e.g., KOD FX Neo, TOYOBO).
3. A pair of oligonucleotides for PCR of 5′ and 3′ junction at the integration site.
4. Cloning vector for sanger DNA sequencing (e.g., TOPO® TA Cloning Kit for Sequencing, Thermo Fisher Scientific).
5. Appropriate *E. coli* hosts for subcloning.
6. Equipment for PCR and electrophoresis.

3 Methods

We here explain MMEJ knock-in in *Xenopus*, showing three examples, one using CRISPR/Cas9 in *X. tropicalis* and the others using TALEN in *X. laevis* (Figs. 1 and 2). If you want to use TALENs, we recommend constructing them using Platinum Gate TALEN kit (Kit # 1000000043: Addgene) [23]. Commonly used methods for RNA synthesis of TALENs, Cas9, and sgRNAs are described elsewhere in this issue and in our previous report [22].

3.1 Construction of Donor Vectors for MMEJ Mediated Knock-in

1. Find the recognition sequence of TALEN and sgRNA for the target chromosomal region into which you want to knock-in the donor vector. You can find it manually or use a web-based predictor such as CRISPR-Direct (https://crispr.dbcls.jp) [24]. If you want to fuse reporter or tag sequences to the N- or C-terminal of the GOI, you should design TALEN and sgRNA target sites adjacent to the start or stop codon.
2. Design appropriate microhomologous sequences in the donor vector. We usually set a single site containing two short microhomologous sequences to join them to the target DSB site via MMEJ repair into a donor vector. Hereinafter, this site is referred to as an "MH site." The MH site also contains a TALEN or Cas9/sgRNA target site (*see* **Note 1**). The lengths of the 5′ and 3′ microhomologous sequences are usually set to 8–16 bp in the MH site. You can employ a fluorescence reporter driven by a tissue-specific promoter/enhancer such as

Fig. 1 (continued) site are split into two and swapped around. Each microhomologous DNA end of the donor vector is annealed corresponding to the 5′ and 3′ resection of DNA ends on the chromosomal DSB site and integrated by MMEJ repair. Underlined letters indicate the start codon of *tyr*.S and mkate2 reporter. pA; poly(A) signal. (**b**) Red fluorescence (mkate2) expression in *tyr*.S promoter trap founders. (**c**) Promoter trap of *npm3*.L founder. Egfp-fused *npm3*.L was driven by the *npm3*.L endogenous promoter and its fluorescence was strongly detected in the central nervous system. (**d**) CMV:Mintbody transgenic founder. Mintbody (EGFP) was ubiquitously expressed by a CMV promoter in the *tyr*.S locus

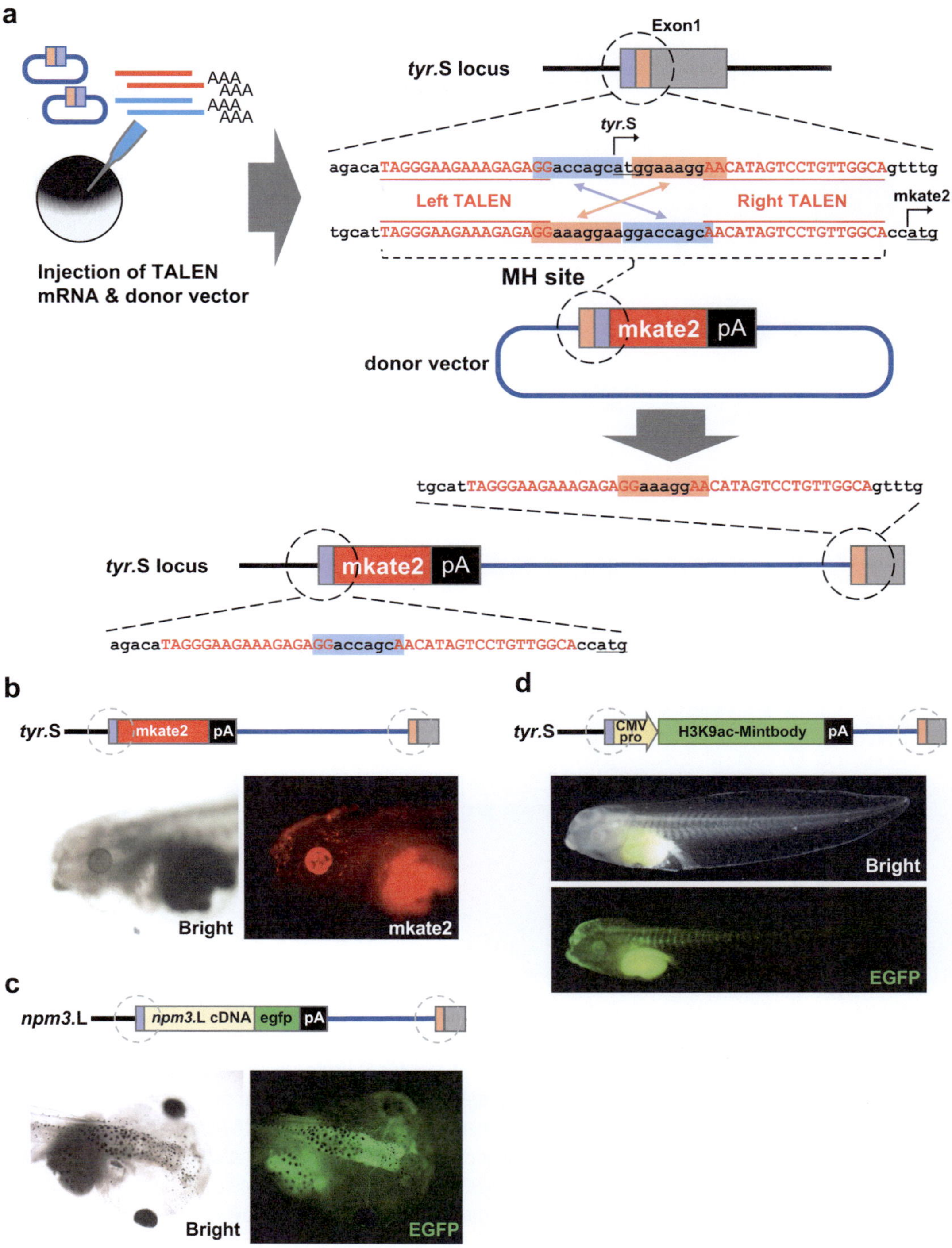

Fig. 1 MMEJ mediated knock-in using TALENs in *X. laevis*. (**a**) A strategy for MMEJ mediated knock-in using TALENs. A pair of TALENs targeting the *tyrosinase* S (*tyr*.S) gene of *X. laevis* is designed at the start codon. Promoter-less donor vector has a single MH site that consists of a TALEN recognition site (red capital letters) and two microhomologous sequences for 5′ and 3′ end joining (blue and orange boxes). These microhomologous sequences based on the original spacer sequence (black small letters) and part of the TALEN recognition

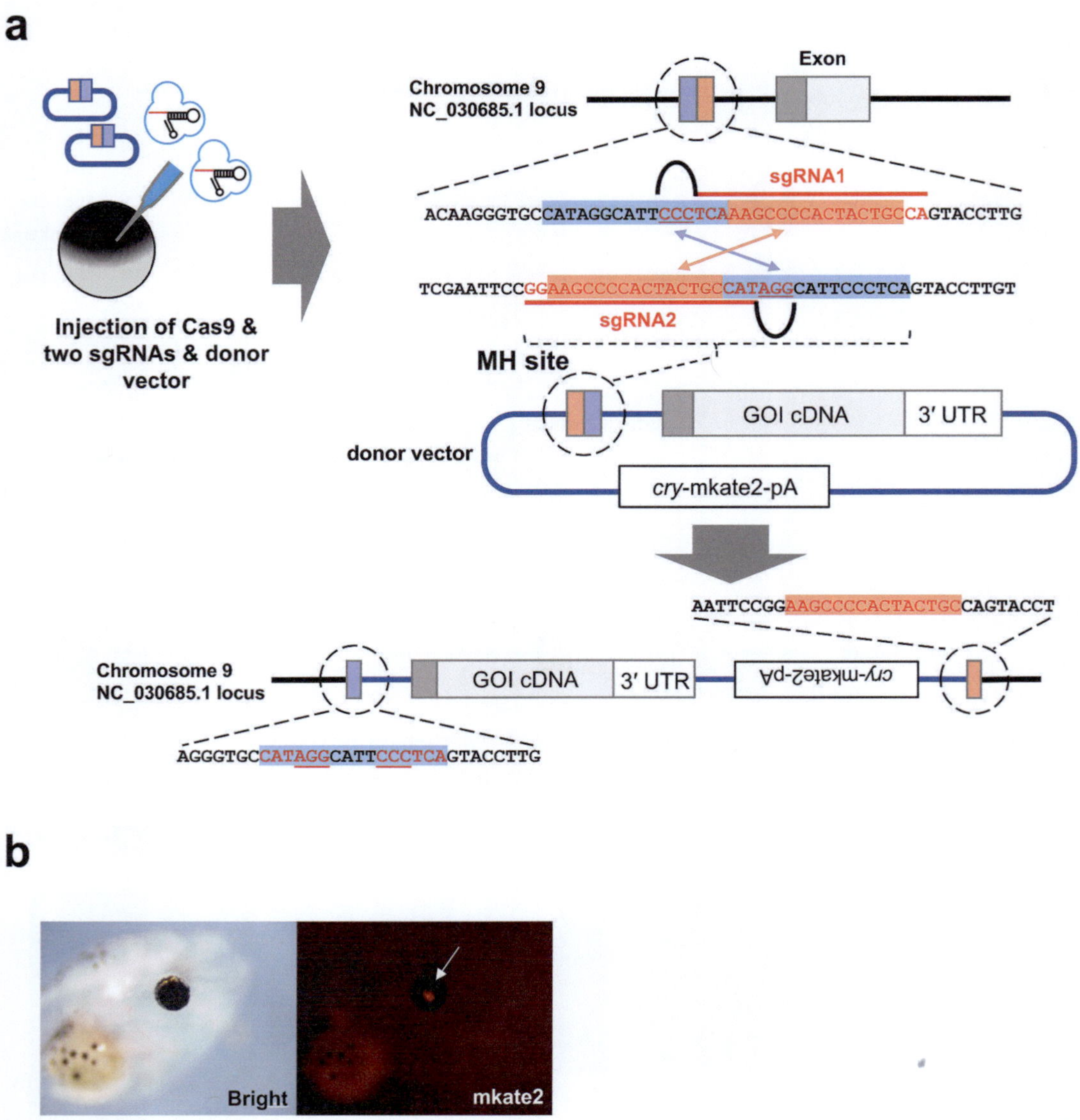

Fig. 2 MMEJ mediated knock-in using CRISPR/Cas9 in *X. tropicalis*. (**a**) A strategy for MMEJ mediated knock-in by CRISPR/Cas9. Two sgRNAs are required for this method: one targets the chromosomal target site (sgRNA1) and the other performs in vivo linearization of the donor vector (sgRNA2). The MH site has three sequences: two microhomologous sequences and a sgRNA recognition site. Blue and orange boxes indicate the microhomologous sequences for 5′ and 3′ end joining, respectively. Two sgRNA recognition sequences are shown in both overlines and red letters. Two protospacer adjacent motifs are indicated by underline. This vector also has a *crystallin* promoter–mkate2 cDNA cassette to screen for lens-specific red fluorescence in vivo. (**b**) Red fluorescence (mkate2) expression in eyes of knocked-in founders. An arrow indicates an eye expressing mkate2

the *crystallin* promoter (lens-specific promoter), for in vivo selection of knock-in embryos (*see* Fig. 2).

3. If you use TALENs, you need only a pair of TALENs targeting the desired site. The spacer sequence between the left and

right TALENs targeting the chromosomal site is used as microhomologous sequences for MMEJ: The spacer sequence is divided into two regions, and the first half of the sequence is swapped for the latter half, as shown in Fig. 1 (see blue and orange highlighted sequences and arrows). Each microhomologous sequence in the MH site can be annealed, corresponding to the 5′ and 3′ resection of DNA ends at the target DSB site. An MH site can be added in the donor vector by inverse PCR (*see* Fig. 3). The length of the MH site is ~60 bp. In this way, a pair of TALENs simultaneously cuts both the target chromosome and the MH site. Both ends of the MH site join to the chromosomal DSB site through microhomology in the right direction. When MMEJ mediated knock-in is completed correctly, the spacer sequence of TALENs at the integration site is shortened and can no longer be cut by the same TALENs again.

4. If you use CRISPR/Cas9, two sgRNAs are required for MMEJ knock-in: one targets the chromosomal target site and the other performs in vivo linearization of donor vector. The MH site in the donor vector is designed as shown in Fig. 2. The target chromosomal region (~40 bp) containing one sgRNA sequence is divided into two. Each microhomologous sequence in the MH site corresponds to the 5′ and 3′ resection of DNA

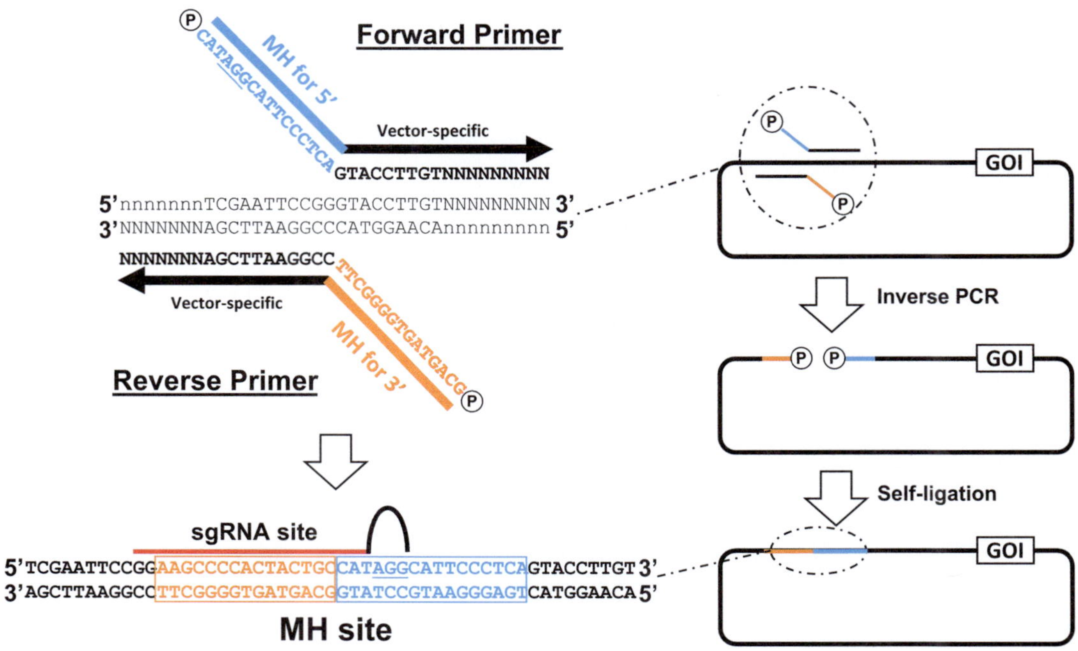

Fig. 3 Workflow for constructing the donor vector for MMEJ mediated knock-in. The MH site consists of three sequences: microhomologous sequences for 5′ and 3′ end joining and a sgRNA site. Overline and underline indicate protospacer sequence and protospacer adjacent sequence of sgRNA for in vivo linearization, respectively. This MH site is referred from Fig. 2

ends on the chromosomal DSB site. The first half is swapped for the latter half and then added into the donor vector by inverse PCR. For in vivo linearization of the donor vector, the other sgRNA should be designed in the MH site. Both ends of the linearized donor vector join to the target DSB site through microhomology in the right direction. When MMEJ mediated knock-in has been completed correctly, the sgRNA sequence at the integration site is altered and can no longer be cut by Cas9 again. There are other designs for MMEJ mediated knock-in possible (*see* **Note 2**).

5. Design and order a pair of oligonucleotides for the insertion of the MH site in accordance with Fig. 3 (*see* **Notes 3** and **4**).
6. Perform phosphorylation of the 5′ ends of oligonucleotides using T4 polynucleotide kinase, in accordance with the supplier's protocol. The final concentration of oligonucleotides is 10 μM in this reaction, after which it can immediately be used in the next step (Fig. 3).
7. Perform inverse PCR with the 5′ phosphorylated forward and reverse oligonucleotides. High-fidelity PCR enzymes should be used with a small number of PCR cycles (<18 cycles).
8. Digest the template plasmid DNA using *Dpn*I. PCR products are not digested because *Dpn*I digests methylated DNA from commonly used *E. coli* hosts.
9. Purified PCR products are self-ligated by DNA ligase or ligation kit and transformed into appropriate *E. coli* hosts.
10. To confirm that the MH site has been inserted, the donor vector should be subjected to DNA sequencing.
11. Prepare the donor vector by small-scale culture. Purify the plasmid using a silica column-based plasmid purification kit. We highly recommend re-purification of the eluted plasmid using a silica column-based DNA purification kit or ethanol precipitation (*see* **Note 5**).

3.2 Preparation of In Vitro Fertilized Eggs and Injection

The use of good-quality eggs is the key to success in knock-in experiments in *X. laevis* and *X. tropicalis* because vector DNA can cause toxicity in the injected embryos. For more details on the handling and injection of *Xenopus* eggs and embryos, we refer to the other chapters in this book or our previous report [22]. Researchers should obtain approval from the ethics committees of their affiliated institutes and perform experiments in accordance with ethics guidelines for animal studies.

1. The amount and ratio of the mixture of TALEN mRNAs (or Cas9 mRNA, protein and sgRNAs) and donor vector should be titrated and optimized for each GOI. We usually test the knock-in under the following conditions in the first experiment:

250 pg each of left and right TALEN mRNA and 100 pg of donor vector in *X. laevis*, 1–2 ng of Cas9 mRNA (protein) and 200 pg each of two sgRNAs and up to 100 pg of donor vector in *X. tropicalis* (*see* **Note 6**).

2. Inject 2.3 and 4.6 nl of the mixture into *X. tropicalis* and *X. laevis* embryos, respectively. To reduce mosaicism, a mixture of TALEN mRNAs, Cas9 mRNA (protein), sgRNAs, and donor vectors should be injected at the one-cell stage.
3. Keep the injected embryos in Ficoll buffer at 26 °C and 18–20 °C for *X. tropicalis* and *X. laevis*, respectively, until the early blastula stage.
4. Transfer the embryos into 0.1× MMR at the blastula stage and then keep the embryos at 26 °C and 18–20 °C for *X. tropicalis* and *X. laevis*, respectively.
5. Perform phenotyping and genotyping at the time point when your target genes should be functioning (*see* **Note 7**). Based on our protocol, up to 30% reporter-positive knock-in founders can be obtained [18, 21]. Note that the efficiency of knock-in is influenced by the chromatin status and sequence context of the GOI loci.

3.3 Genotyping of Knock-In Founders

To confirm the correct integration of exogenous DNA into the target site, genotyping should be performed. Notably, founders exhibit mosaicism to varying degrees. Since both wild type and knock-in alleles are intermingled in a founder, if a pair of PCR primers specific for both outsides of the knock-in site are used, two PCR products of the target regions should be theoretically obtained. The PCR product of the knock-in allele is much larger than that of the wild type allele because the whole donor vector is integrated into the region. However, it is technically difficult to detect a knock-in allele in the founders because small PCR products of the wild type allele are more preferentially amplified than those of the knock-in allele for reasons associated with PCR kinetics. Therefore, we always perform genomic PCR of both junctions (5′ and 3′) between the outside sequence of the target site and the inserted donor vector sequence to check whether knock-in has been successful in the founders.

1. Extract genomic DNA from each sample (whole embryos, tail fin clips, or tips of digits). We recommend using a silica column-based DNA purification kit because the high-quality genomic DNA improves the odds of success in genomic PCR.
2. Perform PCR of the 5′ and 3′ junction of the knock-in site to produce 100- to 300-bp amplicons using each primer set. A PCR primer set for each junction is designed so that one is locus-specific and the other is vector-specific.

3. PCR products of the 5′ and 3′ junctions are electrophoresed on agarose gel. It is expected that only the knock-in allele is amplified and observed in electrophoresis in both junctions (*see* **Note 8**).
4. Subclone the PCR products into the cloning vectors. If necessary, cut out the bands with the expected size.
5. Perform sanger DNA sequencing of plasmids or colony PCR products from the clones. We usually analyze 10 clones from each embryo.
6. Rear genotype-positive founders to adulthood in order to establish knock-in lines for further analysis. Depending on the purpose of the research, knock-in founders could be adequate for analyzing a GOI despite exhibiting mosaicism.

4 Notes

1. Using a donor vector carrying a single MH site, the efficiency of knock-in seems to be higher than that containing two sites flanking the DNA to be integrated, i.e., "gene cassette knock-in." Therefore, we prefer the use of the former. Consequently, the whole donor vector including the backbone sequence is integrated into the target site.
2. In this chapter, we introduce several designs of MH sites for MMEJ mediated knock-in, in accordance with our experimental results. Various designs of microhomologous sequences and applications are also described elsewhere [18, 19]. You can also easily design candidate sgRNAs and microhomologous sequences using a web-based tool (PITCh designer; http://www.mls.sci.hiroshima-u.ac.jp/smg/PITChdesigner/index.html) [25].
3. Each oligonucleotide consists of both a donor vector-specific sequence and a microhomologous sequence for the 5′ and 3′ junctions of the target site (*see* Fig. 3). Donor vector can also be constructed by In-Fusion cloning (Clontech) or using Gibson assembly kit (New England Biolabs), if you add 15-bp overlap sequences to each 5′ end of the oligonucleotides.
4. To fuse reporter genes such as EGFP or tag sequences to the N- or C-terminal of the target genes (proteins), the MH site should be designed to ensure that these sequences and the GOI are in-frame.
5. RNase A treatment is included in the standard plasmid purification (miniprep) process. RNase and other contaminants in plasmid eluate cause decreased viability in the injected embryos; therefore, you should repurify the donor plasmid using a silica column-based DNA purification kit or ethanol precipitation.

6. A large amount of donor vector (over 100 pg) would increase the risk of high toxicity and random integration, so a smaller amount of donor vector is better.
7. If an endogenous GOI is abundantly expressed during embryogenesis, you could observe the reporter fluorescence in the case of a promoter/enhancer trap or in vivo tagging at this point. In the case of transgenesis with a CMV promoter, you could see the reporter fluorescence by the tailbud stage [21].
8. In addition to the expected bands, some extra bands may be detected owing to incorrect knock-in alleles in a founder.

Acknowledgments

This work was supported by JSPS KAKENHI Grant Number 15 K06802 [Grant-in-Aid for Scientific Research (C)] to K.T.S. and 15 J05833 [Grant-in-Aid for JSPS Fellows] to Y.S, NBRP Fundamental Technologies Upgrading Program in AMED to K.T.S. and T.Y., The Narishige Zoological Science Award and The Naito Foundation to K.T.S. We are grateful to Drs. Akihiko Kashiwagi and Keiko Kashiwagi and the National Bio-Resource Project of *X. tropicalis* in the Amphibian Research Center (Hiroshima University) for providing *X. tropicalis*, Golden strain. We also thank Edanz Group (www.edanzediting.com/ac) for editing a draft of this manuscript.

References

1. Hellsten U, Harland RM, Gilchrist MJ, Hendrix D, Jurka J, Kapitonov V, Ovcharenko I, Putnam NH, Shu S, Taher L, Blitz IL, Blumberg B, Dichmann DS, Dubchak I, Amaya E, Detter JC, Fletcher R, Gerhard DS, Goodstein D, Graves T, Grigoriev IV, Grimwood J, Kawashima T, Lindquist E, Lucas SM, Mead PE, Mitros T, Ogino H, Ohta Y, Poliakov AV, Pollet N, Robert J, Salamov A, Sater AK, Schmutz J, Terry A, Vize PD, Warren WC, Wells D, Wills A, Wilson RK, Zimmerman LB, Zorn AM, Grainger R, Grammer T, Khokha MK, Richardson PM, Rokhsar DS (2010) The genome of the Western clawed frog Xenopus tropicalis. Science (New York, NY) 328(5978):633–636. https://doi.org/10.1126/science.1183670
2. Session AM, Uno Y, Kwon T, Chapman JA, Toyoda A, Takahashi S, Fukui A, Hikosaka A, Suzuki A, Kondo M, van Heeringen SJ, Quigley I, Heinz S, Ogino H, Ochi H, Hellsten U, Lyons JB, Simakov O, Putnam N, Stites J, Kuroki Y, Tanaka T, Michiue T, Watanabe M, Bogdanovic O, Lister R, Georgiou G, Paranjpe SS, van Kruijsbergen I, Shu S, Carlson J, Kinoshita T, Ohta Y, Mawaribuchi S, Jenkins J, Grimwood J, Schmutz J, Mitros T, Mozaffari SV, Suzuki Y, Haramoto Y, Yamamoto TS, Takagi C, Heald R, Miller K, Haudenschild C, Kitzman J, Nakayama T, Izutsu Y, Robert J, Fortriede J, Burns K, Lotay V, Karimi K, Yasuoka Y, Dichmann DS, Flajnik MF, Houston DW, Shendure J, DuPasquier L, Vize PD, Zorn AM, Ito M, Marcotte EM, Wallingford JB, Ito Y, Asashima M, Ueno N, Matsuda Y, Veenstra GJ, Fujiyama A, Harland RM, Taira M, Rokhsar DS (2016) Genome evolution in the allotetraploid frog Xenopus laevis. Nature 538(7625):336–343. https://doi.org/10.1038/nature19840
3. Karpinka JB, Fortriede JD, Burns KA, James-Zorn C, Ponferrada VG, Lee J, Karimi K, Zorn AM, Vize PD (2015) Xenbase, the Xenopus model organism database; new virtualized system,

data types and genomes. Nucleic Acids Res 43(Database issue):D756–D763. https://doi.org/10.1093/nar/gku956

4. Young JJ, Cherone JM, Doyon Y, Ankoudinova I, Faraji FM, Lee AH, Ngo C, Guschin DY, Paschon DE, Miller JC, Zhang L, Rebar EJ, Gregory PD, Urnov FD, Harland RM, Zeitler B (2011) Efficient targeted gene disruption in the soma and germ line of the frog Xenopus tropicalis using engineered zinc-finger nucleases. Proc Natl Acad Sci U S A 108(17):7052–7057. https://doi.org/10.1073/pnas.1102030108
5. Lei Y, Guo X, Liu Y, Cao Y, Deng Y, Chen X, Cheng CH, Dawid IB, Chen Y, Zhao H (2012) Efficient targeted gene disruption in Xenopus embryos using engineered transcription activator-like effector nucleases (TALENs). Proc Natl Acad Sci U S A 109(43):17484–17489. https://doi.org/10.1073/pnas.1215421109
6. Ishibashi S, Cliffe R, Amaya E (2012) Highly efficient bi-allelic mutation rates using TALENs in Xenopus tropicalis. Biology Open 1(12):1273–1276. https://doi.org/10.1242/bio.20123228
7. Suzuki KT, Isoyama Y, Kashiwagi K, Sakuma T, Ochiai H, Sakamoto N, Furuno N, Kashiwagi A, Yamamoto T (2013) High efficiency TALENs enable F0 functional analysis by targeted gene disruption in Xenopus laevis embryos. Biology Open 2(5):448–452. https://doi.org/10.1242/bio.20133855
8. Nakayama T, Fish MB, Fisher M, Oomen-Hajagos J, Thomsen GH, Grainger RM (2013) Simple and efficient CRISPR/Cas9-mediated targeted mutagenesis in Xenopus tropicalis. Genesis 51(12):835–843. https://doi.org/10.1002/dvg.22720
9. Blitz IL, Biesinger J, Xie X, Cho KW (2013) Biallelic genome modification in F(0) Xenopus tropicalis embryos using the CRISPR/Cas system. Genesis 51(12):827–834. https://doi.org/10.1002/dvg.22719
10. Guo X, Zhang T, Hu Z, Zhang Y, Shi Z, Wang Q, Cui Y, Wang F, Zhao H, Chen Y (2014) Efficient RNA/Cas9-mediated genome editing in Xenopus tropicalis. Development 141(3):707–714. https://doi.org/10.1242/dev.099853
11. Wang F, Shi Z, Cui Y, Guo X, Shi YB, Chen Y (2015) Targeted gene disruption in Xenopus laevis using CRISPR/Cas9. Cell Biosci 5:15. https://doi.org/10.1186/s13578-015-0006-1
12. Bhattacharya D, Marfo CA, Li D, Lane M, Khokha MK (2015) CRISPR/Cas9: an inexpensive, efficient loss of function tool to screen human disease genes in Xenopus. Dev Biol 408(2):196–204. https://doi.org/10.1016/j.ydbio.2015.11.003
13. Shigeta M, Sakane Y, Iida M, Suzuki M, Kashiwagi K, Kashiwagi A, Fujii S, Yamamoto T, Suzuki KT (2016) Rapid and efficient analysis of gene function using CRISPR-Cas9 in Xenopus tropicalis founders. Genes Cells 21(7):755–771. https://doi.org/10.1111/gtc.12379
14. Sakane Y, Iida M, Hasebe T, Fujii S, Buchholz DR, Ishizuya-Oka A, Yamamoto T, Suzuki KT (2017) Functional analysis of thyroid hormone receptor beta in Xenopus tropicalis founders using CRISPR-Cas. Biology Open. https://doi.org/10.1242/bio.030338
15. Ceccaldi R, Rondinelli B, D'Andrea AD (2016) Repair pathway choices and consequences at the double-strand break. Trends Cell Biol 26(1):52–64. https://doi.org/10.1016/j.tcb.2015.07.009
16. Shi Z, Wang F, Cui Y, Liu Z, Guo X, Zhang Y, Deng Y, Zhao H, Chen Y (2015) Heritable CRISPR/Cas9-mediated targeted integration in Xenopus tropicalis. FASEB J 29(12):4914–4923. https://doi.org/10.1096/fj.15-273425
17. Bae S, Kweon J, Kim HS, Kim JS (2014) Microhomology-based choice of Cas9 nuclease target sites. Nat Methods 11(7):705–706. https://doi.org/10.1038/nmeth.3015
18. Nakade S, Tsubota T, Sakane Y, Kume S, Sakamoto N, Obara M, Daimon T, Sezutsu H, Yamamoto T, Sakuma T, Suzuki KT (2014) Microhomology-mediated end-joining-dependent integration of donor DNA in cells and animals using TALENs and CRISPR/Cas9. Nat Commun 5:5560. https://doi.org/10.1038/ncomms6560
19. Hisano Y, Sakuma T, Nakade S, Ohga R, Ota S, Okamoto H, Yamamoto T, Kawahara A (2015) Precise in-frame integration of exogenous DNA mediated by CRISPR/Cas9 system in zebrafish. Sci Rep 5:8841. https://doi.org/10.1038/srep08841
20. Aida T, Nakade S, Sakuma T, Izu Y, Oishi A, Mochida K, Ishikubo H, Usami T, Aizawa H, Yamamoto T, Tanaka K (2016) Gene cassette knock-in in mammalian cells and zygotes by enhanced MMEJ. BMC Genomics 17(1):979. https://doi.org/10.1186/s12864-016-3331-9
21. Suzuki M, Takagi C, Miura S, Sakane Y, Suzuki M, Sakuma T, Sakamoto N, Endo T, Kamei Y, Sato Y, Kimura H, Yamamoto T, Ueno N, Suzuki KT (2016) In vivo tracking of histone H3 lysine 9 acetylation in Xenopus laevis during tail regeneration. Genes Cells 21(4):358–369. https://doi.org/10.1111/gtc.12349

22. Sakane Y, Suzuki KT, Yamamoto T (2017) A simple protocol for loss-of-function analysis in Xenopus tropicalis founders using the CRISPR-Cas system. Methods Mol Biol 1630:189–203. https://doi.org/10.1007/978-1-4939-7128-2_16
23. Sakuma T, Ochiai H, Kaneko T, Mashimo T, Tokumasu D, Sakane Y, Suzuki K, Miyamoto T, Sakamoto N, Matsuura S, Yamamoto T (2013) Repeating pattern of non-RVD variations in DNA-binding modules enhances TALEN activity. Sci Rep 3:3379. https://doi.org/10.1038/srep03379
24. Naito Y, Hino K, Bono H, Ui-Tei K (2015) CRISPRdirect: software for designing CRISPR/Cas guide RNA with reduced off-target sites. Bioinformatics (Oxford, England) 31(7):1120–1123. https://doi.org/10.1093/bioinformatics/btu743
25. Nakamae K, Nishimura Y, Takenaga M, Nakade S, Sakamoto N, Ide H, Sakuma T, Yamamoto T (2017) Establishment of expanded and streamlined pipeline of PITCh knock-in - a web-based design tool for MMEJ-mediated gene knock-in, PITCh designer, and the variations of PITCh, PITCh-TG and PITCh-KIKO. Bioengineered 8(3):302–308. https://doi.org/10.1080/21655979.2017.1313645

Chapter 8

How to Generate Non-Mosaic CRISPR/Cas9 Mediated Knock-In and Mutations in F0 *Xenopus* Through the Host-Transfer Technique

Emmanuel Tadjuidje and Sang-Wook Cha

Abstract

We have taken advantage of the well-established oocyte host transfer technique to optimize a method for CRISPR editing of *Xenopus* that provides an efficient non-mosaic targeted insertion of small DNA fragment through homology-directed repair mechanism.

Key words *Xenopus*, HDR, CRISPR, Knock-in, F0, Oocyte

1 Introduction

CRISPR/Cas9 genome editing is emerging as a powerful approach for targeted genome modification, including loss-of-function mutations by insertion/deletion or targeted insertion of specific DNA sequences into the genome. The approach is based upon sequence recognition (through base pairing) between a targeting RNA (CRISPR RNA; crRNA) and the target DNA, combined with the ability of the Cas9 enzyme (an RNA-guided DNA endonuclease) to introduce double strand DNA breaks within the target sequences by recognizing a trans-activating CRISPR RNA (tracrRNA) that base pairs with the crRNA. In practice, a portion of the crRNA is fused to the tracrRNA to generate a single guide RNA (sgRNA) that directs the Cas9 enzyme to the target sequence [1–3]. The double strand breaks are often repaired by Non-homologous end-joining (NHEJ), an imprecise pathway that leads to frequent random nucleotides insertion or deletion within the targeted region; this often result in loss-of-function by frame-shift mutations [4–8]. Using a repair construct, the repair pathway can also be engineered to introduce specific nucleotide sequences within the target region through the homology-directed repair (HDR) pathway [9–11].

Kris Vleminckx (ed.), *Xenopus: Methods and Protocols*, Methods in Molecular Biology, vol. 1865,
https://doi.org/10.1007/978-1-4939-8784-9_8, © Springer Science+Business Media, LLC, part of Springer Nature 2018

Despite various reports of the successful application of genome editing technologies (CRISPR/Cas9 and TALENs) in *Xenopus*, the F0 mutants are often mosaic, being a mixture of both mutant and wild type cells; consequently, a lengthy breeding of F1 and F2 mutants is needed to achieve uniform germline transmission [12–14]. Besides, knock-in editing through HDR has proven to be poorly efficient in most animal models [9–11]. We have taken advantage of the well-established oocyte host transfer technique to optimize a method for CRISPR editing of *Xenopus* that not only generates non-mosaic F0 heterozygote INDEL founder mutants, but also provides an efficient non-mosaic targeted insertion of small DNA fragment of 73–104 nucleotides through HDR (Fig. 1a). The method takes advantage of the fact that *Xenopus* oocytes can be isolated, injected, and cultured in vitro for a couple of days before they are fertilized through the host-transfer technique [15, 16]. This approach allows time for the maternal genome to be modified before fertilization [17]. In addition, the *Xenopus* oocytes have a higher HDR activity than the embryos, and the efficiency of HDR is further improved by treating the oocytes with a DNA ligase IV inhibitor (which inhibits the NHEJ repair pathway) [18, 19]. Therefore, while increasing the efficiency of

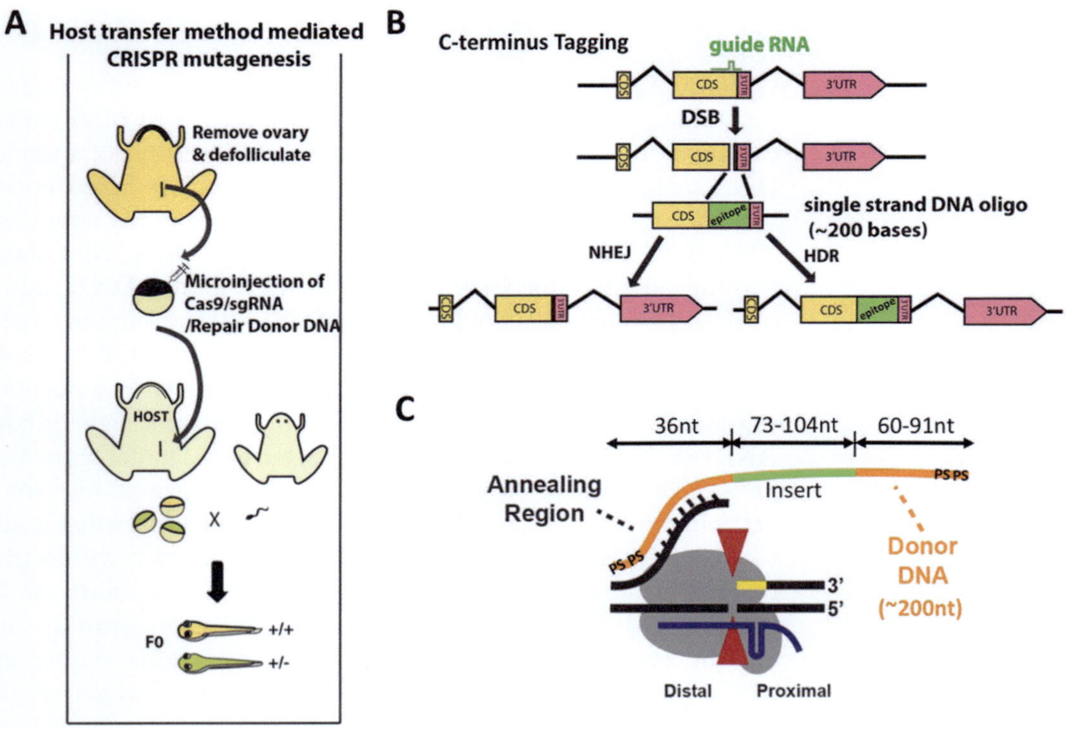

Fig. 1 (**a**) The schematic of CRISPR mutagenesis by host transfer technique. (**b**) The schematic of epitope tagging on C-terminus of a genomic loci. (**c**) A diagram shows an asymmetric DNA donor has two phosphorothioate (PS) modifications on both ends. The lengths of homology arms can be adjusted with the size of the insert

successful knock-in events, this approach also ensures that each mutant embryo from the host-transfer is a non-mosaic heterozygote, since the edited genome in the oocyte is inherited by every single cell of the embryo. The method that we optimized is a straightforward approach to generate non-mosaic F0 heterozygous mutant *Xenopus* with germline transmission, which offers the possibility to obtain homozygous mutants one generation (F1) earlier than previously reported [6, 14, 17].

2 Materials

2.1 Construction of CRISPR/Cas9 Component

1. CRISPRScan (www.crisprscan.org/) and CRISPRdirect (http://crispr.dbcls.jp/) websites for primers design of sgRNA template construction.
2. Primer3 website (http://bioinfo.ut.ee/primer3) for primers design for genomic PCR and sequencing.
3. GGGnome website (https://gggenome.dbcls.jp/) for off-target sites screening.
4. Phusion high-Fidelity PCR kit (NEB), QIAquick PCR purification kit (Qiagen).
5. MegaShortscript Kit (Ambion).
6. Recombinant SpCas9 protein (PNA Bio).
7. Repair oligo (Ultramer™, Integrated DNA technologies).

2.2 Oocyte Collection

1. Ethyl 3-aminobenzoate methanesulfonate salt (tricaine; sigma, MS222) stored in 2 g aliquots at −20 °C and dissolved in 1 L of 0.1× MMR for use.
2. 70% ethanol (for surgical bench sterilization).
3. Surgical instruments: Scissors, surgical blades, needle holders, forceps.
4. Glass bead sterilizer for sterilization of surgical instruments.
5. Oocyte culture media (OCM): 60% L-15 (sigma L4386) supplemented with 0.1 mg/mL BSA and 50 UI penicillin/streptomycin. Complete solution made fresh and stored at 14–16 °C for up to 1 week. The stock solution without BSA or antibiotics is prepared by dissolving the L-15 powder in sterile water (1667 mL of distilled water per a packet of 13.8 g: Sigma L4386), adjusting the pH at 7.6–7.8 (with NaOH) and filter sterilizing using a membrane with a porosity of 0.22 microns (e.g., Steritop Filter Unit, SCGPT10RE).
6. 4–0 silk sutures, black braided c-17, with 12 mm 3/8 circle needle (surgical specialties, cat# 785B).

7. Collagenase VII—Cell culture grade (e.g., sigma, C2799-15KU).
8. Pasteur pipettes.

2.3 Oocyte Injection and Host Transfer

1. DNA ligase IV inhibitor (SCR-7; Selleck chemical).
2. Human Chorionic Gonadotropin (e.g., Chorulon, Intervet).
3. Progesterone (e.g., Sigma), stored at 1 mM in absolute ethanol at −20 °C.
4. Vital dye stock solutions; prepared by vigorous shaking of powder in sterile water for 30 min minimum (2 h for Nile blue) and stored in 1 mL aliquots at −20 °C:
   (a) Red (neutral red, sigma): 2.5 mg/mL.
   (b) Brown (Bismark Brown, sigma): 10 mg/mL.
   (c) Blue (Nile blue, sigma): 1 mg/mL.
5. Marc's Modified Ringer's (MMR). For 1 L of 10× stock in sterile water: 58.44 g NaCl, 1.5 g KCl, 2.94 g $CaCl_2$, 2.03 g $MgCl_2$, 35.7 g HEPES; pH adjusted to 7.6 with 5 M NaOH; filter sterilized and stored at room temperature.
6. High salt solution. For 4 L of 10× stock in sterile water: 280 g NaCl, 7.2 g KCl, 14.1 g $CaCl_2$, 9.7 g $MgCl_2$, 17.12 g HEPES; pH adjusted to 7.6 with NaOH; filter sterilized and stored at room temperature.
7. Dejellying solution: 2% cysteine (Sigma) in 0.1× MMR, adjust to pH 7.6–7.8 with NaOH; prepared fresh each time.
8. For intraperitoneal injection: Shielded IV catheters, BD Insyte Autoguard, (Cat#381457 for *X. laevis* and cat# 330003 for *X. tropicalis*).

2.4 Genotyping

1. Lysis buffer: 50 mM Tris (pH 8.8), 1 mM EDTA, 0.5% Tween 20, 200ug/mL Proteinase K (PCR grade; RPROTK-RO ROCHE; Sigma).
2. Phusion High-Fidelity PCR Kit (NEB).
3. T7E1 endonuclease (NEB).
4. TOPO TA Cloning Kit (Life Technologies).

3 Methods

3.1 Construction of CRISPR/Cas9 Component

1. Potential CRISPR/Cas9 target sites are identified by locating sites adjacent to PAM sequences within the genomic sequence of the gene of interest. This is done using the CRISPRdirect website for *X. laevis* and the CRISPRScan website for *X. tropicalis* (*see* **Note 1**).

2. Off-target sites screening: Two mismatches are allowed in each sgRNA targeting sequence, which is then used to search for potential targets on the GGGenome website. Repeated sequences such as AAAA, TTTT should be avoided. If the mismatch is located 12 bases away from PAM sequence then there is a chance that the sgRNA can still recognize it as a potential target since Cas9 has more tolerance after its 8–12 nucleotides seed sequence. For sgRNA template construction, PCR primers are designed as follows. The forward primer is unique for each target and contains the promoter sequence (for T7 polymerase in the following sequence) followed by an initiator guanine for successful in vitro transcription, then the targeting sequence(N_{18-20}) without the PAM sequence, followed by a portion of the sgRNA backbone: 5'-GAAATTAATACGACTCACTATAGG$\underline{N_{18-20}}$GTTTAAGAGCTATGCTG*GAAACAGCATAGCA-3'.* The reverse primer, which has the sequence 5'-AAAAGCACCGACTCGGTGCCACTTTTTCAAGTTGATAACGGACTAGCCTTATTTAAACT*TGCTATGCTGTTTC*-3′, is the same for all sgRNA templates and is partially complementary (indicated in *italic*) to the sgRNA backbone (*see* **Note 2**).
3. Construction of sgRNA templates: A high fidelity DNA polymerase is used to generate the templates by PCR in a total reaction mixture of 100 μL (10 μL of 10× buffer, 1.2 μL of 25 mM dNTP mix, 2 μL of 50 mM $MgSO_4$, 2 μL each primer at 100 pmol/μL, 1 μL of DNA polymerase, DNase/RNase-free H_2O to 100 μL); the PCR conditions are 94 °C for 5 min and then 20 cycles of 94 °C for 20s, 58 °C for 20s, 68 °C for 15 s followed by a final extension of 5 min at 68 °C. A fraction (2–5 μL) of the reaction is analyzed by agarose gel electrophoresis to confirm the synthesis of the desired product, and the templates are column-purified.
4. Synthesis of sgRNA by in vitro transcription: The MegaShortscript Kit (Ambion) is used according to the manufacturer's recommendations.
5. sgRNA purification: the sgRNA is purified using the RNeasy Kit (Qiagen).
6. Design of the repair construct for knock-in: The repair oligo for tagging of endogenous proteins is a 200 bases single-stranded oligonucleotide with typically 30–50 bases homology that spans the targeted locus, and an epitope tag added just before the stop codon. Two to four silent mutations are introduced in the sgRNA targeting sequence to prevent the repair construct from being cut by the Cas9/sgRNA complex after successful recombination. Also, having phosphorothioate modifications in the last two bases on both ends and having asymmetric homology arms in the donor oligo (which has

different lengths of homology arm) increases the efficiency of the HDR mediated knock-in [20, 21] and unpublished *Xenopus* data (Fig. 1b, c).

3.2 Collection of Oocytes

Except for some minor modifications, the oocyte collection is done following the method described by Mir and Heasman [15].

1. Frog surgeries are performed under aseptic conditions. Scissors, needle holders, and forceps are sterilized using a glass bead sterilizer at 250 °C for 5 min. Surgical blades are single use. The surgical bench is cleaned, thoroughly sprayed with 70% ethanol and wiped down with paper towel.
2. Ovary donor females are submerged in 0.2% Tricaine and closely monitored until they become unresponsiveness to the swallowing reflex.
3. With the animal laying on its back, a small incision is made in the skin on one side of the lower abdomen using a surgical blade. The incision is widened using sterile curved scissors until it is about 1.0 cm in width. The abdominal wall is gently lifted away from underlying organs (to prevent accidental perforation), and a similar incision is made through the muscle layer. A pair of forceps is used to pull a small amount of ovary out from the abdominal cavity. The ovary is cut with scissors, and placed in OCM for evaluation under a dissecting microscope.
4. After ovary collection, the incision is sutured, starting with the muscle layer (three stitches) and then the skin (three stiches).
5. The frog is returned to fresh water and monitored until it is fully awake and alert.
6. The ovary is cut into small-sized pieces and stored in OCM at a density of 6–12 pieces per 90 mm dish.
7. Oocytes isolation:
   (a) For *X. laevis*, oocytes are manually defolliculated in OCM under aseptic conditions (pipettes and forceps are sterilized prior to use). During defolliculation, the dish and medium are changed every time the medium becomes cloudy (**Note 3**).
   (b) For *X. tropicalis*, because of the small size of oocytes, manual defolliculation is often tedious. Instead, small pieces of ovary are incubated at room temperature in OCM containing collagenase VII at 0.1 KU/mL for about 2 h with gentle shaking; individual full-grown oocytes are then sorted and thoroughly rinsed in fresh OCM (without collagenase).
8. Each batch of defolliculated oocytes is tested for maturation before a full experiment is set up. To do so, a couple of defolliculated oocytes are treated with 2 μM progesterone in OCM, for 5–8 h (for *X. laevis*) or 30 min to 1 h (for *X. tropicalis*) at

room temperature. Maturation is confirmed by appearance of a white spot at the animal pole of the oocyte, which marks the germinal vesicle breakdown (GVBD).

3.3 sgRNA Injection and Oocytes Host Transfer

1. For INDEL mutations, mixtures of sgRNA (300 pg for *X. laevis*, 200 pg for *X. tropicalis*) and Cas9 protein (600 pg for *X. laevis*, 300 pg for *X. tropicalis*) are injected into oocytes maintained in OCM (*see* **Note 4 for the injection volume).**
2. For knock-in experiments, the injection cocktail also includes the repair oligo (200 pg for *X. laevis*, 30 pg for *X. tropicalis*) and the post injection culture medium contains the DNA ligase IV inhibitor SCR-7 at a final concentration of 5 μM (*see* **Note 5**).
3. Injected oocytes are cultured in OCM for 72 h at 18 °C (for *X. laevis*) or 23 °C (for *X. tropicalis*); culture medium is changed daily. Before each medium change, dead and unhealthy oocytes are sorted out and discarded (*see* **Note 6**).
4. Oocytes are matured by further culture in 10 mL OCM supplemented with 2 μM progesterone for 12 h at 18 °C (for *X. laevis*) or 3 h a 23 °C (for *X. tropicalis*) (*see* **Note** 7).
5. 12–14 h before the host transfer surgery, host *X. laevis* females are injected with 1000 units of HCG, and kept at 18 °C. It is recommended to prepare more female frogs than needed, to ensure that at least one will be laying eggs within the time available for the surgery to be performed (not later than 16 h from the addition of progesterone to oocytes for both *Xenopus* species). As a general rule, for each host female needed, two extra ones are injected.
6. To distinguish between experimental groups, the oocytes are stained just prior to host transfer by adding vital dye stock solutions to the dish containing the oocytes (in 10 mL OCM); the staining is achieved by a gentle rocking at room temperature. Color combinations are as follows:
   (a) Red: 100 μL of red: Stain for 10 min.
   (b) Brown: 100 μL of Brown: Stain for 10 min.
   (c) Blue: 100 μL of blue: Stain for 10 min.
   (d) Purple: 75 μL of red +100 μL of blue: Stain for 10 min.
   (e) Green: 100 μL of Brown +100 μL of blue: Stain for 10 min (*see* **Note 8**).
7. The stained oocytes are then transferred to a 90 mm dish containing fresh OCM, with care being taken to carry over as little dye as possible (*see* **Note 9**).
8. Midway during the oocyte staining time, a host female that has just begun laying good quality eggs is submerged in 0.2% Tricaine and monitored until it just becomes unresponsive.

3.3.1 Oocytes Transfer by Surgical Implantation

1. A small incision is made in the skin on one side of the midline on the lower abdomen of the frog. Another incision is made through the body wall, with care being taken to raise it away from the underlying organs with forceps to avoid damaging them. The body wall incision should just be large enough to accommodate the tip of the glass pipette (*see* **Note 10**).
2. One edge of the body wall is grasped with forceps and held up, away from the abdomen. To prevent the loss of experimental oocytes, this edge is held up until the first stitch is made following oocytes implantation.
3. An OCM-coated Pasteur pipette is loaded with the treated oocytes by the front filling and is used to deposit the oocytes inside the body cavity, making sure that as little OCM as possible is transferred. Also, the pipette is held at an angle and turned around underneath the incision as the oocytes are expelled; this prevents crowding of the oocytes, by allowing them to spread throughout the area around the incision. The transferred oocytes should mix with the host's own ovulated oocytes, and lie in the space between the ovary and the body wall (*see* **Note 11**).
4. The body wall and skin are sutured as during the ovary collection, and the frog is allowed to recover from anesthesia in water.
5. After the frog is fully awake and alert (within 30 min), it is transferred into 2 L high salt solution.

3.3.2 Oocytes Transfer by Intraperitoneal Injection

1. The anesthetized frog is laid on its back and the needle is inserted in the lower abdomen at low angle (close to parallel to the surgery bench) with the tip facing toward the midline of the frog and a finger applied just above the injection site to provide support.
2. The needle is allowed to insert smoothly through the skin and the muscle layer (The needle will have crossed both layers when resistance against the needle can no longer be felt).
3. The inserted needle is gently lifted (to avoid puncturing internal organs) and carefully pushed until the catheter's head is inside the abdominal cavity.
4. The catheter is stabilized in place by holding the tab with one hand, and the (white) activation button gently pushed to withdraw the needle attached to the safety barrel.
5. Use a glass pipette to push about 500 μL of OCM into the catheter to verify that it is free of obstruction and coat the plastic barrel with BSA.
6. The oocytes are then pipetted through the catheter into the abdominal cavity.

7. The catheter is carefully removed after tilting the frog to help the transferred oocytes spread into the abdominal cavity and the frog is returned to fresh water for recovery. No suture is needed in this case and usually no bleeding has been observed.

3.4 Eggs Collection and Fertilization

1. The host frog is allowed to release the eggs into high salt. 2 h after a the first colored eggs has appeared, the frog is gentle squeezed in high salt from time to time (with at least 30 min intervals) for a maximum of 2 more hours to help release the colored eggs. If no colored eggs are released after 5 h post surgery, the frog is squeezed anyway.
2. The high salt eggs are collected in a Petri dish, and the colored eggs are sorted away from the host eggs into a new dish using a pipette.
3. Excess high salt solution is removed from the eggs by decanting and using a pipette, being careful not to lose any colored eggs, as the high salt prevents them from sticking to the dish.
4. A large piece of testis that matches with transferred species is minced directly in the dish with the eggs, and spread around dish using a plastic transfer pipette.
5. After 5 min, the eggs are flooded with 0.1× MMR.
6. About 1 h after fertilization, the embryos are dejellyed with 2% cysteine and rinsed thoroughly with fresh 0.1 × MMR (*see* **Note 12**).
7. The embryos are sorted by color and allowed to develop to tadpole stage.

3.5 Genotyping

3.5.1 Genomic DNA Extraction from Tadpoles

1. Each tadpoles (NF stage 42–45) is treated with 0.025% Tricaine for 30 s.
2. One fourth of the tail is cut in a petri dish while the tadpole is anesthetized.
3. The tadpole is returned to 0.1 × MMR in a labeled dish and monitored until full recovery. From this point on, each tadpole is assigned an identification number (each been cultured individually in a 24-well plate).
4. Each tail is incubated in lysis buffer (150 μL per tail for *X. laevis*; 100 μL per tail for *X. tropicalis*) for 6 h (up to overnight) at 56 °C.
5. Inactivate Proteinase K activity by incubation at 95 °C for 10 min. Then cool it down to room temperature. Make 1:5 dilution with nuclease-free water for PCR reaction.

3.5.2 Genomic PCR

1. The Phusion High-fidelity Kit is used to perform PCR according to the manufacturer's instructions.
2. For each sample, 1–4 μL of diluted genomic DNA solution is used as template.

3. The general PCR condition is:
   (a) Annealing: 64 °C for 20 s.
   (b) Extension: 72 °C for 20 s.
   (c) Cycles: 40.
4. For each sgRNA, a genomic fragment containing the target regions is amplified by PCR from genomic DNA.
5. A fraction of the PCR reaction is analyzed on a 2% agarose gel for successful amplification.
6. The PCR products are purified using the QIAquick PCR purification.

3.5.3 T7E1 Endonuclease-Based INDEL Mutation Detection

1. 200 ng of PCR product is used to perform the T7 E1 endonuclease essay according to the manufacturer's instructions; two reaction mixtures are prepared for each sgRNA (one where the T7 E1 enzyme is added and one where the enzyme is not added).
2. The samples are then analyzed on a 2% agarose gel: for each sgRNA, the reaction without T7E1 enzyme and the reaction with the enzyme are loaded in adjacent lanes for direct comparison (*see* **Note 13**).

3.6 Sequencing

3.6.1 For General Purpose

1. Direct Sanger sequencing with either one of PCR primers or one nested primer would show the mixed peaks after predicted sgRNA binding site.
2. TIDE or CRISP-ID site can reveal the frequency of the mutations and the nature of mutations [22, 23]. TIDE can tell how many bases are inserted/deleted and the proportion of those mutations among the PCR products and CRISP-ID could tell the unmixed sequences from the mixed peaks up to three sequences.

3.6.2 For Individual Clones

1. To get the actual mutation information for both type and frequency, the PCR products are cloned using the TOPO TA cloning kit according to the manufacturer's instructions.
2. 20 randomly selected clones are sequenced for genotype validation.

4 Notes

1. When designing the primers for the construction of the sgRNA template, it is very important to make sure that a PAM sequence is not present within the sgRNA itself.
2. This design allows the forward and the reverse primers to be complementary at their respective 3′ ends; when the two

primers are mixed, they can base pair and each of them is extended to generate the double stranded sgRNA template.

3. There is no real "one size fits all" formula for successful manual defolliculation, not even for how to sharp or blunt the forceps should be. It takes practice to find a style that works. The key is to hold the stalk that links the oocyte to the ovary with one pair of forceps (preferentially with the weak hand) and use the other pair (in the strong hand) to gently peal away the tissue surrounding the oocyte, without touching the oocyte itself. If a freshly isolated ovary is hard to defolliculate, it is often helpful to leave it overnight (at 14 °C) and defolliculate the following day.
4. Since the sgRNA is intended to target nuclear DNA, the injection is performed in the animal atmosphere of the oocyte where the nucleus is, about midway between the animal pole and the equator. The injection volume for *X. laevis* is up to 10 nl and for *X. tropicalis* is up to 4 nl.
5. SCR-7 inhibits the NHEJ repair pathway and thus increases the frequency of HDR, which favors the insertion of the knock-in (repair) construct.
6. To ease the medium changing process, it is recommended to prepare a new petri dish with fresh OCM and simply transfer the healthy oocytes inside, leaving behind the unhealthy and dead ones.
7. For a host transfer surgery that is scheduled in the morning, progesterone is added to *X. laevis* oocytes the evening before host surgery day and to *X. tropicalis* oocytes the morning of the surgery day (3 h prior to the surgery time).
8. When blue and green are both used in the same experiment, it might be difficult to tell a light blue embryo from a green one. Increasing the blue dye to 150 μL/10 mL might be helpful. Alternatively, when staining the green group, the oocytes can be incubated in Brown alone for about 5 min before the green is added; this will give a brown foundation to the vegetal mass making the green oocyte easy to distinguish from blue ones for which the vegetal mass will keep its white background color.
9. To limit dye carry-over, the oocytes are collected in a glass Pasteur pipette; with the pipette tip facing down, the bulb is gently released to allow oocytes to settle near the pipette opening; the oocytes are then emptied in the new dish with one quick but gentle squeeze on the bulb.
10. To make sure that the incision has gone through the body wall, a glass pipette containing a bit of fresh OCM is inserted through the incision to pipette out a few host eggs.

11. Before loading the oocytes into the host, it is critical to make sure that all the groups are homogeneously mixed to ensure that they will be laid at the same rate.
12. If no early analysis is to be performed on the embryos, they can be allowed to develop until the blastula stage before dejellying. This tends to improve the survival rate of host transfer embryos.
13. If the CRISPR editing of the oocytes DNA was successful, the host transfer embryo should be heterozygous. Therefore, some hybrid DNA duplex should form between the wild type (paternal) fragment and the mutated (maternal) one after denaturation and renaturation of the PCR product. Since these hybrids contain mismatches at the mutation sites, they will be cut by the T7E1 endonuclease, generating fragments that are smaller than the full length PCR product. If there was no mutation, there will be no mismatch and thus no cut.

References

1. Mali P, Yang L, Esvelt KM, Aach J, Guell M, DiCarlo JE, Norville JE, Church GM (2013) RNA-guided human genome engineering via Cas9. Science 339(6121):823–826. https://doi.org/10.1126/science.1232033
2. Wang H, Yang H, Shivalila CS, Dawlaty MM, Cheng AW, Zhang F, Jaenisch R (2013) One-step generation of mice carrying mutations in multiple genes by CRISPR/Cas-mediated genome engineering. Cell 153(4):910–918. https://doi.org/10.1016/j.cell.2013.04.025
3. Nakayama T, Blitz IL, Fish MB, Odeleye AO, Manohar S, Cho KW, Grainger RM (2014) Cas9-based genome editing in Xenopus tropicalis. Methods Enzymol 546:355–375. https://doi.org/10.1016/B978-0-12-801185-0.00017-9
4. Blitz IL, Biesinger J, Xie X, Cho KW (2013) Biallelic genome modification in F(0) Xenopus tropicalis embryos using the CRISPR/Cas system. Genesis 51(12):827–834. https://doi.org/10.1002/dvg.22719
5. Kotani H, Taimatsu K, Ohga R, Ota S, Kawahara A (2015) Efficient multiple genome modifications induced by the crRNAs, tracrRNA and Cas9 protein complex in Zebrafish. PLoS One 10(5):e0128319. https://doi.org/10.1371/journal.pone.0128319
6. Nakayama T, Fish MB, Fisher M, Oomen-Hajagos J, Thomsen GH, Grainger RM (2013) Simple and efficient CRISPR/Cas9-mediated targeted mutagenesis in Xenopus tropicalis. Genesis 51(12):835–843. https://doi.org/10.1002/dvg.22720
7. Wang F, Shi Z, Cui Y, Guo X, Shi YB, Chen Y (2015) Targeted gene disruption in Xenopus laevis using CRISPR/Cas9. Cell Biosci 5:15. https://doi.org/10.1186/s13578-015-0006-1
8. Xue W, Chen S, Yin H, Tammela T, Papagiannakopoulos T, Joshi NS, Cai W, Yang G, Bronson R, Crowley DG, Zhang F, Anderson DG, Sharp PA, Jacks T (2014) CRISPR-mediated direct mutation of cancer genes in the mouse liver. Nature 514(7522):380–384. https://doi.org/10.1038/nature13589
9. Aslan Y, Tadjuidje E, Zorn AM, Cha SW (2017) High-efficiency non-mosaic CRISPR-mediated knock-in and indel mutation in F0 Xenopus. Development 144(15):2852–2858. https://doi.org/10.1242/dev.152967
10. Miyaoka Y, Chan AH, Judge LM, Yoo J, Huang M, Nguyen TD, Lizarraga PP, So PL, Conklin BR (2014) Isolation of single-base genome-edited human iPS cells without antibiotic selection. Nat Methods 11(3):291–293. https://doi.org/10.1038/nmeth.2840
11. Yu Z, Chen H, Liu J, Zhang H, Yan Y, Zhu N, Guo Y, Yang B, Chang Y, Dai F, Liang X, Chen Y, Shen Y, Deng WM, Chen J, Zhang B, Li C, Jiao R (2014) Various applications of TALEN- and CRISPR/Cas9-mediated homologous recombination to modify the Drosophila genome. Biol Open 3(4):271–280. https://doi.org/10.1242/bio.20147682
12. Harrison MM, Jenkins BV, O'Connor-Giles KM, Wildonger J (2014) A CRISPR view of

development. Genes Dev 28(17):1859–1872. https://doi.org/10.1101/gad.248252.114

13. Peng Y, Clark KJ, Campbell JM, Panetta MR, Guo Y, Ekker SC (2014) Making designer mutants in model organisms. Development 141(21):4042–4054. https://doi.org/10.1242/dev.102186
14. Guo X, Zhang T, Hu Z, Zhang Y, Shi Z, Wang Q, Cui Y, Wang F, Zhao H, Chen Y (2014) Efficient RNA/Cas9-mediated genome editing in Xenopus tropicalis. Development 141(3):707–714. https://doi.org/10.1242/dev.099853
15. Mir A, Heasman J (2008) How the mother can help: studying maternal Wnt signaling by antisense-mediated depletion of maternal mRNAs and the host transfer technique. Methods Mol Biol 469:417–429. https://doi.org/10.1007/978-1-60327-469-2_26
16. Olson DJ, Hulstrand AM, Houston DW (2012) Maternal mRNA knock-down studies: antisense experiments using the host-transfer technique in *Xenopus laevis* and Xenopus tropicalis. Methods Mol Biol 917:167–182. https://doi.org/10.1007/978-1-61779-992-1_10
17. Ratzan W, Falco R, Salanga C, Salanga M, Horb ME (2017) Generation of a *Xenopus laevis* F1 albino J strain by genome editing and oocyte host-transfer. Dev Biol 426(2):188–193. https://doi.org/10.1016/j.ydbio.2016.03.006
18. Chu VT, Weber T, Wefers B, Wurst W, Sander S, Rajewsky K, Kuhn R (2015) Increasing the efficiency of homology-directed repair for CRISPR-Cas9-induced precise gene editing in mammalian cells. Nat Biotechnol 33(5):543–548. https://doi.org/10.1038/nbt.3198
19. Maruyama T, Dougan SK, Truttmann MC, Bilate AM, Ingram JR, Ploegh HL (2015) Increasing the efficiency of precise genome editing with CRISPR-Cas9 by inhibition of nonhomologous end joining. Nat Biotechnol 33(5):538–542. https://doi.org/10.1038/nbt.3190
20. Renaud JB, Boix C, Charpentier M, De Cian A, Cochennec J, Duvernois-Berthet E, Perrouault L, Tesson L, Edouard J, Thinard R, Cherifi Y, Menoret S, Fontaniere S, de Croze N, Fraichard A, Sohm F, Anegon I, Concordet JP, Giovannangeli C (2016) Improved genome editing efficiency and flexibility using modified oligonucleotides with TALEN and CRISPR-Cas9 nucleases. Cell Rep 14(9):2263–2272. https://doi.org/10.1016/j.celrep.2016.02.018
21. Richardson CD, Ray GJ, DeWitt MA, Curie GL, Corn JE (2016) Enhancing homology-directed genome editing by catalytically active and inactive CRISPR-Cas9 using asymmetric donor DNA. Nat Biotechnol 34(3):339–344. https://doi.org/10.1038/nbt.3481
22. Brinkman EK, Chen T, Amendola M, van Steensel B (2014) Easy quantitative assessment of genome editing by sequence trace decomposition. Nucleic Acids Res 42(22):e168. https://doi.org/10.1093/nar/gku936
23. Dehairs J, Talebi A, Cherifi Y, Swinnen JV (2016) CRISP-ID: decoding CRISPR mediated indels by sanger sequencing. Sci Rep 6:28973. https://doi.org/10.1038/srep28973

Chapter 9

Targeted Electroporation in the CNS in *Xenopus* Embryos

Hovy Ho-Wai Wong and Christine E. Holt

Abstract

Electroporation allows targeting of genetic materials (e.g., DNA, RNA, antisense morpholinos) to the tissue of interest with high spatial and temporal specificity. Here, we describe a highly efficient and reproducible electroporation strategy for targeting the central nervous system in *Xenopus*. This versatile approach can be combined with live imaging or other existing experimental procedures to aid the investigation of different research questions.

Key words Axon navigation, Electroporation, Cell Imaging, Fluorescence, In vivo imaging, Living imaging, Neurodevelopment, Neural wiring, Optic tract, Xenopus

1 Introduction

Transfection has been a crucial tool for biologists to answer fundamental questions in complex systems. It allows the labeling of cells for visualization, overexpression of proteins, as well as knockdown/knockout of gene expression and translation. Two major considerations in the choice of transfection methods are the spatial and temporal selectivity. Blastomere microinjection [1], for example, allows bulk targeting of cells from early stages as young as the 1-cell stage. This method has been widely used and is often preferred in biochemical experiments or in population-level imaging experiments. However, it is not ideal in scenarios where the labeling of individual cells is needed or when the manipulation of protein expression would result in embryogenesis defects or even lethality during early development [2–5]. An approach to circumvent this problem is to obtain restricted ectopic protein expression by the use of cDNA under control of cell type-specific or inducible promoters, although these promoters tend to be leaky and show only modest levels of expression. This could arise, in part, from the temporal expression patterns of the promoters and the dilution of cDNA by the many cell divisions during embryogenesis. Moreover, the use of promoters to restrict or induce ectopic protein expression

Kris Vleminckx (ed.), *Xenopus: Methods and Protocols*, Methods in Molecular Biology, vol. 1865,
https://doi.org/10.1007/978-1-4939-8784-9_9,

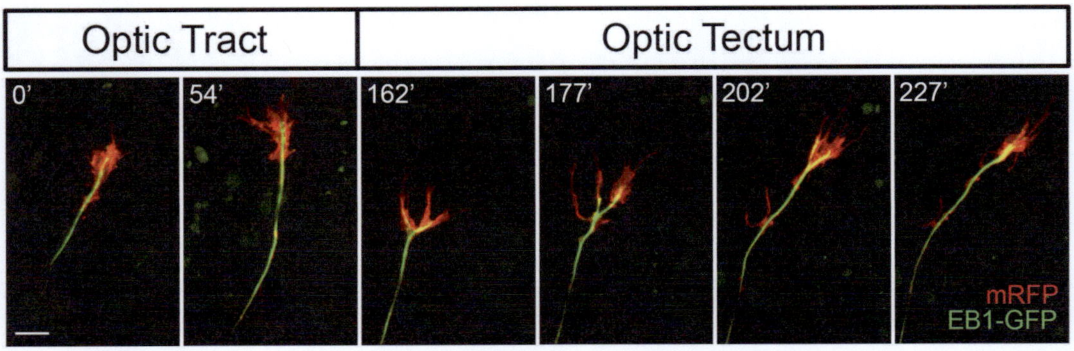

Fig. 1 In vivo time-lapse images showing the navigation of a retinal axon navigating through the optic tract and arborizing after reaching the optic tectum. The axon is labeled by mRFP and expresses GFP-tagged microtubule plus-end-tracking protein EB1. Scale bar, 10 μm

is not possible with mRNA, antisense oligonucleotides, or morpholino injection.

Here, we give a detailed description of a highly reproducible electroporation strategy using an electroporation chamber [6], with a particular focus on the practical aspects. This method allows efficient and spatially restricted targeting of cells in the central nervous system, in particular the retina, during the critical developmental periods of stages 21–32. Moreover, we have adapted the method for local delivery of genetic materials into the optic tract and into the optic tectum (stages 33/34–43) in the intact brain [7]. Combining the careful selection of stages and the two methods described above, it allows selective targeting of different cell types (e.g., cells in the retina vs. cells in the brain) and different neuronal compartments (e.g., cell body/dendrites vs. axons of retinal ganglion cells). This strategy can be used for validating and enhancing our understanding of the different steps of neural wiring in the visual circuit. In addition to cell labeling and protein expression manipulations [2, 4–6, 8–13], this method has been successfully implemented to label and track the dynamics of organelles and macromolecules such as microtubule plus-end-tracking protein EB1 (Fig. 1), mitochondria [3, 7], RNA binding proteins [14], and endogenous RNA in vivo [7]. Electroporation of post-mitotic cells also allows high transfection efficiency at an individual cell level, which could be critical for certain experiments such as protein expression knockdown and live imaging [7, 8, 15, 16]. The electroporated cells, like their non-electroporated counterparts, are also amenable to normal in vitro culture systems [15–19]. Therefore, it is highly adaptable to different research questions and experimental setups.

2 Materials

2.1 Electroporation Chamber

1. Sylgard dish: Mix the base and curing agents from the Sylgard 184 Silicone Elastomer Kit in a 10:1 ratio. Pour the mixture into a 35 mm dish to create a 0.8 cm layer of silicone elastomer at the bottom. Leave the Sylgard to polymerize at room temperature for 24 h.
2. Design for Stage 21–23 embryos: A "T" shape chamber with a length (*L*) of 2.5 mm × width (*W*) of 1 mm × depth (*D*) of 1 mm in the longitudinal channel, and 2 mm (*L*) × 0.5 mm (W) × 0.25 mm (*D*) in the transverse channel (Fig. 2a).
3. Design for Stage 24–32 embryos: A "†" shape chamber with a dimension of 7 mm (*L*) × 1 mm (W) × 1 mm (*D*) in the longitudinal channel and 4 mm (*L*) × 0.8 mm (*W*) × 0.25 mm (*D*) in the transverse channel, which crosses at 1 mm from the end of longitudinal channel (Fig. 2a).
4. Design for Stage 33/34–43 embryos: A "†" shape chamber with a dimension of 9 mm (*L*) × 1 mm (*W*) × 1 mm (*D*) in the longitudinal channel and 4 mm (*L*) × 0.8 mm (*W*) × 0.25 mm (*D*) in the transverse channel, which crosses at 1 mm from the end of longitudinal channel (Fig. 2a).
5. Millimetric grid graph papers.
6. No. 15A surgical scalpel blade.
7. 27 G Needle.
8. Fine forceps (#5).

2.2 Solutions

1. 10× Modified Barth's saline (MBS) pH 7.6: 880 mM NaCl, 10 mM KCl, 24 mM $NaHCO_3$, 100 mM HEPES, 8.2 mM $MgSO_4$, 3.3 mM $Ca(NO_3)_2$, 4.1 mM $CaCl_2$. Add water to a volume of 900 ml with double distilled water (ddH_2O). Mix and adjust pH with about 7 ml of 4 M NaOH. Make up to 1 L with ddH_2O. Filter and store at 4 °C.
2. Electroporation solution: 1× MBS, 0.4 mg/ml of tricaine methanesulfonate (MS-222), pH 7.6. Make 50 ml 1× MBS by mixing 5 ml of 10× MBS with 45 ml of ddH_2O. Weigh 20 mg of MS-222 and dissolve in 50 ml of 1× MBS. Mix and adjust pH with 1 M NaOH.
3. Rearing solution: 0.1× MBS. Mix 0.5 ml of 10× MBS with 49.5 ml of ddH_2O.
4. Injection mixture (*see* **Note 1**): DNA (1–2 μg/μl); RNA (1–1.5 μg/μl); Morpholino (0.5–1 mM) (*see* **Note 2**).

2.3 Brain Exposure

1. Sylgard dish.
2. 0.1 mm insect pins.
3. Insect pin holders.

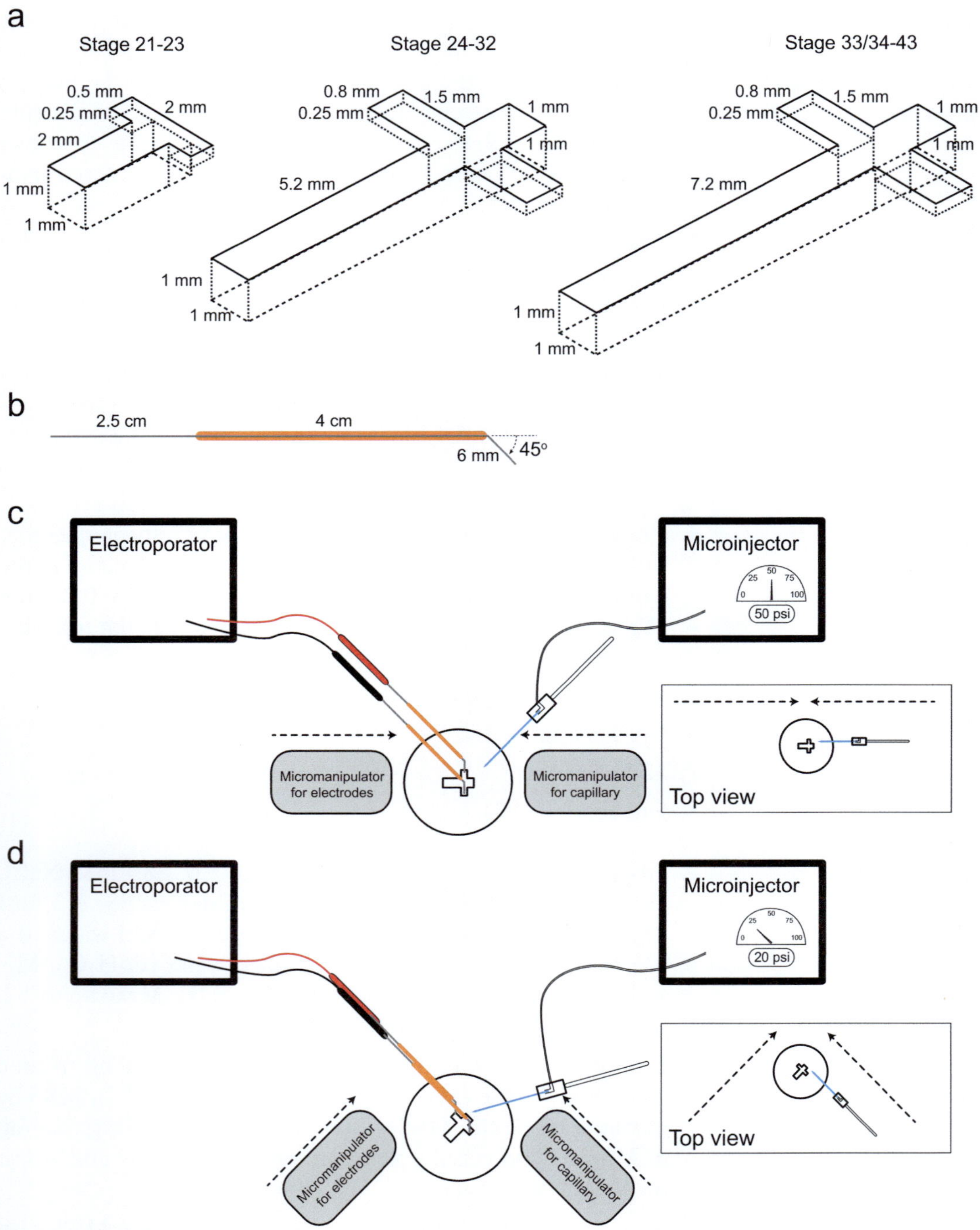

Fig. 2 Electroporation setup. (**a**) Electroporation chamber designs for embryos of different stages. (b) Electrode design. The orange segment denotes the insulated part of the electrode. The electrode tip can be flattened and polished with electric sander and sharpening stones in a rotary tool kit. The 45° bend facilitates the positioning of electrodes at the ends of the transverse channel. (**c**) Setup for targeted retina electroporation. The micromanipulator-held electrode adaptor, chamber, and the micromanipulator-held capillary holder are arranged from left to right in a straight horizontal line. (**d**) Setup for targeted brain electroporation. The micromanipulator-held electrode adaptor and chamber are arranged from left to right in a straight line with a 45° angle toward the chamber. The micromanipulator-held capillary holder and chamber from right to left in a straight line with a 45° angle toward the chamber

2.4 Electroporation Setup

1. Pressurized microinjector with adjustable pulse duration and pressure.
2. Standard 1 mm outer diameter (OD) glass capillary holder.
3. Pulled 1 mm OD borosilicate glass capillary with 0.78 mm internal diameter (ID) (*see* **Note 3**).
4. Electroporator with square wave pulse generation capability (*see* **Note 4**).
5. Electrode adaptor.
6. Electrodes: A flat-ended 0.5 mm diameter platinum electrode consisting of three segments. (a) A 2.5 cm unmodified segment for insertion into the electrode adaptor. (b) A 4 cm insulated segment to allow a convenient working distance (*see* **Note 5**). (c) A 6 mm flattened electrode tip bent with a 45° angle from segment ii (*see* **Note 6**) (Fig. 2b).
7. Micromanipulators for glass capillary holder and electrode adaptor (*see* **Note 7**).
8. Steering glass rod with rounded end (*see* **Note 8**): Carefully heat the middle part of the thinner segment of a borosilicate glass Pasteur pipette with a Bunsen burner until malleable (*see* **Note 9**). Gently pull apart the two ends of the pipette to create a 5 mm segment angled at 45° on the end of the pipette tip. Wait for the angled glass rod to cool down and solidify, heat the tip of the 5 mm segment until malleable, and push gently against a hard surface to create a rounded end with 0.5 mm diameter. Clip and break off the excessive end to create an angled glass rod of 3.5 cm long (*see* **Note 10**). Seal the clipped end with tapes (*see* **Note 11**).
9. Stereo microscope with light source.

3 Methods

All procedures are carried out at room temperature. Chamber carving and electroporation are both done under a stereo microscope.

3.1 Electroporation Chamber Carving

1. Draw the dimensions of the desired electroporation chamber design on millimetric grid graph paper to create a blueprint (Fig. 2a).
2. Place the blueprint underneath a Sylgard dish.
3. Sculpt the channel edges with a No. 15A scalpel blade by following the blueprint.
4. Carve the z-dimension of the chamber using a 27 G needle and fine forceps.

5. Place an anesthetized embryo of desired stage in the chamber to check chamber fitting.
6. Repeat **steps 3–5** until a good generic fit is achieved.

3.2 Preparation

1. Remove vitelline membrane of the embryos with fine forceps (*see* **Note 12**).
2. Wash the embryos in 0.1× MBS and transfer to a dish filled with 0.1× MBS (*see* **Note 12**).
3. Prepare the injection mixture.
4. Clean the electrodes and electroporation chamber with 70% ethanol.
5. Rinse the electrodes and electroporation chamber with 1× MBS.
6. Clamp the electrode adaptor and capillary holder separately onto the micromanipulators.
7. Use the manipulator to place and secure the electrodes at the two ends of the transverse channel in the electroporation chamber (*see* **Note 13**).
8. Attach the cables of the electrode adaptor to the electroporator to determine polarity (*see* **Note 14**).
9. Set the electroporator parameters—voltage, number, duration, and intervals of electric pulses. Generally, 18 V, 8× 50 ms pulses with 1 s intervals is used (*see* **Note 15**).
10. Transfer 1 ml of electroporation solution to cover the chamber channels and remove any trapped air bubbles with the steering rod (*see* **Note 16**).
11. Set the pressure output of the microinjector to 50 pound-forces per square inch (psi) for retina electroporation or 20 psi for brain electroporation.
12. Back-fill glass capillary with injection mixture and attach it to the capillary holder.
13. Break the capillary tip with fine forceps and adjust the pulse duration of the microinjector to deliver 5 nl of injection mixture per pulse (*see* **Note 17**).
14. Anesthetize the embryos in electroporation solution.

3.3 Targeted Retina Electroporation

1. Line up the micromanipulator-held electrode adaptor, chamber and the micromanipulator-held capillary holder from left to right in a straight horizontal line.
2. Transfer an anesthetized embryo with a drop of electroporation solution to the electroporation chamber (*see* **Note 18**) (Fig. 2c).

3. With the help of the steering rod, position the embryos in the channels such that the dorsal side is facing up and the retina/retina primordia are between the electrodes in the transverse channel (*see* **Note 18**) (Fig. 3a, b).

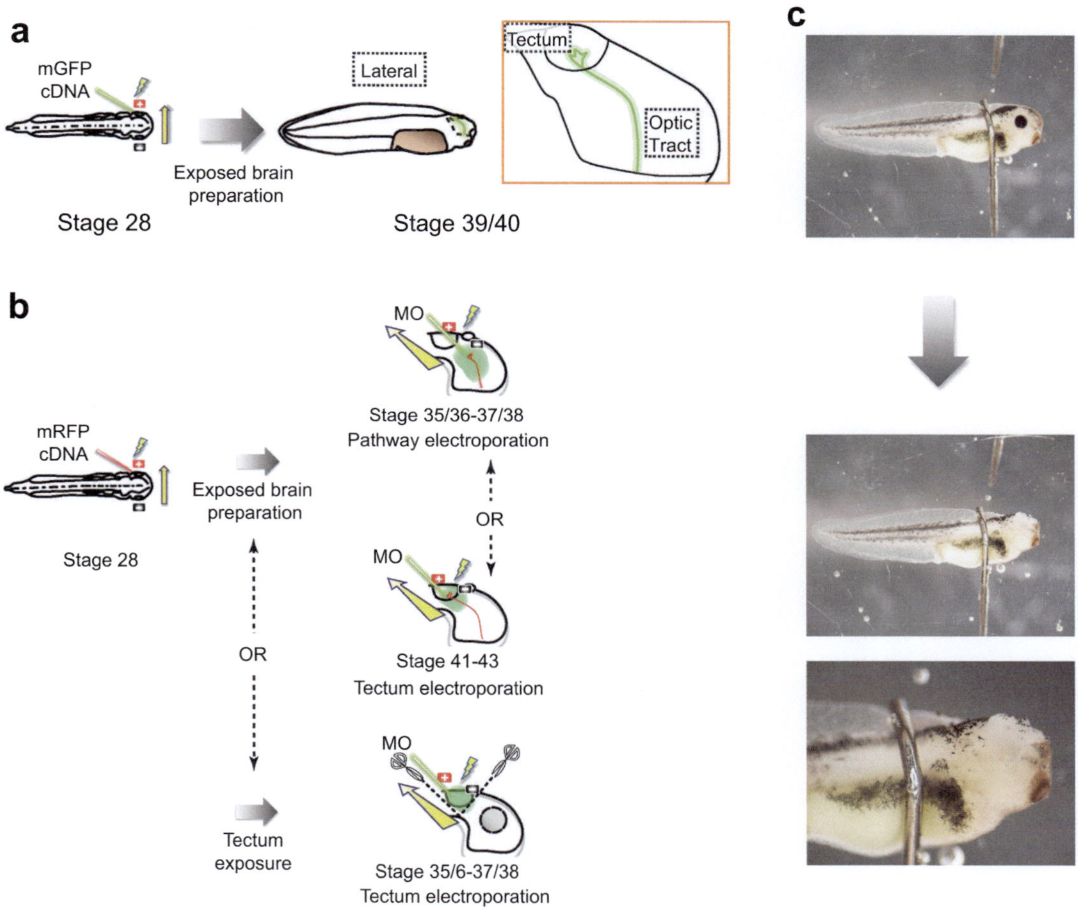

Fig. 3 Examples of electroporation and imaging experiments. (**a**) Targeted retina electroporation of membrane targeted GFP (mGFP) is carried out at Stage 28. mGFP-labeled retinal axons can be observed to arrive the optic tectum at Stage 39/40. (**b**) Targeted retina electroporation of membrane targeted RFP (mRFP) is carried out at Stage 28. Pathway electroporation and tectum electroporation can be carried out at different stages to target the retinal axons or brain cells. The examples shown here involve the local delivery of antisense morpholino (MO) to dissect the role of local protein synthesis in axon navigation in the optic tract and axon arborization in the optic tectum [7]. Pathway electroporation at Stage 35/36–37/38 allows the targeting of RGC axons which are navigating in the optic tract. Tectum electroporation at Stage 41–43 allows the targeting of RGC axons which are arborizing in the optic tectum. Tectum electroporation at Stage 35/36–37/38 allows the targeting of the tectal cells before the entry of RGC axons into the optic tectum. The two tectum electroporation procedures can be used in conjunction to test if the observed effects are RGC axons autonomous or resulted indirectly from the tectal cells. (**c**) An example of an exposed brain preparation for targeted brain electroporation or live in vivo imaging

4. Leave only sufficient volume of electroporation solution to cover the embryo in the channels and remove the excess (*see* **Note 18**).
5. Position and insert the capillary into the retina of interest (*see* **Note 19**).
6. Deliver 1–8 injection pulses into the retina (*see* **Note 20**).
7. Withdraw the capillary away from the retina.
8. Deliver the electric pulses immediately (*see* **Note 21**).
9. Use a plastic Pasteur pipette to gently pipette rearing solution up and down near the chamber channels in order to flush the embryo out (*see* **Note 18**).
10. Transfer the embryo to a dish filled with rearing solution (*see* **Note 18**).
11. Remove bubbles, mucus, and debris from the chamber channels by rinsing with electroporation buffer using a plastic Pasteur pipette.
12. Check to see if capillary is clogged before starting the subsequent round (*see* **Note 22**).

3.4 Targeted Brain Electroporation

1. Line up the micromanipulator-held electrode adaptor, chamber, and micromanipulator-held capillary holder from left to right in a triangular formation, with the chamber being further away from the user (Fig. 2d).
2. Line up the micromanipulator-held electrode adaptor and chamber from left to right in a straight line with a 45° angle toward the chamber (Fig. 2d).
3. Line up the micromanipulator-held capillary holder and chamber from right to left in a straight line with a 45° angle toward the chamber (*see* **Note 23**) (Fig. 2d).
4. Expose the brain hemisphere of interest in a Sylgard dish by removing the overlying epidermis (optional: and eye) of an anesthetized embryo with the help of insect pins (*see* **Note 24**) (Fig. 3c).
5. Transfer an anesthetized embryo with a drop of electroporation solution to the electroporation chamber (*see* **Note 18**).
6. With the help of the steering rod, position the embryos in the channels such that the dorsal side is facing up and the brain area of interest is between the electrodes in the transverse channel (*see* **Note 18**).
7. Leave only sufficient volume of electroporation solution to cover the embryo in the channels and remove the excess (*see* **Note 18**).
8. Position the capillary adjacent to the area of interest (*see* **Note 25**).

9. Deliver eight injection pulses into the vicinity of the area of interest (*see* **Note 25**).
10. Withdraw the capillary (*see* **Note 25**).
11. Deliver 8× 18 V, 50 ms electric pulses with 1 s intervals immediately.
13. Use a plastic Pasteur pipette to gently pipette rearing solution up and down near the chamber channels in order to flush the embryo out (*see* **Note 18**).
14. Transfer the embryo to a dish filled with rearing solution (*see* **Note 18**).
15. Remove bubbles, mucus, and debris from the chamber channels.
16. Check to see if capillary is clogged before starting the subsequent round (*see* **Note 22**).

4 Notes

1. Higher concentrations of injection mixture help to increase electroporation efficiency at the expense of a higher occurrence of capillary clogging. The suggested concentrations provide a compromise between the two parameters. For practicing purpose, dyes such as fast green can be added to the injection mixture for easy visualization.
2. The phosphate groups in the phosphodiester backbones confer the negatively charged nature of DNA/RNA and make electroporation possible. As morpholino has an uncharged backbone, a charged fluorescent tag such as carboxyfluorescein or lissamine is needed. If co-electroporation of morpholino with DNA/RNA is intended, carboxyfluorescein tagged morpholino should be used as it possesses negative charge and would thus be carried to the anode (positive electrode) together with DNA/RNA. Lissamine, on the other hand, is positively charged and would be electroporated in the opposite direction to DNA/RNA.
3. Settings for parameters can differ between different glass capillary pullers, but one should aim for long and sharp capillaries. This prevents tissue or skin from getting trapped in the capillaries and thus reduces the occurrence of capillary clogging. The parameters should therefore be set to low tension, high heat, and high delay, if available.
4. Square wave pulse allows easy manipulation of the electroporation efficiency by adjusting electric pulse parameters—voltage, duration, number of pulses, and time intervals.

5. A simple insulation method is to apply nail polish onto the electrodes. However, the major drawbacks are the low durability of nail polish and the tendency of uneven application. These could increase the difficulties when adjusting the orientation and position of the electrodes. Alternatively, a plastic sheath with an ID of 0.5 mm can be used for insulation. To prevent the bending of platinum wire, the plastic-insulated section can then be surrounded by a metallic sheath with 1 mm ID and sealed with nail polish at both ends.
6. The electrode tip can be flattened and polished with electric sander and sharpening stones in a rotary tool kit. The 45° bend facilitates the positioning of electrodes at the ends of the transverse channel.
7. Although a clamp holder can be used in place of a micromanipulator for the electrode adaptor, micromanipulator allows easy adjustment for fine tuning the placement of electrodes.
8. The steering glass rod provides a useful tool to steer the orientation and position of embryos in the electroporation chamber. This is crucial for the reproducibility of electroporation efficiency. The rounded end prevents the embryos and the electroporation chambers from being damaged during the steering process. Since each steering rod is likely to slightly differ from another, it is advisable to make multiple steering rods to find the most suitable one for the experiment.
9. Be cautious with the Bunsen burner. One hand should be used to hold the thicker end of the Pasteur glass pipette and forceps for the thinner end.
10. Eye protection should be worn to prevent the unlikely chance of being injured by small pieces of broken glass.
11. This prevents injuries arising from sharp glass ends.
12. The left over vitelline membrane or unclean embryos are the major sources of capillary clogging and should therefore be avoided.
13. The flat surface of the electrodes should be orientated in parallel to maximize electroporation efficiency.
14. The cathode (negative electrode) and the anode should be placed in a direction according to the charge of the injection mixture and intended direction of the current, where the cathode repels and the anode attracts a negatively charged mixture and vice versa for positively charged mixture.
15. The suggested parameters aim to provide high efficiency electroporation. If single cell labeling of retinal ganglion cell is intended, a single 18 V, 50 ms pulse is recommended. If labeling of multiple individual cells is intended, the setting of 18 V, 4× 50 ms pulses with 1 s intervals is recommended.

16. Try to have the solution droplet covering only the channels as the spreading of solution on the rest of the chamber would make the subsequent positioning of embryos more difficult, due to surface tension. It is also important to remove trapped air bubbles as it can reduce or prevent current being passed through the electroporation solution. The electroporation setup can be tested by doing test runs with only sufficient volume of electroporation solution covering the channels, a successful setup is characterized by the appearance of bubbles adjacent to the electrodes during and after electroporation.
17. Set a long pulse duration (~100 ms) to begin with and break a small bit of the capillary at a time in order to obtain an opening as small as possible. This can reduce the occurrence of capillary clogging. The injection volume can be calibrated to 5 nl by pulsing into mineral oil and adjusting to a drop size of 214 μm diameter. For eye electroporation, try to create a capillary opening with an angle to facilitate skin penetration.
18. Cautions should be taken at all time to prevent exposing the embryos outside of the solution as the air pressure would result in instant disintegration and death of the embryos.
19. If the correctness of the capillary insertion location is uncertain, deliver an injection pulse to check for eye bulging.
20. The electroporation efficiency increases with the number of injection pulses and should therefore be optimized based on experimental design. Not only could too many injection pulses waste reagents, it could also rupture the retina. As a rule of thumb, matching the number of injection pulses to the number of electric pulses is a good starting point to adjust efficiency.
21. Due to diffusion of the injection mixture, a longer lapse time between injection and electroporation often results in more cells being electroporated, but with lower efficiency at the level of individual cells. A successful electroporation is characterized by the formation of bubbles adjacent to the electrodes and slight twitching of the embryo when the electric pulses are delivered.
22. Clogged needles can sometimes be cleaned be gently rubbing the capillary tip against the sides of the chamber channels and/or by pulsing out with longer pulse duration. If these fail, a new capillary is needed for the subsequent injection.
23. This setup lines up the capillary holder to be perpendicular to the longitudinal channel of the chamber and facilitates easy visualization of the brain electroporation procedures.
24. Securing the embryos to the Sylgard dish helps the brain exposure procedure. This can be done by using insect pin traps,

which consists of a 1.5 mm segment with the blunt end angled at 45° to the 3.5 mm segment with the sharp end. The bend can be made with the use of two pairs of forceps and the excessive segment on the blunt end can be removed by scissors. The traps can then be used by applying the sharp end to the Sylgard dish at 45°, with the space created by the bend of the traps used for accommodating the embryos in lateral orientation.

25. Do not allow the capillary to touch the brain at any time as this could lead to damage to the brain tissue and cells.

Acknowledgment

We thank R. Cagnetta and M. Eldred for critical reading of the manuscript. This work was supported by Cambridge Trust, Croucher Foundation, Sir Edward Youde Memorial Fund (H.H.W), Wellcome Trust Programme Grant (085314/Z/08/Z), and ERC Advanced Grant (322817) (C.E.H).

References

1. Vize PD, Melton DA, Hemmati-Brivanlou A, RM H (1991) Assays for gene function in developing Xenopus embryos. Methods Cell Biol 36:367–387
2. Konopacki FA, Wong HH, Dwivedy A, Bellon A, Blower MD, Holt CE (2016) ESCRT-II controls retinal axon growth by regulating DCC receptor levels and local protein synthesis. Open Biol 6(4):150218. https://doi.org/10.1098/rsob.150218
3. Roque CG, Wong HH, Lin JQ, Holt CE (2016) Tumor protein Tctp regulates axon development in the embryonic visual system. Development 143(7):1134–1148. https://doi.org/10.1242/dev.131060
4. Zhang S, Li J, Lea R, Vleminckx K, Amaya E (2014) Fezf2 promotes neuronal differentiation through localised activation of Wnt/beta-catenin signalling during forebrain development. Development 141(24):4794–4805. https://doi.org/10.1242/dev.115691
5. Falk J, Konopacki FA, Zivraj KH, Holt CE (2014) Rab5 and Rab4 regulate axon elongation in the Xenopus visual system. J Neurosci 34(2):373–391. https://doi.org/10.1523/JNEUROSCI.0876-13.2014
6. Falk J, Drinjakovic J, Leung KM, Dwivedy A, Regan AG, Piper M, Holt CE (2007) Electroporation of cDNA/morpholinos to targeted areas of embryonic CNS in Xenopus. BMC Dev Biol 7:107. https://doi.org/10.1186/1471-213X-7-107
7. Wong HH, Lin JQ, Strohl F, Roque CG, Cioni JM, Cagnetta R, Turner-Bridger B, Laine RF, Harris WA, Kaminski CF, Holt CE (2017) RNA docking and local translation regulate site-specific axon remodeling in vivo. Neuron 95(4):852–868 e8. https://doi.org/10.1016/j.neuron.2017.07.016
8. Leung LC, Urbancic V, Baudet ML, Dwivedy A, Bayley TG, Lee AC, Harris WA, Holt CE (2013) Coupling of NF-protocadherin signaling to axon guidance by cue-induced translation. Nat Neurosci 16(2):166–173. https://doi.org/10.1038/nn.3290
9. Yoon BC, Jung H, Dwivedy A, O'Hare CM, Zivraj KH, Holt CE (2012) Local translation of extranuclear lamin B promotes axon maintenance. Cell 148(4):752–764. https://doi.org/10.1016/j.cell.2011.11.064
10. Baudet ML, Zivraj KH, Abreu-Goodger C, Muldal A, Armisen J, Blenkiron C, Goldstein LD, Miska EA, Holt CE (2011) miR-124 acts through CoREST to control onset of Sema3A sensitivity in navigating retinal growth cones. Nat Neurosci 15(1):29–38. https://doi.org/10.1038/nn.2979
11. Bonev B, Pisco A, Papalopulu N (2011) MicroRNA-9 reveals regional diversity of neural progenitors along the anterior-posterior axis. Dev Cell 20(1):19–32. https://doi.org/10.1016/j.devcel.2010.11.018
12. Drinjakovic J, Jung H, Campbell DS, Strochlic L, Dwivedy A, Holt CE (2010) E3 ligase

Nedd4 promotes axon branching by downregulating PTEN. Neuron 65(3):341–357. https://doi.org/10.1016/j.neuron.2010.01.017

13. Agathocleous M, Iordanova I, Willardsen MI, Xue XY, Vetter ML, Harris WA, Moore KB (2009) A directional Wnt/beta-catenin-Sox2-proneural pathway regulates the transition from proliferation to differentiation in the Xenopus retina. Development 136(19):3289–3299. https://doi.org/10.1242/dev.040451

14. Kalous A, Stake JI, Yisraeli JK, Holt CE (2014) RNA-binding protein Vg1RBP regulates terminal arbor formation but not long-range axon navigation in the developing visual system. Dev Neurobiol 74(3):303–318. https://doi.org/10.1002/dneu.22110

15. Strohl F, Lin JQ, Laine RF, Wong HH, Urbancic V, Cagnetta R, Holt CE, Kaminski CF (2017) Single molecule translation imaging visualizes the dynamics of local beta-actin synthesis in retinal axons. Sci Rep 7(1):709. https://doi.org/10.1038/s41598-017-00695-7

16. Shigeoka T, Jung H, Jung J, Turner-Bridger B, Ohk J, Lin JQ, Amieux PS, Holt CE (2016) Dynamic axonal translation in developing and mature visual circuits. Cell 166(1):181–192. https://doi.org/10.1016/j.cell.2016.05.029

17. Bellon A, Iyer A, Bridi S, Lee FC, Ovando-Vazquez C, Corradi E, Longhi S, Roccuzzo M, Strohbuecker S, Naik S, Sarkies P, Miska E, Abreu-Goodger C, Holt CE, Baudet ML (2017) miR-182 regulates Slit2-mediated axon guidance by modulating the local translation of a specific mRNA. Cell Rep 18(5):1171–1186. https://doi.org/10.1016/j.celrep.2016.12.093

18. Yoon BC, Zivraj KH, Strochlic L, Holt CE (2012) 14-3-3 proteins regulate retinal axon growth by modulating ADF/cofilin activity. Dev Neurobiol 72(4):600–614. https://doi.org/10.1002/dneu.20955

19. Lin AC, Tan CL, Lin CL, Strochlic L, Huang YS, Richter JD, Holt CE (2009) Cytoplasmic polyadenylation and cytoplasmic polyadenylation element-dependent mRNA regulation are involved in Xenopus retinal axon development. Neural Dev 4:8. https://doi.org/10.1186/1749-8104-4-8

Chapter 10

Conditional Chemogenetic Ablation of Photoreceptor Cells in *Xenopus* Retina

Albert Chesneau, Odile Bronchain, and Muriel Perron

Abstract

Xenopus is an attractive model system for regeneration studies, as it exhibits an extraordinary regenerative capacity compared to mammals. It is commonly used to study body part regeneration following amputation, for instance of the limb, the tail, or the retina. Models with more subtle injuries are also needed for human degenerative disease modeling, allowing for the study of stem cell recruitment for the regeneration of a given cellular subtype. We present here a model to ablate photoreceptor cells in the *Xenopus* retina. This method is based on the nitroreductase/metronidazole (NTR/MTZ) system, a combination of chemical and genetic tools, allowing for the conditional ablation of targeted cells. This type of approach establishes *Xenopus* as a powerful model to study cellular regeneration and stem cell regulation.

Key words *Xenopus*, Cell ablation, Regeneration, Retina, Photoreceptors, NTR-MTZ

1 Introduction

In contrast to mammals, the amphibian *Xenopus* has the exceptional capacity during larval stages to regenerate body parts, including entire appendages such as limbs and tail [1, 2]. Despite the longstanding interest in this remarkable ability, we are still in the early stages of understanding the cellular and molecular mechanisms underlying regeneration processes. Given the explosion of research in the stem cell field, and the potential therapeutic impact of stem cell-dependent tissue repair, there is a need to develop suitable animal models to study regeneration. *Xenopus* is thus re-emerging as a foremost model for regeneration studies [1, 3, 4].

Retinal regeneration is of particular interest: (1) eye development has been studied intensively in various animal models including *Xenopus* and this knowledge is key to study regeneration, (2) retinal tissue repair holds enormous promise for therapeutic approaches

Albert Chesneau and Odile Bronchain contributed equally to this work.

Kris Vleminckx (ed.), *Xenopus: Methods and Protocols*, Methods in Molecular Biology, vol. 1865,
https://doi.org/10.1007/978-1-4939-8784-9_10, © Springer Science+Business Media, LLC, part of Springer Nature 2018

to treat patients with retinal degenerative diseases, which affect millions of people worldwide. *Xenopus laevis* was thought to lose its eye regeneration ability after metamorphosis. Ten years ago, the re-examination of this issue revealed that adult frogs are still able to regenerate their retina [5]. Complete or partial retinectomy contributed to highlight the cellular sources involved in retinal neuron replacement, i.e., stem cells of the ciliary marginal zone and cells of the retinal pigmented epithelium [6–10]. More recently, genetic models allowing for targeted cell ablation of a particular retinal cell type have also been established as powerful models to study regeneration [11–13]. One of the techniques relies on the nitroreductase (NTR)/metronidazole (MTZ) system. The NTR enzyme is able to convert the prodrug MTZ into a cytotoxic DNA cross-linking agent, which specifically ablates cells expressing NTR. This method has initially been developed in zebrafish [14, 15]. It has been successfully implemented in *NTR* transgenic zebrafish models to study regeneration of different organs [16–22], including the retina [17, 23, 24]. This approach has been used in *Xenopus* to target oligodendrocytes [25] or photoreceptors in the retina [11, 13]. Here, we present a protocol based on the use of the *Tg(Rho:GFP-NTR) Xenopus* transgenic line, in which the *Rhodopsin* promoter (*Rho*) drives the expression of NTR fused to GFP (GFP-NTR) in rod photoreceptors. MTZ treatments lead to the conditional ablation of these cells. Such chemogenetic system offers the opportunity to further our understanding of retinal regeneration in *Xenopus* following retinal injury. For instance, this approach revealed that Müller cells are able to reenter into the cell cycle following photoreceptor loss and to contribute to photoreceptor regeneration in *Xenopus* [13]. This type of approach can be applied to any other types of cells to study the regeneration of other organs or specific cellular subtypes. As recently reported, it can also be associated to large-scale chemical screens, for which *Xenopus* is well suited [26]. This is of importance as elucidating the mechanisms of regeneration in *Xenopus* may help the design of new strategies in regenerative medicine to restore damaged tissues using the patient's own cells.

2 Materials

2.1 Plasmid Preparation

1. PCR Thermocycler.
2. Set up to perform agarose gel electrophoresis.
3. Set up to perform bacteria transformation and growth through ampicillin resistance.
4. Forward oligonucleotide: 5′-ATG <u>CTC GAG</u> CCA TGG ATA TCA TTT CTG TCG CCT TA-3′ (XhoI site underlined), at a 10 μM stock solution in distilled water.

5. Reverse oligonucleotide: 5′-GGG GAT CCG ATC GAT CTC AAT ACC CGC TAA ATA-3′ (BamHI site underlined), at a 10 μM stock solution in distilled water.
6. *Escherichia coli* genomic DNA purified as previously described [14].
7. High-fidelity, proofreading DNA polymerase (e.g., AccuSure DNA Polymerase).
8. 2 mM dNTPs.
9. 25 mM $MgCl_2$.
10. *pEGFP-C1* cloning vector (Clontech, GenBank: U55763.1), containing the Green Fluorescent Protein (GFP) open reading frame downstream of the *cytomegalovirus* (*CMV*) promoter.
11. Appropriate restriction enzymes for cloning: XhoI, BamHI, AseI, AgeI.
12. *pXOP(−508/+41)GFP* plasmid containing a fragment of the *Xenopus Rhodopsin* promoter, which drives expression in rod photoreceptors of the retina [27] (*see* **Note 1**).
13. Purification kits (NucleoSpin gel and PCR cleanup, Nucleobond Xtra Midi/Maxi, Macherey-Nagel).
14. T4 DNA polymerase and accompanying reaction buffer.
15. T4 DNA ligase and accompanying reaction buffer.
16. Competent bacteria cells (DH5α).

2.2 Generation of Transgenic Animals Using the Restriction Enzyme Mediated Integration Method (REMI)

The protocol is based on a previously described procedure (*see* [28, 29]) with minor modifications. Sperm nuclei were frozen just after preparation and kept as small aliquots at −80 °C. The REMI reaction was made by just adding the linearized transgene to sperm nuclei (*see* **Note 2**).

1. Puller for capillary micropipettes (e.g., Sutter instrument Co, model p-97).
2. Capillary micropipettes (Drummond, # 1-000-1000) clipped with forceps to produce a beveled tip of 60 μm diameter.
3. Microinjector (e.g., Havard apparatus PHD 2000 infusion, set at a slow flow speed of about 20–25 nl/s with mineral oil).
4. Micromanipulator (e.g., Drummond).
5. Tygon tubing.
6. Incubator (16–18 °C).
7. Aquarium air pump with air stone diffuser.
8. Stereo microscope.
9. Fluorescent stereo microscope using a GFP filter.
10. Human chorionic gonadotropin.

11. 10× Marc's Modified Ringer's (MMR) stock solution: 1 M NaCl, 20 mM KCl, 10 mM $MgCl_2$, 20 mM $CaCl_2$, 50 mM HEPES, pH 7.5.
12. 2.5% L-cysteine in 0.1× MMR, adjust pH to 7.8 with NaOH just before use.
13. 60 mm Petri dishes coated with 1% agarose in 0.1× MMR.
14. Ficoll injection solution (to be stored at 4 °C): 3% Ficoll 400 dissolved in 0.1× MMR.
15. Gentamicin.
16. Sperm dilution buffer: 250 mM sucrose, 75 mM KCl, 0.5 mM spermidine trihydrochloride, 0.2 mM spermine tetrahydrochloride, 5 mM $MgCl_2$; titrated to pH 7.3–7.5 with NaOH.
17. Frozen Sperm nuclei preparation at 10^5 nuclei/μl prepared as in [28, 30].
18. Appropriate restriction enzyme for linearization: ApaLI.
19. Nettle and spirulina mixture: ¾ nettle, ¼ spirulina.
20. Artemia nauplii, bloodworms, fish pellets (e.g., neo supra S-AL2G, neo start semi F3, B-Nature grower-semi F5, from Le Gouessant).

2.3 Metronidazole Treatment

1. Incubator at 20 °C protected from light.
2. Magnetic stir bars and a magnetic agitator.
3. Metronidazole (MTZ, Sigma).
4. Dimethyl sulfoxyde (DMSO).
5. Tadpoles above NF45 stage, when the photoreceptors are mature. Animals are staged according to the Normal table of *Xenopus laevis* [31].
6. Food (*see* Subheading 2.2 **steps 10** and **19**).

2.4 Histological Processing

1. Roller shaker.
2. Incubator at 60 °C.
3. Microtome (e.g., Microm HM340E).
4. Microscope slides and coverslips.
5. Disposable plastic embedding molds.
6. Heating block.
7. Microwave.
8. Parafilm roll.
9. 0.04% tricaïne methane sulfonate solution (MS222, Sigma-Aldrich) buffered to pH 7.0–7.5 with sodium bicarbonate.
10. 1× PBS: 140 mM NaCl, 2.7 mM KCl, 10 mM Phosphate buffer (adjust pH to 7.4 by adding dilute (1 N) HCl if necessary).

11. 4% Paraformaldehyde (PFA) in 1× PBS.
12. 50%, 70%, 90%, and 100% ethanol in 1× PBS.
13. Paraffin wax (e.g., 00403 HISTOWAX®, Histolab).
14. Freshly made 3% glycerin albumen (e.g., VWR) in distilled water from a 100% stock solution stored at −20 °C.
15. Xylene.
16. 1-Butanol.
17. PBT solution: 1× PBS containing 0.2% Triton X-100.
18. Unmasking citrate solution: 10 mM sodium citrate, 0.05% Tween 20 in distilled water.
19. Blocking solution: 10% goat serum in PBT.
20. Primary antibody: mouse anti-rhodopsin (EMD Millipore).
21. Secondary antibody: goat anti-mouse (Alexa fluor 594 or 488; Thermo Fisher).
22. 5 μg/mL Hoechst solution.
23. FluoSave reagent (e.g., Calbiochem).

3 Methods

3.1 Preparation of the Rho:GFP-NTR Vector, Encoding a GFP-Nitroreductase Fusion Protein Downstream the Rhodopsin Promoter (Fig. 1)

1. To amplify the bacterial *Nitroreductase* gene (*NfsB,* thereafter named *NTR*) by PCR, mix all PCR components in a 25 μl total volume: 2 μl dNTP, 2.5 μl 10× Accusure buffer, 2.5 μl of 25 mM $MgCl_2$, 1 μl of each oligonucleotide, 2 U of AccuSure DNA polymerase, 100 ng of *Escherichia coli* genomic DNA, H_2O to a final volume of 25 μl (*see* **Note 3**). Then perform the PCR reaction using the cycling conditions as follows: 95 °C for 5 min, 35 cycles of 95 °C for 1 min, 50 °C for 1 min, 72 °C for 1 min, and then 72 °C for 10 min.
2. Assess the PCR reaction on an agarose gel: the reaction should provide a ~700-bp product that can be directionally inserted in frame with the eGFP cassette into the *peGFP-C1* vector following XhoI/BamHI double digestion.
3. Purify the PCR product using the DNA purification kit.
4. Double digest the insert and the *peGFP-C1* vector by XhoI and BamHI.
5. Clean up the digested PCR product using the DNA purification kit.
6. Isolate the digested vector on a 1% agarose gel and purify it with the DNA purification kit.
7. Ligate 10 ng of the digested *NTR* amplicon with 30 ng of digested *peGFP-C1* vector using 10 U of T4 DNA ligase

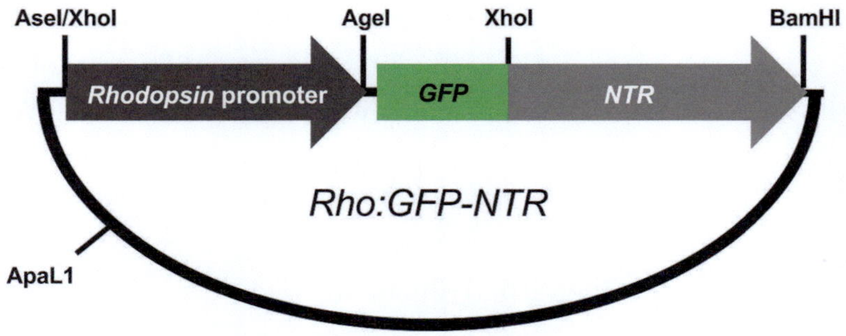

Fig. 1 Schematic representation of the transgene. The PCR amplified bacterial *Nfs*B gene (*NTR*) has been cloned in frame to *GFP* in the *pEGFP-C1* vector using the XhoI and BamHI restricted enzymes. The original *cytomegalovirus* (*CMV*) promoter present in this vector was removed by digestion with AseI (blunted with T4 polymerase) and AgeI. It was then replaced by a minimal version of the *Rhodopsin* promoter, known as XOP−508/+41 [27], previously digested with XhoI (blunted) and AgeI, creating the vector used to generate the *Tg(Rho:GFP-NTR) Xenopus* transgenic line

overnight at 13 °C in a 10 μl final volume of 1× T4 DNA ligase buffer to produce the *CMV-eGFP-NTR* plasmid.

8. Transform the *CMV-eGFP-NTR* plasmid into competent bacterial cells and plate onto antibiotic selective plates using standard techniques.
9. Select individual colonies for DNA preparation using a DNA purification kit.
10. Validate the construct by double XhoI/BamHI digestion and sequencing.
11. Digest the *CMV-eGFP-NTR* plasmid with AseI to remove the original *CMV* promoter cassette.
12. Purify the digested product using a DNA purification kit. DNA can be eluted in the elution buffer from the kit or in water.
13. Blunt the extremities using 1 U T4 DNA polymerase per μg of DNA for 15 min at 12 °C in 1× reaction buffer supplemented with 100 μM dNTPs.
14. Purify the plasmid with a DNA purification kit.
15. Digest the resulting vector with AgeI and isolate on a 1% agarose gel the ~4.8 kbp band corresponding to the vector without the CMV promoter (593 bp) and purify it with a DNA purification kit.
16. Digest the *XOP(−508/141)GFP* plasmid with XhoI (*see* **Note 1**).
17. Blunt the extremities using T4 DNA polymerase as above.

18. Purify the linearized plasmid with a DNA purification kit.
19. Digest with AgeI and isolate the promoter fragment (~650 bp fragment, thereafter named *Rho*) from the rest of the vector on a 1% agarose gel and purify it with a DNA purification kit.
20. To generate the final *Rho:GFP-NTR* vector, mix 10 ng of the *Rho* fragment with 30 ng of the vector and add 10 U of T4 DNA ligase in a 10 μl final reaction and incubate overnight at 13 °C.
21. Transform the *Rho:GFP-NTR* vector and prepare DNA as above.
22. Control the final construct by double ApaLI/AgeI enzymatic digestion. Two fragments of ~900 and ~3.7 bp should be obtained. Sequence the final construct.
23. Linearize the plasmid with ApaLl to perform the REMI reaction.

3.2 Generation of Tg(Rho:GFP-NTR) Transgenic Animals Using the REMI Method (Fig. 2)

The method below is for *Xenopus laevis* (*see* **Note 4**).

1. Females are induced to lay eggs by dorsal lymph sac injection of 4 U per g of hCG (1.5 U/μl) the evening before the experiment. A medium size female is estimated to weigh about 150 g. Females should begin laying eggs about 12 h after hCG injection.
2. Keep frogs at 18 °C until morning.
3. Fresh eggs are obtained by gently massaging the abdomen and sides of the frogs. Collect the eggs into a dish. Keep eggs from different females separately as egg quality and amount of jelly coat may vary extensively from one female to the other.
4. Remove the jelly coat of the eggs in 2% cysteine, pH 7.8 by gently swirling the eggs around until they roll freely and start touching each other.
5. Wash many times with 0.1× MMR.
6. Place the dejelled eggs in agarose coated Petri dishes containing Ficoll injection solution supplemented with 50 μg/ml gentamicin. They should be used for nuclear transplantation within half an hour.
7. The injection procedure largely follows the REMI transgenic method described previously [30]. Briefly, mix 5 μl of the ApaLI linearized vector (at a concentration of 0.1 μg/μl) with 5 μl of frozen sperm preparation (at a concentration of 10^5 nuclei/μl) and incubate at room temperature for 5 min. Use cut yellow tips whenever pipetting sperm nuclei. The resulting solution is thereafter named REMI reaction (*see* **Note 5**).

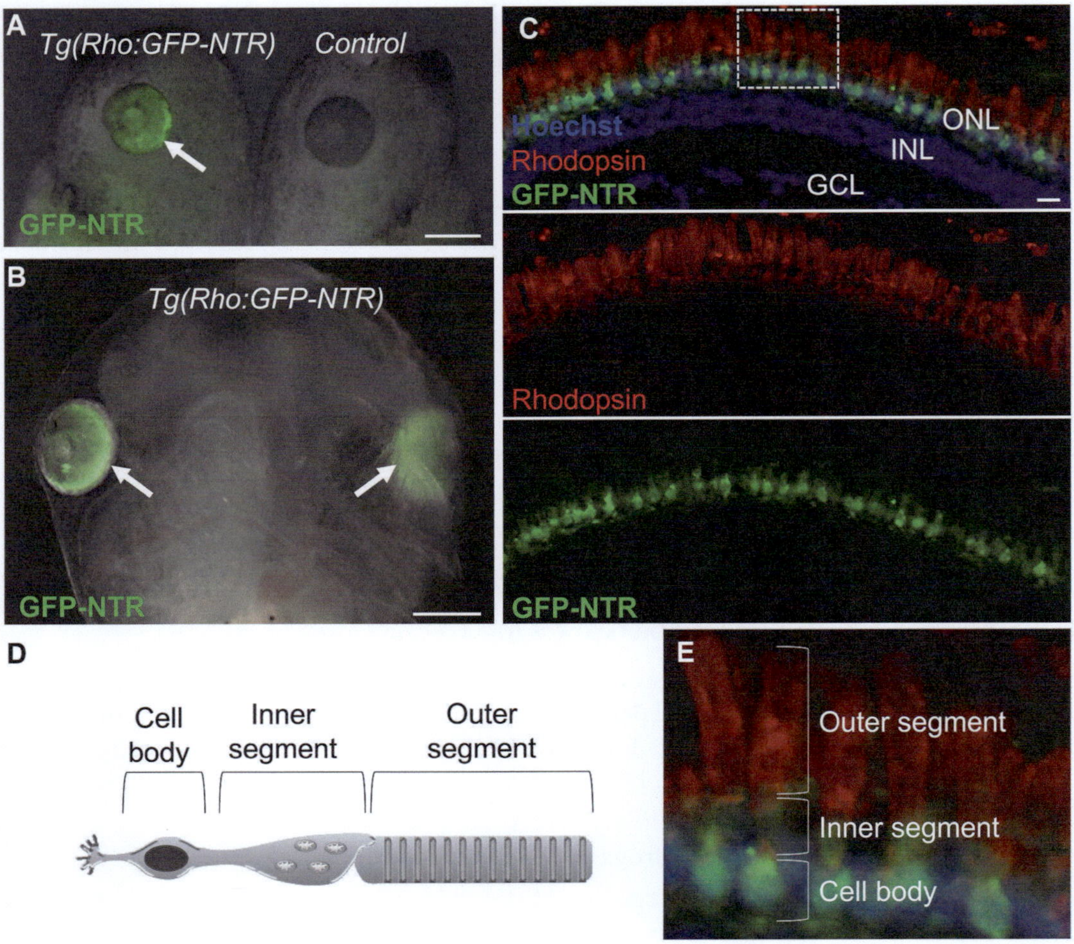

Fig. 2 Rod-specific expression of NTR in *Tg(Rho:GFP-NTR)* tadpoles. (**a**, **b**) Specific eye expression (arrows) of the transgene in *Tg(Rho:GFP-NTR)* tadpoles as assessed by GFP fluorescence (green) at stage 38 (**a**) and 47 (**b**). A non-transgenic tadpole is also shown in (**a**) as a negative control. All tadpoles are shown with the anterior pole to the top, (**b**) is a ventral view while in (**a**) ventral sides are respectively on the left for the transgenic tadpole and on the right for the control one. The experiment in B has been performed on an albino animal in order to easily detect the GFP. (**c**) A retinal section from a stage 54 *Tg(Rho:GFP-NTR)* tadpole, immunostained for GFP (to label *GFP-NTR* positive cells) and Rhodopsin (to label rod cells). (**d**) Schematic of a rod photoreceptor cell. (**e**) Enlargement of the area delineated with white dashes in (**c**) showing that Rhodopsin mainly labels outer segments of rods while the GFP mainly labels cell bodies and inner segments. Scale bars: 0.5 mm in (**a**, **b**); 100 μm in (**c**)

8. Dilute the REMI reaction to reach a concentration of 1–2 sperm nuclei per 10–15 nl (~1/500) and mix thoroughly with a cut 200 μl tip.
9. Attach a piece of tygon tubing to a cut yellow tip to draw up the diluted sperm suspension and back load a capillary micropipette.
10. Connect the capillary micropipette to the injection apparatus and transplant nuclei by injecting each unfertilized egg with a volume of about 10 nl (1 s inside the egg). A hole should be

visible for several seconds indicating that the solution has been delivered.

11. Keep injected eggs at 16–18 °C.
12. Sort fertilized embryos at the 4-cell stage (about 3–4 h after injection at 16 °C). Gently transfer normally dividing embryos into Petri dishes containing 0.1× MMR supplemented with 50 μg/ml gentamicin and place in an incubator at 16–18 °C.
13. Sorting transgenic embryos is based on the expression of the *GFP-NTR* transgene. The expression thus relies on the promoter used which requires a prior knowledge of its activity in living animals. In our case, we expect to see GFP expression in rod photoreceptor cells of the retina from stage NF37/38 onward. Transgenic tadpoles are thus separated from non-transgenic tadpoles according to the expression of the GFP in the retina under a fluorescent stereo microscope (*see* **Note 6**).
14. Tadpoles are raised in tanks at 18–22 °C with filtered tap water aerating with air bubbles. Try to keep a density of 3–6 tadpoles per liter.
15. Tadpoles are fed every day with the spirulina/nettle mixture until they reach the metamorphic stage (NF62–63).
16. Weekly remove waste and uneaten food and provide new filtered tap water to compensate for water evaporation.
17. During growth, separate tadpoles according to their size. This seems to improve the overall raising. When transgenic individuals complete metamorphosis, transfer young froglets into a new tank.
18. Feeding regimen changes during metamorphosis from "powder/vegetarian" to pellets/carnivore type food. We perform a gradual food replacement as follows: from stage NF65, individuals are fed every day with artemia nauplii for a week. From the second week to one month, artemia nauplii and bloodworms are alternatively given every other day. From 1 to 5 months, fish pellets (neo supra S-AL2G) and bloodworms are alternatively given every other day. From 6 months to 2 years, individuals are fed with fish pellets (neo start semi F3) every 2 days and bloodworms once a week. From 2 years of age, frogs are fed with fish pellets (B-Nature grower-semi F5), every other day.

3.3 Rod Cell Ablation by MTZ Treatment

1. Prepare a fresh MTZ solution by dissolving the MTZ powder in a dark bottle (MTZ is light sensitive) in 0.1× MMR containing 0.1% DMSO (*see* **Note 7**), using a magnetic stir-bar and a magnetic agitator to obtain a 10 mM MTZ working solution (*see* **Note 8**). Complete dissolution may take up to 10 min.
2. Transfer transgenic and non-transgenic siblings (controls) separately in the MTZ solution and raise the tadpoles at 20 °C

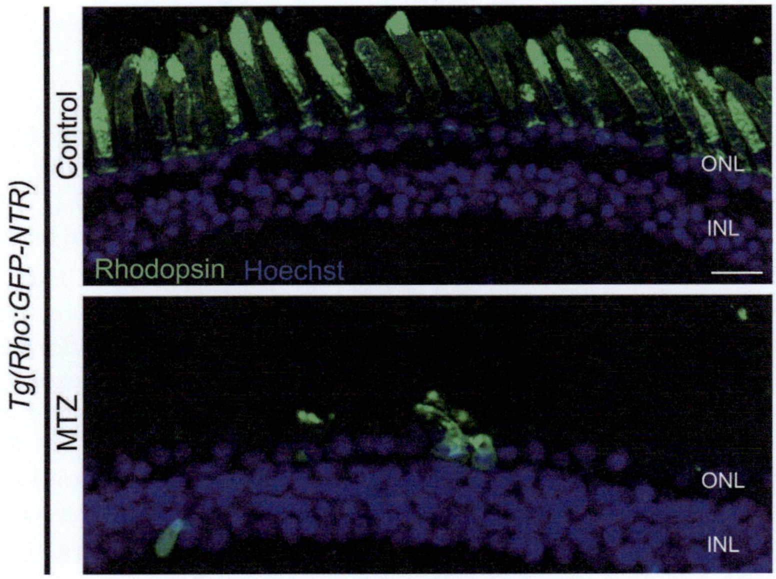

Fig. 3 Targeted rod photoreceptor degeneration. Rhodopsin immunostaining on retinal sections from stage 54 *Tg(Rho:GFP-NTR)* tadpoles treated or not with MTZ for 1 week. Scale bar: 100 μm

in complete darkness for a week. Another useful control are transgenic siblings raised in 0.1× MMR containing 0.1% DMSO without MTZ (*see* **Note 9**).

3. Change the MTZ working solution every other day.
4. During the MTZ treatment, tadpoles should be fed normally (*see* Subheading 3.2, **steps 15–18**).

3.4 Histological Analysis to Confirm Rod Cell Ablation (Fig. 3)

1. Anesthetize tadpoles in 0.04% MS222 solution.
2. Transfer the anesthetized tadpoles in 4% PFA solution. We use about ten times more volume of PFA than the approximate volume of the tadpoles. Incubate for 1–2 h with agitation using a roller shaker. For tadpoles above stage NF54, only the head is kept for fixation.
3. Wash fixed tadpoles 3 × 5 min in 1× PBS and successively transfer them 3 × 5 min in 70% ethanol, 30 min in 90% ethanol, 3 × 1 h in 100% ethanol, and overnight in 100% 1-Butanol (all steps at room temperature).
4. Transfer tadpoles in disposable plastic embedding molds and cover them with melted paraffin. Incubate at 60 °C for ~6 h. Change the paraffin at least twice.
5. Transfer the molds at room temperature and let them harden before sectioning.

6. Remove the paraffin blocks from the molds.
7. Cut paraffin blocks at 10–12 μm thickness using a microtome.
8. Mount the sections on slides in a 3% glycerin albumen.
9. Transfer slides for 20 s on a heating block (50 °C) to allow the sections to extend, then remove the excess of glycerin albumen.
10. Dry slides for 10 s on a heating block (50 °C) to prevent their detachment or sliding and stand them up to let any remaining liquid run down the slides.
11. Completely dry the slides by incubating at 37 °C overnight.
12. Slides can be stored at room temperature in a dry place.
13. For the immunostaining, transfer the slides 2 × 10 min in 100% xylene, then successively for 5 min in 100%, 90%, 70% and 50% ethanol, and finally 3 × 3 min in distilled water.
14. Transfer the slides in the unmasking citrate solution in a glass jar and incubate in a microwave at high power (800 W) until the solution starts boiling. Then continue heating at low power (thaw mode) for 10 min.
15. Let the glass jar and solution cool down for 20 min at room temperature.
16. Wash the slides 3 × 5 min in PBT.
17. Cover the slides with 800 μl of blocking solution and incubate for 1 h at room temperature in a box with a cover to prevent evaporation.
18. Discard the blocking solution from the slides and replace it with the primary antibody diluted at 1/500 in the blocking solution (add enough to nicely cover the sections, typically 200 μl per slide).
19. Cover each slide with a parafilm band to prevent them from drying out and incubate overnight at 4 °C in a humid chamber.
20. Remove parafilm bands and wash the slides 3 × 5 min in 1× PBS.
21. Add the anti-mouse secondary antibody diluted at 1/1000 in the blocking buffer (typically 400 μl per slide) and incubate for 2 h in a dark box.
22. Wash the slides 3 × 5 min in 1× PBS.
23. Stain the nuclei by incubating the slides for 10 min in Hoechst solution and then 3 × 5 min in 1× PBS.
24. Mount slides with coverslips using a fluorsave reagent.
25. Let the slides dry for at least 1 h at room temperature before further observation using a fluorescent microscope.

4 Notes

1. Any other specific promoter can be used instead of the *Rhodopsin* promoter by using the AseI and AgeI cloning sites within the *CMV-eGFP-NTR* vector.
2. Alternative methods to generate transgenic founders may also be used (e.g., the meganuclease I-SceI, the sleeping Beauty transposon, the phi31 integrase, etc.). We chose the REMI method since fewer mosaic founders are generated making it better adapted to work in *F0* individuals. Yet, we still recommend generating transgenic *F1* progenies to select lines with a homogenous transgene expression.
3. This is the specific protocol for AccuSure DNA polymerase but other proofreading polymerases can be used instead.
4. For this part of the procedure, researchers should get approval from their affiliated institutes and perform experiments according to their local ethics guidelines for animal studies.
5. Unlike the initial protocol was recommending, restriction enzyme is no longer added to the mixture.
6. The intensity of the GFP florescence depends on both the number of integrated copies and the insertion sites in the genome.
7. DMSO is not essential, but it helps solubilize MTZ and may facilitate permeation of the drug into the animals.
8. The concentration of MTZ solution should not exceed 10 mM because we observed toxic effects at 20 mM [25].
9. For the design of the experiment, it is important to anticipate inter-individual variability in the response to MTZ treatment. Indeed, although early development in *Xenopus* is highly synchronous, allowing to work on homogeneous batches of embryos, this is not the case for tadpoles over NF45. Indeed, important variations in tadpole growth can be observed, although raised in the same conditions [32]. Therefore, in order to have enough individuals at the same stage, anticipate to generate excess transgenic animals. Besides, these variations can explain subsequent differences in cell ablation efficiency and in the regeneration process.

Acknowledgments

We thank B.E. Knox for providing the *Rhodopsin* promoter plasmid. Our lab is supported by grants from the Fondation pour la Recherche Médicale (FRM), Association Retina France and Fondation Valentin Haüy.

References

1. Li J, Zhang S, Amaya E (2016) The cellular and molecular mechanisms of tissue repair and regeneration as revealed by studies in *Xenopus*. Regeneration 3:198–208. https://doi.org/10.1002/reg2.69
2. Tseng a-S, Levin M (2008) Tail regeneration in *Xenopus laevis* as a model for understanding tissue repair. J Dent Res 87:806–816. https://doi.org/10.1177/154405910808700909
3. Beck CW, Izpisúa Belmonte JC, Christen B (2009) Beyond early development: *Xenopus* as an emerging model for the study of regenerative mechanisms. Dev Dyn 238:1226–1248. https://doi.org/10.1002/dvdy.21890
4. Slack JMW, Lin G, Chen Y (2008) The Xenopus tadpole: a new model for regeneration research. Cell Mol Life Sci 65:54–63. https://doi.org/10.1007/s00018-007-7431-1
5. Yoshii C, Ueda Y, Okamoto M, Araki M (2007) Neural retinal regeneration in the anuran amphibian *Xenopus laevis* post-metamorphosis: transdifferentiation of retinal pigmented epithelium regenerates the neural retina. Dev Biol 303:45–56. https://doi.org/10.1016/j.ydbio.2006.11.024
6. Araki M (2007) Regeneration of the amphibian retina: role of tissue interaction and related signaling molecules on RPE transdifferentiation. Develop Growth Differ 49:109–120. https://doi.org/10.1111/j.1440-169X.2007.00911.x
7. Araki M (2014) A novel mode of retinal regeneration: the merit of a new Xenopus model. Neural Regen Res 9:2125–2127. https://doi.org/10.4103/1673-5374.147942
8. Chiba C (2014) The retinal pigment epithelium: an important player of retinal disorders and regeneration. Exp Eye Res 123:107–114. https://doi.org/10.1016/j.exer.2013.07.009
9. Miyake A, Araki M (2014) Retinal stem/progenitor cells in the ciliary marginal zone complete retinal regeneration: a study of retinal regeneration in a novel animal model. Dev Neurobiol. https://doi.org/10.1002/dneu.22169
10. Ail D, Perron M (2017) Retinal degeneration and regeneration—lessons from fishes and amphibians. Curr Pathobiol Rep 5:67–78. https://doi.org/10.1007/s40139-017-0127-9
11. Choi RY, Engbretson GA, Solessio EC et al (2011) Cone degeneration following rod ablation in a reversible model of retinal degeneration. Invest Ophthalmol Vis Sci 52:364–373. https://doi.org/10.1167/iovs.10-5347
12. Hamm LM, Tam BM, Moritz OL (2009) Controlled rod cell ablation in transgenic *Xenopus laevis*. Invest Ophthalmol Vis Sci 50:885–892. https://doi.org/10.1167/iovs.08-2337
13. Langhe R, Chesneau A, Colozza G et al (2017) Müller glial cell reactivation in Xenopus models of retinal degeneration. Glia 65:1333–1349. https://doi.org/10.1002/glia.23165
14. Curado S, Anderson RM, Jungblut B et al (2007) Conditional targeted cell ablation in zebrafish: a new tool for regeneration studies. Dev Dyn 236:1025–1035. https://doi.org/10.1002/dvdy.21100
15. Curado S, Stainier D (2008) Nitroreductase-mediated cell/tissue ablation in zebrafish: a spatially and temporally controlled ablation method with applications in developmental and regeneration studies. Nat Protoc 3: 948–954. https://doi.org/10.1038/nprot.2008.58
16. White DT, Mumm JS (2013) The nitroreductase system of inducible targeted ablation facilitates cell-specific regenerative studies in zebrafish. Methods 62:232–240. https://doi.org/10.1016/j.ymeth.2013.03.017
17. White DT, Sengupta S, Saxena MT et al (2017) Immunomodulation-accelerated neuronal regeneration following selective rod photoreceptor cell ablation in the zebrafish retina. Proc Natl Acad Sci 114:E3719–E3728. https://doi.org/10.1073/pnas.1617721114
18. Pipalia TG, Koth J, Roy SD et al (2016) Cellular dynamics of regeneration reveals role of two distinct Pax7 stem cell populations in larval zebrafish muscle repair. Dis Model Mech 9:671–684. https://doi.org/10.1242/dmm.022251
19. Ariga J, Walker SL, Mumm JS (2010) Multicolor time-lapse imaging of transgenic zebrafish: visualizing retinal stem cells activated by targeted neuronal cell ablation. J Vis Exp. https://doi.org/10.3791/2093
20. Ohnmacht J, Yang Y, Maurer GW et al (2016) Spinal motor neurons are regenerated after mechanical lesion and genetic ablation in larval zebrafish. Development 143:1464–1474. https://doi.org/10.1242/dev.129155
21. Oosterhof N, Kuil LE, van Ham TJ (2017) Microglial activation by genetically targeted conditional neuronal ablation in the zebrafish. Meth Mol Biol (Clifton, NJ) 1559:377–390

22. White YAR, Woods DC, Wood AW (2011) A transgenic zebrafish model of targeted oocyte ablation and de novo oogenesis. Dev Dyn 240:1929–1937. https://doi.org/10.1002/dvdy.22695
23. Fraser B, DuVal MG, Wang H, Allison WT (2013) Regeneration of cone photoreceptors when cell ablation is primarily restricted to a particular cone subtype. PLoS One 8:e55410. https://doi.org/10.1371/journal.pone.0055410
24. Montgomery JE, Parsons MJ, Hyde DR (2010) A novel model of retinal ablation demonstrates that the extent of rod cell death regulates the origin of the regenerated zebrafish rod photoreceptors. J Comp Neurol 518:800–814. https://doi.org/10.1002/cne.22243
25. Kaya F, Mannioui A, Chesneau A et al (2012) Live imaging of targeted cell ablation in Xenopus: a new model to study demyelination and repair. J Neurosci 32:12885–12895. https://doi.org/10.1523/JNEUROSCI.2252-12.2012
26. Mannioui A, Vauzanges Q, Fini JB et al (2017) The *Xenopus* tadpole: an in vivo model to screen drugs favoring remyelination. Mult Scler J. https://doi.org/10.1177/1352458517721355
27. Knox BE, Schlueter C, Sanger BM et al (1998) Transgene expression in Xenopus rods. FEBS Lett 423:117–121
28. Ishibashi S, Kroll KL, Amaya E (2012) Generating transgenic frog embryos by restriction enzyme mediated integration (REMI). Methods Mol Biol 917:185–203. https://doi.org/10.1007/978-1-61779-992-1_11
29. Chesneau A, Sachs LM, Chai N et al (2008) Transgenesis procedures in Xenopus. Biol Cell 100:503–521. https://doi.org/10.1042/BC20070148
30. Kroll KL, Amaya E (1996) Transgenic Xenopus embryos from sperm nuclear transplantations reveal FGF signaling requirements during gastrulation. Development 122:3173–3183
31. Nieuwkoop P, Faber J (1994) Normal table of Xenopus laevis. In: Garland, TX
32. Nye HLD, Cameron JA (2005) Strategies to reduce variation in Xenopus regeneration studies. Dev Dyn 234:151–158. https://doi.org/10.1002/dvdy.20508

Chapter 11

Cancer Models in *Xenopus tropicalis* by CRISPR/Cas9 Mediated Knockout of Tumor Suppressors

Thomas Naert and Kris Vleminckx

Abstract

The recent advent of CRISPR/Cas9 as a straightforward genome editing tool has allowed the establishment of the first bona fide genetic cancer models within the diploid aquatic model organism *Xenopus tropicalis (X. tropicalis)*. Within this chapter, we demonstrate the methods for targeting tumor suppressors with the CRISPR/Cas9 system in the developing *X. tropicalis* embryo. We further illustrate genotyping and phenotyping of the resulting tumor-bearing F0 mosaic mutant animals (crispants). We focus in detail on the histopathological analysis of cancer neoplasms, the methodology to illustrate high proliferative index by proliferation marker immunofluorescence and how to isolate specific (tumor) cell populations by laser capture microdissection. As such, the described pipeline allows for rapid establishment of novel cancer models by CRISPR/Cas9 targeting of established tumor suppressor genes, or novel candidates obtained from clinical data. In conclusion, we thus provide the methodology for modeling human cancer with the highly efficient CRISPR/Cas9 system in F0 *X. tropicalis*.

Key words Cancer, CRISPR/Cas9, *Xenopus tropicalis*, Tumor, Suppressor, Disease models

1 Introduction

With the advent of targeted genome editing technologies, in particular CRISPR/Cas9, it has recently become possible to perform true reverse genetics in *Xenopus tropicalis* (*X. tropicalis*) [1, 2]. As such, we are standing at a crossroad in time in which knowledge amassed over decades of basic research, such as the *X. tropicalis* genome sequencing efforts and the fate map, can now be employed to generate models for human disease [3]. This allows *X. tropicalis* to enter the realm of truly clinically relevant animal models due to possibilities to cost-effectively, rapidly, and efficiently generate knockouts within specific genes of interest [4]. Capitalizing on these unprecedented possibilities we turned to modeling cancer within *X. tropicalis* and have generated, over the past few years, penetrant, and clinically relevant disease models for familial adenomatous polyposis (FAP) and retinoblastoma [5, 6]. Within these

Kris Vleminckx (ed.), *Xenopus: Methods and Protocols*, Methods in Molecular Biology, vol. 1865,
https://doi.org/10.1007/978-1-4939-8784-9_11,

efforts, we observed desmoid tumors, limb tumors, retinoblastoma, lung tumors, choroid plexus tumors, and medulloblastoma indicating that modeling a diverse spectrum of human malignancies in *X. tropicalis* is within arm's reach. Moreover, *X. tropicalis* has unique advantages to disease modeling when compared to other non-mammalian models [7]. Namely, unlike zebrafish, *X. tropicalis* has a true diploid genome and hence gene disruption studies are not suffering from redundancy [8]. In addition, we have demonstrated the possibilities to perform multiplex CRISPR/Cas9 experiments, easily targeting three genes or more within an F0 mosaic mutant animal (crispant) allowing for targeting of modifier genes within developing tumors (*see* Chapter 3 for methodology) [6].

Over the past years, due to rapidly declining costs of sequencing, an increasing number of cancer types, metastatic profiles, and recurrence profiles are seeing their genome fully sequenced in high impact papers leading to identification of more and more rare tumor suppressor genes and driver mutations with profound clinical significance [9–12]. We believe that, by combining *X. tropicalis*, the CRISPR/Cas9 revolution, and the ever increasing rate of cancer genotyping information, profound discoveries will be made in the field over the next couple of years. As such, we provide in this chapter the guide for employing the CRISPR/Cas9 technology to generate F0 tumor suppressor mutant *X. tropicalis* crispants, in order to model in a clinically relevant manner cancer syndromes in the frog.

2 Materials

2.1 General

1. Benzocaine: 100× stock solution: 5 g benzocain dissolved in EtOH.
2. 0.66× (amphibian) PBS: 91 mM NaCl, 1.8 mM KCl, 2.9 mM Na_2HPO_4, 1.15 mM KH_2PO_4 adjusted to pH 7.4.

2.2 Tumor Genotyping

1. Lysis buffer: 50 mM Tris–HCl pH 8.8, 1 mM EDTA, 0.5% (v/v) tween-20, 200 μg/mL Proteinase K.

2.3 Histopathological Assessment of Tumors

1. 4% paraformaldehyde (PFA) in 1× PBS. Stored in aliquots at −20 °C, after thawing can be used for a couple of days when kept at 4 °C.
2. Morse's acid decalcification solution: 10% (w/v) sodium citrate and 22.5% formic acid.
3. Ethanol (absolute).
4. Xylene.
5. Paraffin wax (e.g., HistoLab HistoWax).

6. Automated tissue processor (e.g., Thermo-Shandon Citadel 2000).
7. Paraffin embedding station (e.g., Thermo-Scientific HistoStar).
8. Microtome (e.g., Microm HM360).
9. Klinipath Plus slides (Klinipath).
10. Automated slide stainer (e.g., ThermoFisher VariStain V24-4).
11. Eosine (1%) in BiDi water.
12. HCl.
13. Hematoxylin Solution, Mayer's.
14. Xylene-based mounting solution (e.g., Entellan).
15. 20-slide Glass staining dish and rack.
16. Histology cassettes.

2.4 Immunofluorescence of Proliferating Cell Nuclear Antigen (PCNA) to Demonstrate High Proliferative Index of Tumor Tissue

1. Tissue retriever (e.g., PickCell 2100-Retriever; ProteoGenix).
2. Citrate buffer: 10 mM citric acid, 0.1% Tween-20, pH 6.0.
3. PAP pen.
4. 1× PBS (prepared from 10× stock): 137 mM NaCl, 2.7 mM KCl, 4.3 mM Na_2HPO_4, 1.47 mM KH_2PO_4; Adjust to a final pH of 7.4.
5. 1× PBST: 1× PBS with 0.1% (v/v) Tween-20.
6. Blocking buffer: 3% natural goat serum, 1% bovine serum albumin (BSA) in 1× PBST.
7. Slide staining tray.
8. PCNA antibody (Clone PC10; Agilent-Dako; M0879).
9. Goat anti-mouse IgG dylight-594 (ThermoFisher) (or similar).
10. Hoechst-33342.
11. Immunofluorescence mounting medium.
12. Coverslip.
13. Nail polish.

2.5 Laser Capture Microdissection to Isolate Pure Tumor Tissue

1. Automated slide stainer and/or the reagents for (manual) processing of H&E stain: Xylene, Ethanol (absolute), Hematoxylin, Eosin.
2. Laser capture microdissection apparatus (e.g., P.A.L.M. MicroBeam laser microdissection system).
3. QIAamp DNA Micro Kit (Qiagen).

3 Methods

3.1 Design Considerations for gRNA Targeting Tumor Suppressor Genes

Best practice for design of gRNA, and the pitfalls associated, are described in detail in Chapter 3. However we would like to stress additionally here that it can be rewarding to identify, if possible, an (catalytically) important protein domain of your tumor suppressor to target by the CRISPR/Cas9 system. Namely, we have demonstrated that targeting the catalytical site of the *rbl1* tumor suppressor leads to an increased frequency of retinoblastoma tumorigenesis within *rb1/rbl1* double crispants [6]. This, as small in-frame INDELs in regions of catalytic importance, has a much higher chance to abrogate protein function (*see* **Notes 1** and **2**). Furthermore, in specific cases it may be required to target the last exon in order to generate a truncated protein—possibly acting as a dominant-active or dominant-negative—to mimic a specific clinical situation. This is the case for targeting of the tumor suppressor gene *apc* [5, 13].

3.2 Genome Editing of Tumor Suppressors in X. tropicalis

Be aware that the choice of blastomere, injected with the gRNA/Cas9 ribonucleoprotein (RNP), can be of paramount importance in obtaining the desired tumor phenotype (*see* Chapter 3 for details on tissue-restrictive CRISPR/Cas9 delivery in the developing *X. tropicalis* embryo). We have shown that targeting of distinct blastomeres within an 8-cell *X. tropicalis* embryo with a RNP targeting the same tumor suppressor can culminate in tumorigenesis within different organs (unpublished data). As such, we recommend employing the *X. tropicalis* fate maps to inject specifically those blastomeres within an 8-cell embryo from which the lineage is expected to give rise to the tumor precursor cell pool (*see* **Note 3**) [14]. When unsure of which lineage to specifically target, unilateral injection for comparison across midline can be employed, or even whole embryo injections can be performed. However, this dramatically increases the chances for at least some developmental side effects of tumor suppressor loss of function during early embryogenesis. Additionally, it is possible to titrate down gRNA concentration in your injection mixture in order to reduce genome editing efficiencies and associated embryolethality. Note that this associated reduction in efficiencies might also impact the penetrance of your phenotype; however, this is of lesser concern when doing cancer modeling as there is always a positive selection mechanism in place (*see* below).

Microinjection of the pre-complexed gRNA/Cas9 ribonucleoprotein (RNP) should be performed as described in Chapter 3. In essence the RNP will edit the genomic target site within a subset of cells such that the CRISPR/Cas9 editing efficiencies will dictate the level of mosaicity within the developing embryo. By employing gRNA designed as described before, followed by quantitative assessment of genome editing (*see* Chapter 6), we routinely obtain genome editing in between 50% and 100% of the cells within the

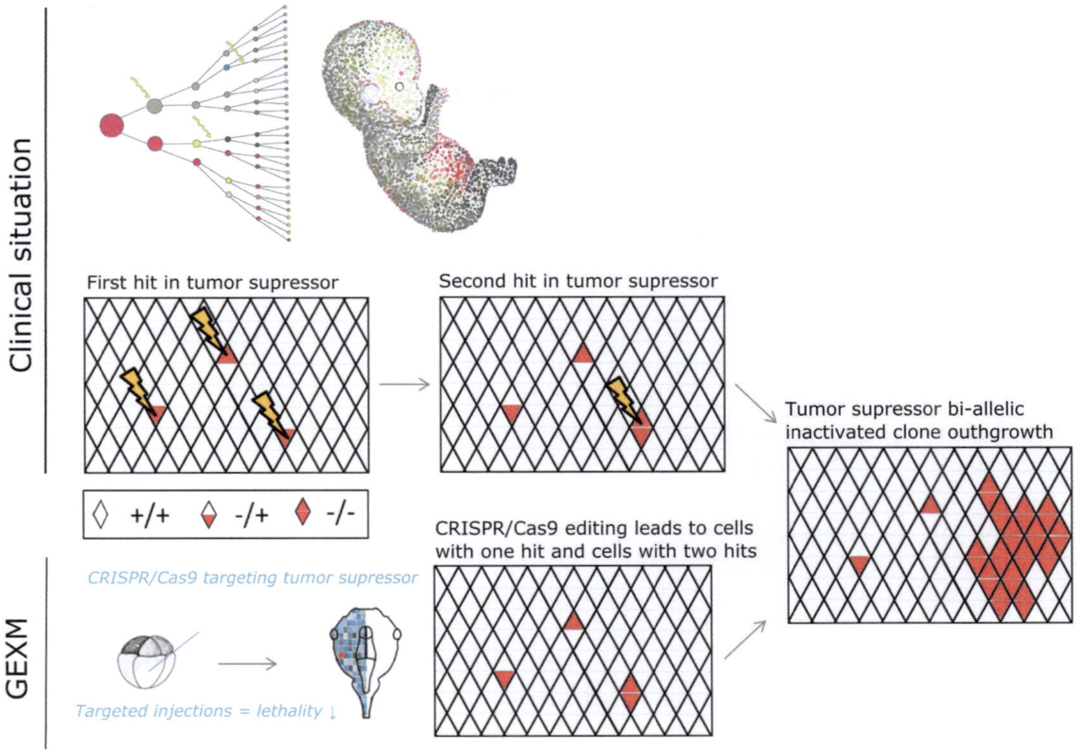

Fig. 1 Schematic comparison of mutational events occurring prior to tumor initiation between the sporadic clinical situation and the CRISPR/Cas9 mediated genetically engineered ***Xenopus*** models (GEXM). (Clinical situation) Mutational insults (UV, carcinogen exposure, etc.) can lead to deactivation of the first allele of a tumor suppressor. Subsequently, only these cells in which a concurrent mutational insult will inactivate the second tumor suppressor allele have the ability to undergo neoplastic transformation. ***(GEXM)*** By microinjection in a ***Xenopus tropicalis*** embryo of Cas9 recombinant protein and gRNAs targeting tumor suppressors, an F0 mosaic crispant is generated in which cells can be either wild type, heterozygous mutant, or homozygous mutant for the tumor suppressor gene. ***(Conclusion)*** In both situations, the result is a clone of cells containing bi-allelic inactivating mutations in the tumor suppressor, which will concurrently obtain a proliferative growth advantage and has the potential to further progress to a fully malignant macroscopically detectable outgrowth

(tissue-restricted) targeted cells. Within these cells, some will be bi-allelic edited in a way that a resulting tumor suppressor null situation will provide a Darwinian advantage and the positive selection pressure will result in tumorigenesis originating from this original clone (Fig. 1). Please note that these tumors will thus develop within a context of surrounding cells which are not necessarily CRISPR/Cas9 edited, and this is much akin to the human clinical situation.

3.3 Phenotyping of Tumor Suppressor Crispants

There will be inevitably some lag (on average between 50 and 150 days within our established models) between microinjection and tumor development within the developing tadpole or froglet, and as such tadpoles will have to be raised. After tadpoles have become somewhat established and mortality due to developmental defects becomes unlikely (±2 weeks post-fertilization), one should closely investigate the clutch for sudden unexplained deaths. This can be indicative for cancer-related mortality and, if and when observed, several crispants should be sampled to verify tumor development within your setup. If tumor development can be observed the whole clutch can be sacrificed and tumor incidence can be determined. Alternatively, if tumor development can be observed by external signs (e.g., retinoblastoma), the above becomes evidently less relevant. Crispants should then be sacrificed as soon as they show signs of tumorigenesis or upon reaching of the ethically determined humane end-point.

3.4 Tumor Genotyping

Genotyping of the tumor is quintessential in providing evidence that the CRISPR/Cas9 mediated genome editing of your tumor suppressor(s) has been the driving factor in tumorigenesis. This is best practice in the field and should be performed [15]. Within tumors induced by CRISPR/Cas9 genome editing, tumor suppressors should show clear positive selection for bi-allelic mutations. Fortunately, when carefully dissecting out monoclonal tumor, relatively straightforward genotyping can be performed as the tumor should only show two distinct mutational events for each tumor suppressor and this reduction in genomic complexity (as compared to the rest of the mosaic mutant crispant) allows genotyping by sanger sequencing methodologies. If these methodologies fail due to tumor heterogeneity we recommend performing laser capture microdissection of tumor tissues (*see* Subheading 3.7).

1. Euthanize the frog in a sufficient volume of 0.05% benzocaine diluted in 0.66× PBS.
2. Move the specimen to a petri dish containing 0.05% benzocaine diluted in 0.66× PBS (*see* **Note 4**).
3. Document tumor occurrence and gross pathology by photographing using a binocular stereomicroscope setup with a camera.
4. Isolate the tumor. For this perform macrodissection of the neoplasm and cut away as much non-transformed tissue as possible (*see* **Notes 5** and **6**).
5. Transfer the tumor to 100 μL lysis buffer and incubate overnight at 56 °C in a thermoshaker with gently agitation (500 rpm).
6. Boil the sample to 98 °C in a thermoshaker for 5 min to inactivate Proteinase K.

7. Spin down the samples for 2 min at a benchtop centrifuge set to full speed (> 14,000 x *g*).
8. Tumors can be further genotyped by
   (a) Subcloning and sanger sequencing (*see* Chapter 5),
   (b) Sanger sequencing and TIDE-analysis (*see* Chapter 5),
   (c) Next-generation sequencing (NGS) followed by BATCH-GE analysis (*see* Chapter 6).

3.5 Histopathology Assessment of Tumors

Histopathological analysis of the crispants can be done as soon as macroscopic tumor development becomes apparent or at time-points of your specific interest. For this, the crispant is humanely euthanized and the tissues of interest are further processed for histopathology.

1. Euthanize the frog in a sufficient volume of 0.05% benzocaine diluted in 0.66× PBS.
2. Move the frog to a petri dish containing 0.05% benzocaine diluted in 0.66× PBS (*see* **Note 4**).
3. Isolate the organs, tumors, or tissue of interest by macrodissection under a binocular stereomicroscope (*see* **Note 5**). Alternatively, the entire crispant can be fixed for downstream analysis leaving all the organ systems within their native context. When performing this, be sure to open up the belly of the crispant slightly to allow the PFA to immediately reach and thus correctly fixate the internal organs (*see* belly of animal in Fig. 2c). This can as thus be as either very specific (e.g., isolation of organs) or more broad (e.g., isolation of the entire head region or even the entire crispant). Carefully document any aberrant findings by photographing.

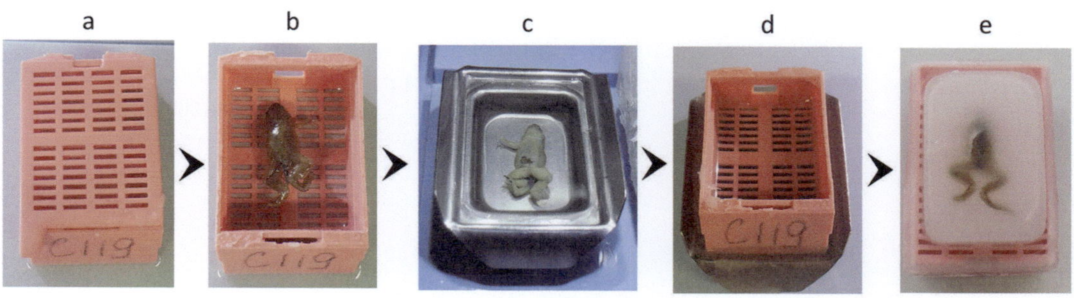

Fig. 2 Processing of a PFA fixed crispant to paraffin and the subsequent imbedding procedure to generate paraffin blocks for microtomy cutting. (**a** → **b**) Place the PFA fixed sample within the tissue cassette, dehydrate and process to hot paraffin (*see* **step 7** of Subheading 3.5). (**b** → **c**) Open up to tissue cassette and transfer the specimen to a mold containing melted paraffin. (**c** → **d**) Place the bottom of the tissue cassette on the top of this mold, then place the mold on the cold plate of the imbedding station and allow it to fully solidify. (**d** → **e**) Remove the paraffin block, now attached to the tissue cassette from the mold and archive until microtomy cutting

4. Immerse isolated tissue in a solution of 4% PFA and fix overnight at 4 °C. As a rule of thumb have at least 5 volumes of PFA per volume of tissue.
5. If tissue contains bone, decalcification is necessary to avoid issues during microtomy procedure (*see* **Note 7**). For this immerse tissue section for 5–7 h within Morse's acid decalcification solution. Wash the tissues once in 70% EtOH.
6. Move tissue to histology cassette (*see* **Note 8**) and immerse in 70% EtOH solution (Fig. 2a).
7. Process tissue to paraffin. Employ an automated tissue processor (e.g., Citadel 2000 Tissue Processor) or perform manual processing, respecting the immersion times, as follows (Fig. 2b).

Reagent	Immersion time, h
Ethanol (70%)	2
Ethanol (90%)	7
Xylene	3
Paraffin (heated)	5

8. Imbed the tissue within histology molds with hot paraffin in an imbedding station and put the top of the cassette above the tissue. Let it cool down on the cold plate and remove the paraffin block from the mold (Fig. 2c–e).
9. Cut paraffin tissue sections at 5 μm on a microtome and collect on slides suitable for downstream immunostaining procedures (e.g., Klinipath Plus).
10. Stain (a subset—keep some slides for immunohistochemistry) the slides by hematoxylin and eosin (H&E) staining. Employ an automated slide stainer or perform manual processing, respecting the immersion times, as follows:

Reagent	Immersion time
Xylene	15 min
Ethanol (99–100%)	2 min
Ethanol (90%)	2 min
Ethanol (70%)	1 min
Bidi	1 min
Hematoxyline	8 min
Rinse with running tap water	1 min
Bidistilled water + HCl	5 s

(continued)

Reagent	Immersion time
Rinse with running tap water	2 min and 30 s
Eosine (1% in Bidi)	3 min
Rinse with running tap water	2 min and 30 s
Tap water	5 min
Ethanol (70%)	50 s
Ethanol (90%)	1 min and 40 s
Ethanol (99–100%)	1 min and 40 s
Xylene	4 min

11. Mount the slides with a coverslip using a xylene-based mounting solution (e.g., Entellan).
12. Perform, using a standard microscope, histopathological assessment of slides. Ideally, an additional assessment should be made by a clinical pathologist with oncology experience.

3.6 Immunofluorescence of Proliferating Cell Nuclear Antigen (PCNA) to Demonstrate High Proliferative Index of Tumor Tissue

The most straightforward way of validating a neoplasm by demonstrating increased cellular proliferation. We describe here an established immunofluorescence protocol for performing proliferating cell nuclear antigen (PCNA) staining on sections from *X. tropicalis* tumors in order to demonstrate extensive hyperproliferation within tumor sections. Note that this protocol can be adapted in order to perform immunofluorescence on *X. tropicalis* sections with other antibodies, assuming that the epitope is sufficiently conserved in *X. tropicalis.*

1. Deparaffinize and rehydrate tissue sections with an automatic slide processor or perform manual processing, respecting the immersion, times as follows:

Reagent	Immersion time
Xylene	15 min
Ethanol (99–100%)	2 min
Ethanol (90%)	2 min
Ethanol (70%)	1 min
Bidistilled water	1 min

2. Perform antigen retrieval with a tissue retriever in citrate buffer (*see* **Note 9**), then wash slide(s) once in a staining container containing 100 mL 1× PBS.
3. Aspirate excess fluid of slide(s) by vacuum pump aspiration and draw a circle with a PAP pen around tissues of interest (*see* **Note 10**). Note that you need at least two tissues circled (tissue to be incubated with and without primary antibody—as control).

4. Apply blocking buffer to the sections within the constraints of the PAP pen and incubate for 2 h at room temperature in a humidified slide staining tray (*see* **Note 11**).
5. Dilute anti-PCNA antibody 1/500 in blocking buffer. Keep one section per slide in blocking buffer as a "no primary antibody" control tissue. For the other section(s), aspirate by vacuum pump the blocking buffer from within the constraints of the PAP pen circle (*see* **Note 10**). Apply primary antibody mix to the sections to be stained, by gently pipetting within the PAP pen circle. Then, incubate the tissues overnight at 4 °C in a humidified slide staining tray.
6. Wash the slide(s) four times 5 min and one time 30 min in 1× PBST. This is achieved by transferring the slides through a series of staining containers containing 100 mL 1× PBST.
7. Dilute secondary antibody (DyLight 594) 1/1000 in blocking buffer. Aspirate excess fluid of slide(s) and from within the PAP pen circles by vacuum pump aspiration (*see* **Note 10**). Apply secondary mix to *all* sections (including the "no primary antibody"), by gently pipetting within the PAP pen circle. Then, incubate the tissues for 2 h at room temperature in a humidified slide staining tray.
8. Wash the slide(s) four times 5 min in 1× PBST and one time 30 min in 1× PBS, as described in **step 6** of this section.
9. Dilute nuclear counterstain (Hoechst 33342) 1/1000 in 1× PBS. Aspirate excess fluid of slide(s) and from within the PAP pen circles by vacuum pump aspiration (*see* **Note 10**). Apply nuclear counterstain to *all* sections (including the "no primary antibody"), by gently pipetting within the PAP pen circle. Then, incubate the tissues for 30 min at room temperature in a humidified slide staining tray.
10. Wash the slide(s) four times 5 min in 1× PBS and one time 30 min in 1× PBS, as described in **step 6** of this section.
11. Aspirate excess fluid of slide(s) and from within the PAP pen circles by vacuum pump aspiration. Apply three drops of immunofluorescence mounting medium, apply a coverslip, and seal with nail polish.
12. Examine sections by fluorescence microscopy. PCNA immunoreactivity should be observed within actively proliferating tumors. *See* Fig. 3 for a representative imaging.

3.7 Laser Capture Microdissection to Isolate Pure Tumor Tissue

Laser capture microdissection is a straightforward and rapid way of isolating tumor tissue from sections. This isolation can be extremely useful in order to gain insight into the mutational INDEL pattern occurring within the tumor suppressors targeted by the CRISPR/Cas9 system.

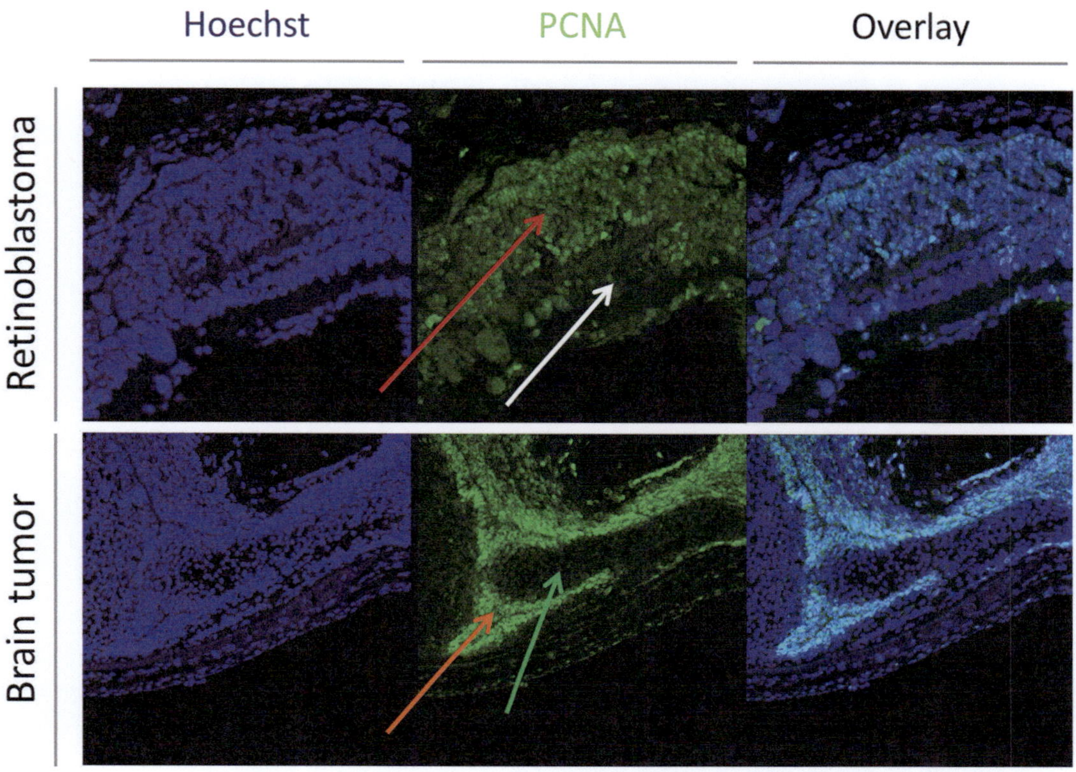

Fig. 3 Representative imaging of a PCNA staining on a retinoblastoma and a brain tumor exhibiting a high proliferative index as imaged by immunofluorescence techniques. (**a**) Retinoblastoma shows clear PCNA immunostaining (red arrow) while the normal retinal layers remain quiescent (white arrow). (**b**) Brain tumor shows clear PCNA immunostaining (orange arrow) while the surrounding normal brain tissue remains quiescent (green arrow)

1. Rehydrate, perform hematoxylin and eosin staining, and dehydrate the sections containing the tumors up to 100% ethanol (employ same staining procedure as **step 10** in Subheading 3.5, but omit the final xylene step).
2. Allow sections to briefly air-dry.
3. Perform microdissection by laser pressure catapulting (LPC) methodologies. *See* Fig. 4 for a representative microdissection.
4. Capture microdissected tissue within the caps of a 200 μL microcentrifuge tube and process to isolate DNA by using the QIAamp DNA Micro Kit per manufacturer's instructions (partum: Isolation of Genomic DNA from Laser-Microdissected Tissues).
5. Amplify the genome edited region by PCR and further genotype by
   (a) Subcloning and Sanger sequencing (*see* Chapter 5).
   (b) Sanger sequencing and TIDE-analysis (*see* Chapter 5).

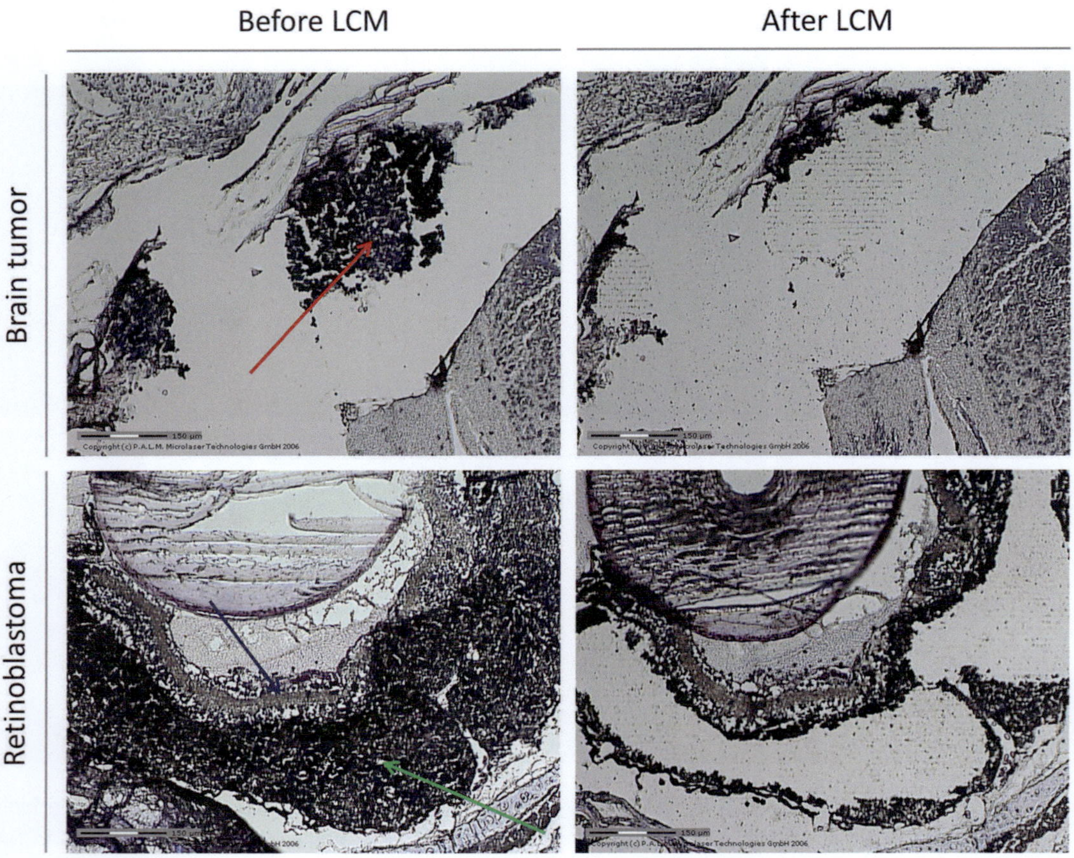

Fig. 4 Laser capture micro dissection (LCM) to isolate specific tumor cells from a H&E stained histological section. (Top) Isolation of a pure tumor cell population from an intracranial brain tumor (red arrow). (Bottom) Isolation of a pure tumor cell population (green arrow) from a retinoblastoma developing within the retina of an ***X. tropicalis***. Appreciate that the surrounding normally appearing retina (blue arrow) remains on the section post-LCM, highlighting the possibility for rapid, spatially restricted isolation of a specific pathological cell population from a section

(c) Next-generation sequencing (NGS) followed by BATCH-GE analysis (*see* Chapter 6).

4 Notes

1. CRISPR/Cas9 induced in-frame mutation (insertion or deletion of $(n) \times 3$ bp) can lead to production of protein missing or gaining $(n) \times 1$ amino acid. Whether or not this protein will be functional is a priori very hard to determine. One can make an educated guess on known protein 3D structure from structural biology work (if available) but in either way it remains hard to determine, based on DNA sequence information, the specific effect of the protein perturbation of the protein function.

2. It has to be noted that targeting a structurally important domain will result in mosaic crispants by which per definition some of the cells will express an in-frame variant of the protein product from your targeted gene (losing or gaining one or more amino acids). It is possible that these variants might acquire positive or negative dominant functions (gain-of-function) and as such some cells composing your mosaic crispant might not represent a full loss-of-function experiment. While still valuable in order to increase penetrance of resulting F0 phenotypes (due to loss of several evolutionary conserved amino acids in in-frame edited cells within the crispant probably leading to protein null in the majority of the situations), it does serve you well to be at least aware of this possibility. A signal that this might be occurring is positive selection pressure within the spectrum of CRISPR/Cas9 induced mutations to certain specific in-frame mutations within a large number of tumors developing within your crispants.
3. The *Xenopus* fate map can be consulted on Xenbase: http://www.xenbase.org/anatomy/static/xenbasefate.jsp

   One should identify the tissues/organs one wishes to engineer by CRISPR/Cas9 and subsequently employ the reverse fate map to determine the blastomere to inject in order to ensure major contribution (red) to the tissue lineage you wish to affect.
4. Benzocaine should be added to the dissection dish to achieve complete palliative surgical anesthesia throughout the autopsy procedure. Frogs should remain immersed in 0.05% benzocaine solution permanently during dissection procedures and during the disposal in order to avoid any chance of regaining consciousness.
5. When needing to isolate tumor tissues, or in fact any very small tissue, it becomes a worthwhile investment to purchase micro scissors (e.g., Vannas Spring Scissors—2.5 mm Blades; Fine Science Tools) and small watchmaker tweezers.
6. Note that extremely small pieces of tumor (akin to the size of putting a ballpoint pen on a paper) should still yield a sufficient amount of DNA for successful PCR post-lysis. As such focusing on isolating "pure tumor" is more important than isolating "a lot of" tumor.
7. Failure to decalcify bone prior to further downstream processing can lead to rapid microtome blade dulling during cutting of the sample. This will lead to curled sections and tearing of the tissue culminating in generally uninterpretable tissue sections due to artifacts. We have previously tried different methodologies to perform decalcification such as EDTA which, while functional, take a much longer time to completely decalcify

the tissue. In our experience the fastest and most straightforward way is as described here using Morse's solution.

8. When choosing a size of cassette, make sure to choose one in which the tissue is at least two times larger in diameter than the holes, because during processing the tissue will still shrink.
9. We employ a pressure cooker retriever as this allows standardization of the antigen retrieval procedure. Each cycle through this pressure cooker should, in theory, be identical to one another. If one does not have access to this equipment, it is possible to use alternative antigen retrieval methodologies such as but not limited to heat-induced antigen retrieval [16].
10. Tissue sections should be *left standing dry* (after aspiration of fluid) as *short a time as possible*. Work rapidly to rapidly apply new solution to the tissue sections after aspiration. Allowing a tissue section to stand dry for longer than a couple of minutes will lead to a substantial increase in background signal.
11. A slide chamber can be commercially bought (e.g., Sigma-Aldrich; H6644-1EA) or home-made by constructing plastic holders that can carry slides and that fit in a larger metal container. In either way the slide-trays need to be humidified in order to avoid evaporation of sample when incubating the sections. This is achieved by placing a stack (±10 pieces) of 1× PBS soaked Tork tissue within the bottom of the container. Upon closure of the container, the evaporation of this 1× PBS will humidify the chamber leading to a reduced evaporation of the antibody containing solution.

Acknowledgments

The authors would like to acknowledge Tarryn Porter (tarryn.porter@sanger.ac.uk) for allowing reuse of a figure by her hand (top part of Fig. 1). Furthermore, the authors would like to thank Dr. Tom van Nieuwenhuysen for the collaborative effort in setting up the PCNA immunofluorescence methodologies. Furthermore, the authors would like to thank Trees Lepez and prof. Dieter Deforce for the initial tutorial on how to use the PALM LCM system. Finally, the authors would like to thank Dieter Tulkens and Marjolein Carron for critical proof-reading of this chapter. Research in the authors' laboratory is supported by the Research Foundation—Flanders (FWO-Vlaanderen) (grants G0A1515N and G029413N), by the Belgian Science Policy (Interuniversity Attraction Poles—IAP7/07) and by the Concerted Research Actions from Ghent University (BOF15/GOA/011). Further support was obtained by the Hercules Foundation, Flanders (grant AUGE/11/14), and the Desmoid Tumor Research Foundation.

References

1. Bhattacharya D, Marfo CA, Li D et al (2015) CRISPR/Cas9: an inexpensive, efficient loss of function tool to screen human disease genes in Xenopus. Dev Biol 408:196–204. https://doi.org/10.1016/j.ydbio.2015.11.003
2. Nakayama T, Fish MB, Fisher M et al (2013) Simple and efficient CRISPR/Cas9-mediated targeted mutagenesis in Xenopus tropicalis. Genesis 51:835–843. https://doi.org/10.1002/dvg.22720
3. Hellsten U, Harland RM, Gilchrist MJ et al (2010) The genome of the western clawed frog Xenopus tropicalis. Science (80) 328:633–636. https://doi.org/10.1126/science.1183670
4. Guo X, Zhang T, Hu Z et al (2014) Efficient RNA/Cas9-mediated genome editing in Xenopus tropicalis. Development 141(3):707–714
5. Van Nieuwenhuysen T, Naert T, Tran HT et al (2015) TALEN-mediated apc mutation in Xenopus tropicalis phenocopies familial adenomatous polyposis. Oncoscience 2. https://doi.org/10.18632/oncoscience.166
6. Naert T, Colpaert R, Van Nieuwenhuysen T et al (2016) CRISPR/Cas9 mediated knockout of rb1 and rbl1 leads to rapid and penetrant retinoblastoma development in Xenopus tropicalis. Sci Rep 6. https://doi.org/10.1038/srep35264
7. Naert T, Van Nieuwenhuysen T, Vleminckx K (2017) TALENs and CRISPR/Cas9 fuel genetically engineered clinically relevant *Xenopus tropicalis* tumor models. Genesis 55:e23005. https://doi.org/10.1002/dvg.23005
8. Howe K, Clark MD, Torroja CF et al (2013) The zebrafish reference genome sequence and its relationship to the human genome. Nature 496:498–503. https://doi.org/10.1038/nature12111
9. Robinson DR, Wu Y-M, Lonigro RJ, et al (2017) Integrative clinical genomics of metastatic cancer. doi:https://doi.org/10.1038/nature23306
10. Northcott PA, Buchhalter I, Morrissy AS et al (2017) The whole-genome landscape of medulloblastoma subtypes. Nature 547:311–317. https://doi.org/10.1038/nature22973
11. Scarpa A, Chang DK, Nones K et al (2017) Whole-genome landscape of pancreatic neuroendocrine tumours. Nature 543:65–71. https://doi.org/10.1038/nature21063
12. Morrissy AS, Garzia L, Shih DJH et al (2016) Divergent clonal selection dominates medulloblastoma at recurrence. Nature 529:351–357. https://doi.org/10.1038/nature16478
13. Kwong LN, Dove WF (2009) APC and its modifiers in colon cancer. Adv Exp Med Biol 656:85–106
14. Moody SA (1987) Fates of the blastomeres of the 32-cell-stage Xenopus embryo. Dev Biol 122:300–319
15. Weber J, Öllinger R, Friedrich M et al (2015) CRISPR/Cas9 somatic multiplex-mutagenesis for high-throughput functional cancer genomics in mice. Proc Natl Acad Sci U S A 112:13982–13987. https://doi.org/10.1073/pnas.1512392112
16. Hussaini SMQ, Jun H, Cho CH et al (2013) Heat-induced antigen retrieval: an effective method to detect and identify progenitor cell types during adult hippocampal neurogenesis. J Vis Exp. https://doi.org/10.3791/50769

Chapter 12

CRISPR/Cas9 F0 Screening of Congenital Heart Disease Genes in *Xenopus tropicalis*

Engin Deniz, Emily K. Mis, Maura Lane, and Mustafa K. Khokha

Abstract

In the US and Europe, birth defects are the leading cause of infant mortality. Among birth defects, Congenital Heart Disease (CHD) occurs in approximately 8 out of 1000 live births, affects 1.3 million newborns per year worldwide, and has the highest mortality rate. While there is evidence to indicate that CHD does have a genetic basis, most of the CHD burden remains unexplained genetically. Fortunately, new genomics technologies are enabling genetic analyses of CHD patients. Whole exome sequencing of trios as well as copy number variations assayed by high-density SNP arrays can now be obtained at high efficiency and relatively low cost. These efforts are identifying a number of sequence variations in patients with CHD, but only a small percentage have second unrelated alleles to validate them as disease causing. Importantly, most of these candidate genes do not have an identified molecular mechanism implicating them in cardiac development. Therefore, there is a pressing need to develop rapid functional assays to evaluate candidate genes for a role in cardiac development, and then to investigate the underlying developmental mechanisms. Most recently, the advent of CRISPR/Cas9 genome editing technology has greatly enhanced the ability to manipulate and observe the function of the genome in model systems and cell culture. Incorporating the power of a developmental system such as *Xenopus tropicalis* with the CRISPR/Cas9 system and the microscale imaging modality optical coherence tomography (OCT), the analysis of thousands of different genes in cardiac development becomes possible.

Key words CRISPR/Cas9, Congenital heart disease (CHD), Human disease modeling, Optical coherence tomography (OCT), Microinjection

1 Introduction

In order to investigate gene function of candidate genes identified in congenital heart disease patients, animal models are essential to examine the 3D structure of complex organs such as the heart. Ideally, we would use mammalian models that closely resemble the human heart; however, hundreds of candidate genes have been identified and only a few have multiple alleles to confirm pathogenicity[1–4]. Given the high cost of mouse knockouts, testing hundreds of candidate genes is simply not feasible. For this reason, rapid, inexpensive animal models as closely related to human as

Kris Vleminckx (ed.), *Xenopus: Methods and Protocols*, Methods in Molecular Biology, vol. 1865,
https://doi.org/10.1007/978-1-4939-8784-9_12,

possible are attractive. *Xenopus*, a frog model, is particularly well suited.

Xenopus laevis has been used as a model system to study developmental biology since the 1950s. *X. laevis* could be induced to ovulate and produce hundreds to thousands of embryos whenever needed, revolutionizing the field of experimental embryology [5].

Subsequently, *Xenopus tropicalis* has emerged as a powerful amphibian genetic system. Its genome is relatively small and diploid (as opposed to *X. laevis*' allo-tetraploid genome), a well-annotated genome sequence is available, and many of the methods already optimized for *X. laevis* work well in *X. tropicalis* [6, 7]. CRISPR/Cas9 genetic manipulations are ideally suited for *X. tropicalis* due to its diploid genome, efficient genomic mutation rate when using Cas9 protein, and ability to screen complex structural phenotypes within 3–5 days of development [8–10]. Specifically, in studying candidate CHD genes, the *Xenopus* cardiovascular system develops within 72 h and remains optically accessible throughout the early stages of development [11]. Most importantly, oxygen delivery in tadpoles relies on simple diffusion rather than cardiac output. Even in the presence of a non-beating heart, tadpoles survive for several days, which is a major advantage when studying severely disrupted cardiogenesis in some forms of CHDs.

Here, we describe the design and production of CRISPR sgRNA targets, microinjection, and subsequent phenotypic analysis of cardiac phenotypes using OCT. Design and production of sgRNA targets is limited by the specific sequence of the gene of interest; however, new Cas9-like proteins such as Cpf1 with different target sequence requirements are being developed and tested to enhance the number of potential genomic targets in any organism [12].

2 Materials

All solutions are prepared using ultrapure water (prepared by purifying deionized water to attain a sensitivity of 18 MΩ cm at 25 °C) and analytical grade reagents. Follow all waste disposal regulations when disposing of waste materials.

2.1 Reagents and Buffers

1. Sequence specific target primer for sgRNA at 1 μM (sgRNA oligo).
2. EnGen sgRNA synthesis kit (NEB).
3. Nuclease-free water.
4. Zymo RNA Clean and Concentrator™-5 kit.
5. Cas9 protein (PNA Bio CP02) reconstituted at 2 mg/ml in nuclease-free water (*see* **Note 1**).

6. Dextran Tetramethylrhodamine lysine fixable, or another fluorescent tracer.
7. 1/9× MR (Modified Ringer's solution): We keep a stock of 1× MR in distilled water and dilute ninefold.
8. 100× MR salts: 180 mM KCl, 200 mM $CaCl_2$, 100 mM $MgCl_2$.
9. 1× MR: To make 20 l, add 117 g NaCl, 200 ml of 100× MR salts and 100 ml of 1 M HEPES, pH 7.4.
10. 1% low-melt agarose.

2.2 Supplies

1. Microinjector (e.g., Parker-Picospritzer).
2. 3D micromanipulator (e.g., Narishige MM-3 Micromanipulator).
3. Thorlabs-OCT-XYR1-XY Linear Translation and Rotation Stage with Removable Solid Top Plate.
4. Thorlabs Telesto 1325 nm spectral domain-OCT System.

3 Methods

All the steps of the protocol should be carried out at room temperature unless otherwise noted.

3.1 sgRNA Synthesis

1. Design sgRNA targeting primer (sgRNA oligo) for your gene of interest using CRISPRscan.org [13] and the UCSC genome browser [14] (*see* **Note 2**).
   - On CRISPRscan.org, click on UCSC tracks, then *Xenopus*. Search for your gene of interest by entering the gene name in the browser window. If your gene is not well annotated (*see* **Note 3**), find your gene using BLAT and the sequence for your gene of interest. Select a bright green labeled CRISPR that targets an early exon with a high score (>60 if possible) (Fig. 1) (*see* **Note 4**).
   - Click on the sgRNA to reveal the target sequence and the oligo to order. Add CTAGC to the 5′ end of the oligo sequence to increase efficiency of the T7 promoter, and order the sgRNA oligo (*see* Fig. 2).
2. Follow the protocol for EnGen sgRNA synthesis (NEB) (*see* **Note 5**).
   - Thaw EnGen 2× reaction mix and sgRNA oligo (1 μM). Mix and pulse spin before setting up reaction.
   - Assemble reaction at room temperature in the order listed, avoiding master mixes per the instructions for the EnGen sgRNA synthesis kit, and adding the enzyme last (*see* Table 1).

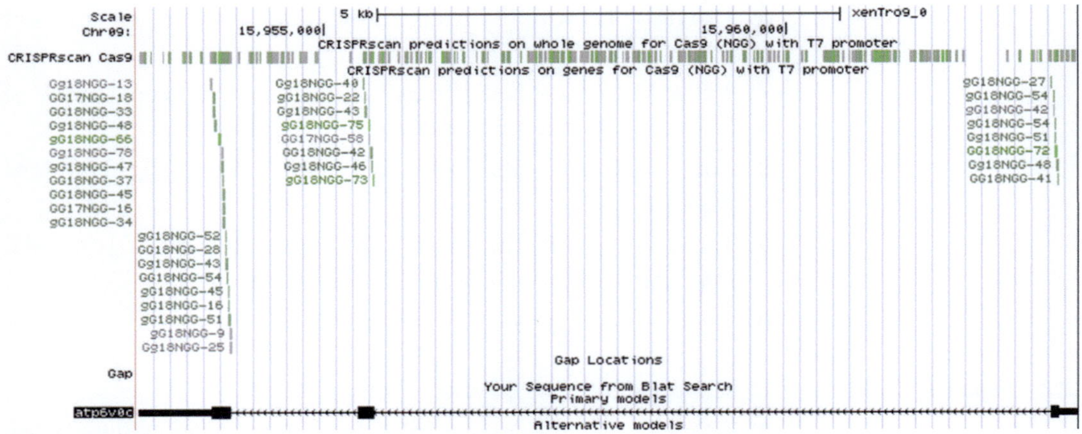

Fig. 1 Screenshot of CRISPRscan targets in the UCSC genome browser. Image acquired from *http://genome.ucsc.edu* on *X. tropicalis* Aug 2015 assembly (xenTro9)

Item: GG18NGG-72
Score: 72
Position: Chr09:15962905-15962927
Genomic Size: 23
Strand: +
View DNA for this feature (xenTro9_0/X. tropicalis)

Target site sequence (with PAM)	GGAGTACTCGGGCGCGCTTGCGG
Oligo sequence	taatacgactcactataGGAGTACTCGGGCGCGCTTGgttttagagctagaa

Number of off-targets

CFD score	0.5144674
All	0
Seed	0

Off-targets (max 30)

	CFD score	All	Seed	Gene
Chr06:14626643[X....X.XX...........]+	0.32			Xetrov90015635m.g-5UTR
scaffold_603:62072[......X.....X..XX...]-	0.15			
Chr09:9429300[........X...X..X..X.]-	0.02			
Chr07:1345567[XX.....X..........X.]+	0.00			Xetrov90017392m.g-CDS
Chr08:64095225[.........X...X.X....X]+	0.00			
Chr03:51811970[.X...X..X..........X]+	0.00			
Chr04:86443059[.........X..X.....XX]+	0.00			
Chr02:77782637[.XX...X............X]+	0.00			
Chr03:94860631[X.X........XX.......]+	0.00			
Chr07:17856419[....X...X...X.....X.]+	0.00			
scaffold_351:34952[......X.X...X.X.....]+	0.00			
Chr01:169091215[....XX......X.....X.]+	0.00			

Fig. 2 Detailed view of sgRNA information from CRISPRscan and UCSC genome browser. Includes score, target sequence, sgRNA oligo sequence, and off-target sites within the *X. tropicalis* genome. Image acquired from *http://genome.ucsc.edu* on *X. tropicalis* Aug 2015 assembly (xenTro9)

Table 1
Protocol for EnGen sgRNA synthesis (NEB)

Reagent	For 20 μl	For 10 μl
Nuclease-free water	3 μl	1.5 μl
EnGen 2× reaction mix	10 μl	5 μl
sgRNA oligo (1 μM)	5 μl	2.5 μl
EnGen enzyme mix	2 μl	1 μl

- Mix thoroughly and pulse-spin in a microcentrifuge. Incubate at 37 °C for 1 h.
- Bring volume to 50 μl with nuclease-free water. Add 2 μl of DNAse I and incubate at 37 °C for 15 min.

3. Purify RNA using the Zymo RNA Clean and Concentrator™-5 kit, then resuspend in a final volume of 13.5 μl.
4. Measure concentration and purity using a nanodrop.
5. Immediately aliquot and freeze at −80 °C until ready to use. Avoid excessive freeze-thaw cycles (*see* **Note 6**).

3.2 Prepare Reaction Mixes

1. Retrieve Cas9 protein aliquot and sequence-specific sgRNA from −80 °C.
2. Combine 1 μl sequence-specific sgRNA at 600 ng/μl, 1.2 μl of cas9 protein at 2 mg/ml, and 0.8 μl of fluorescent tracer for *Xenopus tropicalis* injections.
3. Vortex and pulse-spin in a microcentrifuge.
4. Leave at room temperature until ready to inject (*see* **Note 7**).
5. Inject 2 nl per one cell embryo.

3.3 Scoring Cardiac Phenotypes

3.3.1 Optical Coherence Tomography (OCT) Image Acquisition and Imaging Planes

The Thorlabs Telesto 1325 nm spectral domain-OCT System has a 7 mm imaging depth in air and 12 μm axial resolution (in water, the axial imaging resolution is 9 μm). In *Xenopus*, the cardiac cycle varies between 110 and 140 beats/min. Therefore, to capture the in vivo cardiac cycle, obtain cardiac 2D cross sectional movies at high-speed mode-91 kHz (91,000 A-scan per second) with 91 dB sensitivity.

1. Embed stage 45 (day 5 post fertilization, raised at 26 °C) tadpoles in a 1% agarose gel (Low-melt agarose gel warmed >60 °C and stirred until agarose melts). Pour 2.5 ml of melted agarose into a 35 × 10 mm polystyrene petri dish, allow the agarose to cool down (1–2 min, alternatively an infrared thermometer can be used to precisely track agarose temperature and cool to 25–30 °C, which is suitable for embedding), and place the tadpole in the middle of the dish before the agarose

is solid. While the agar solidifies (~1 min), manually align the tadpole(s) while viewing under a stereomicroscope so the ventral body and cardiac sac are exposed en-face to the lens. Importantly, refrain from using paralytic agents (such as benzocaine, Tricaine-MS222), as these cause alterations in cardiac function with prolonged exposure.

2. Using Thor-Image OCT Software (Thorlabs-version 4.4.6) to acquire optical cross sections of the tadpole heart to analyze cardiac structure and function. For a full cardiac assessment, follow the following steps (https://www.thorlabs.com):

   - Determine a reference plane (XZ-plane) as the plane intersecting the mid-eye point (*see* Fig. 3). This plane is defined as 0 degrees. Relative to this reference plane, adjust the imaging plane to capture three imaging views: (1) VTCV (Ventral Two-Chamber View), (2) OTV (Outflow track view), and (3) LV (Lateral View).
   - *VTCV (Ventral Two-Chamber View)* (*see* Fig. 4)**:** This imaging plane is captured as the OCT beam travels posteriorly along the *y*-axis at 0° (based on the reference plane).

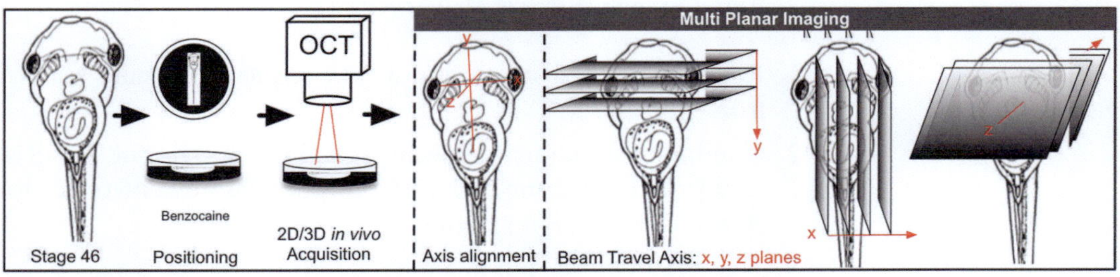

Fig. 3 Stage 46 tadpole embedded in 1% agarose and positioned so the ventral side faces the OCT lens. Next, using a 2-axis stage, the embryo is oriented to the reference axis-mid eye plane, and the heart is scanned along the *x*, *y*, and *z* axes to capture the three imaging planes: Ventral Three Chamber View (VTCV) and Outflow Tract View (OTV) and Lateral View (LV). Image acquired from Deniz et al. 2017 [11]

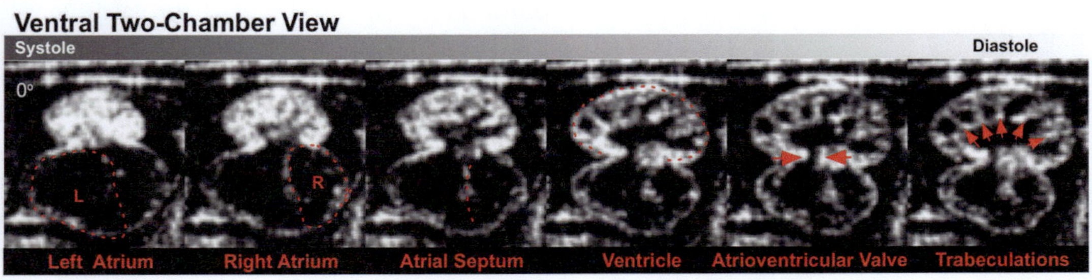

Fig. 4 Imaging stage manually moved along the cardiac sac posteriorly toward the tail on the y-axis until the ventricle captured. Ventral Three Chamber View: Visualizes the single ventricle and both atria. Red dashed line marks the left atrium, right atrium, atrial septum, ventricle, atrioventricular valve, and trabecula in different stages of the cardiac cycle from systole to diastole (left to right). Image acquired from Deniz et al. 2017 [11]

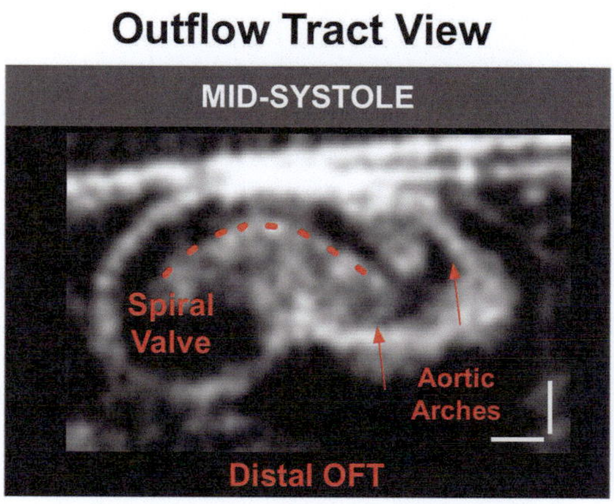

Fig. 5 Outflow tract view: by moving the imaging plane anteriorly to the cranium from the VTCV, OCT captures the distal outflow tract including the spiral valve (red dashed line) and systemic arteries. Image acquired from Deniz et al. 2017 [11]

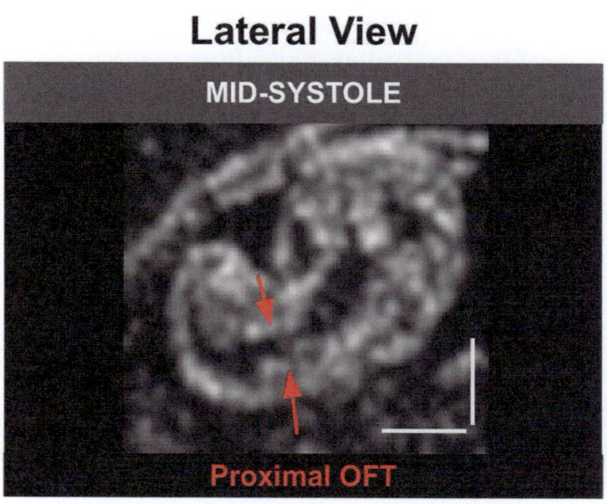

Fig. 6 Lateral View: Rotating the OCT beam angle 30–50° reveals the proximal outflow tract and the ventricle visualized during mid-systole. Image acquired from Deniz et al. 2017 [11]

The left atrium, right atrium, atrial septum, single ventricle, atrioventricular valve, and the trabeculations are captured with this view.

- *OTV (Outflow Tract View)* (*see* Fig. 5): After capturing the VTCV view, adjust the OCT beam anteriorly on the *y*-axis toward the cranium to capture the truncus arteriosus, spiral valve, and the distal systemic vessels.
- *LV (Lateral View)* (*see* Fig. 6): Next, adjust the OCT imaging plane 40–50° clockwise, from the reference axis, and move the embryo along the *y*- and *x*-axis posteriorly toward the tail to visualize most of the proximal outflow tract, as

well as the ventricle. This rotational adjustment can be achieved either using the OCT-Software (which allows 0–360° rotation of the optical cross-section) or the imaging stage (We use Thorlabs—OCT-XYR1—XY linear translation and rotation stage).

3. Acquire 200–400 frames for each imaging plane in order to record 2–3 cardiac cycles. Play each acquisition frame by frame using Thorlabs Software to identify maximal contraction (end-systole) and maximal dilation (end-diastole). For accurate measurements, at least three full cardiac cycles should be obtained. Depending on the pathology (i.e., in cases where the cardiac rate may be slower), frame number should be adjusted to cover 3 cardiac cycles at a minimum.
4. Following acquisition, immobilized tadpoles can be recovered from the 1% agarose with multiple 1/9× MR washes. They can be reimaged if needed over time.

3.3.2 Optical Coherence Tomography (OCT) Cardiac Measurements

OCT, as we outlined above, provides cardiac phenotyping data similar to clinical echocardiography, allowing investigators to visualize detailed internal structural anatomy that is undetectable by simple inspection. We analyze each movie frame by frame (200–400 frames—2/3 cardiac cycles) and obtain measurement using Thorlabs Software provided with the OCT setup. (https://www.thorlabs.com) (*see* Fig. 7 and Table 2).

3.3.3 Xenopus tropicalis Wild type Cardiac Measurements by OCT

Xenopus heart is three-chambered, with two atria, an atrial septum, a highly trabeculated single ventricle, an atrioventricular valve, and an outflow tract. Using OCT measurements, we have delineated the normative data for the wild type stage 46 *Xenopus tropicalis* hearts (*see* Table 3). All measurements obtained at room temperature. We determined normative data and variance based on three observers and accepted variance <20%. Based on these results, only outflow tract proximal end measurements seemed to vary among the observers.

Fig. 7 (continued) wall-to-wall. End diastolic diameter (EDD) and End systolic diameter (ESD) recorded as the average of these three measurements. (**b**) Outflow tract shown during a cardiac cycle. (**b1**) Represent the mid-systole phase where outflow tract is at the maximal stretch. (**b2**–**b3**) Enlarged images from distal OFT where "distal outflow tract is at maximal diameter" measured with the wall-to-wall distance presented as green-a line. The "distal outflow tract maximal excursion distance" measured as the distance between the leaflets presented as yellow-b line. (**c**) Shows the tadpole heart from the lateral view during a cardiac phase. (c1–c2) Focused on the proximal outflow tract during maximal diastole and systole. Inner wall-to-wall measurements made during systole presented as yellow-a line. Image acquired from Deniz et al. 2017 [11]

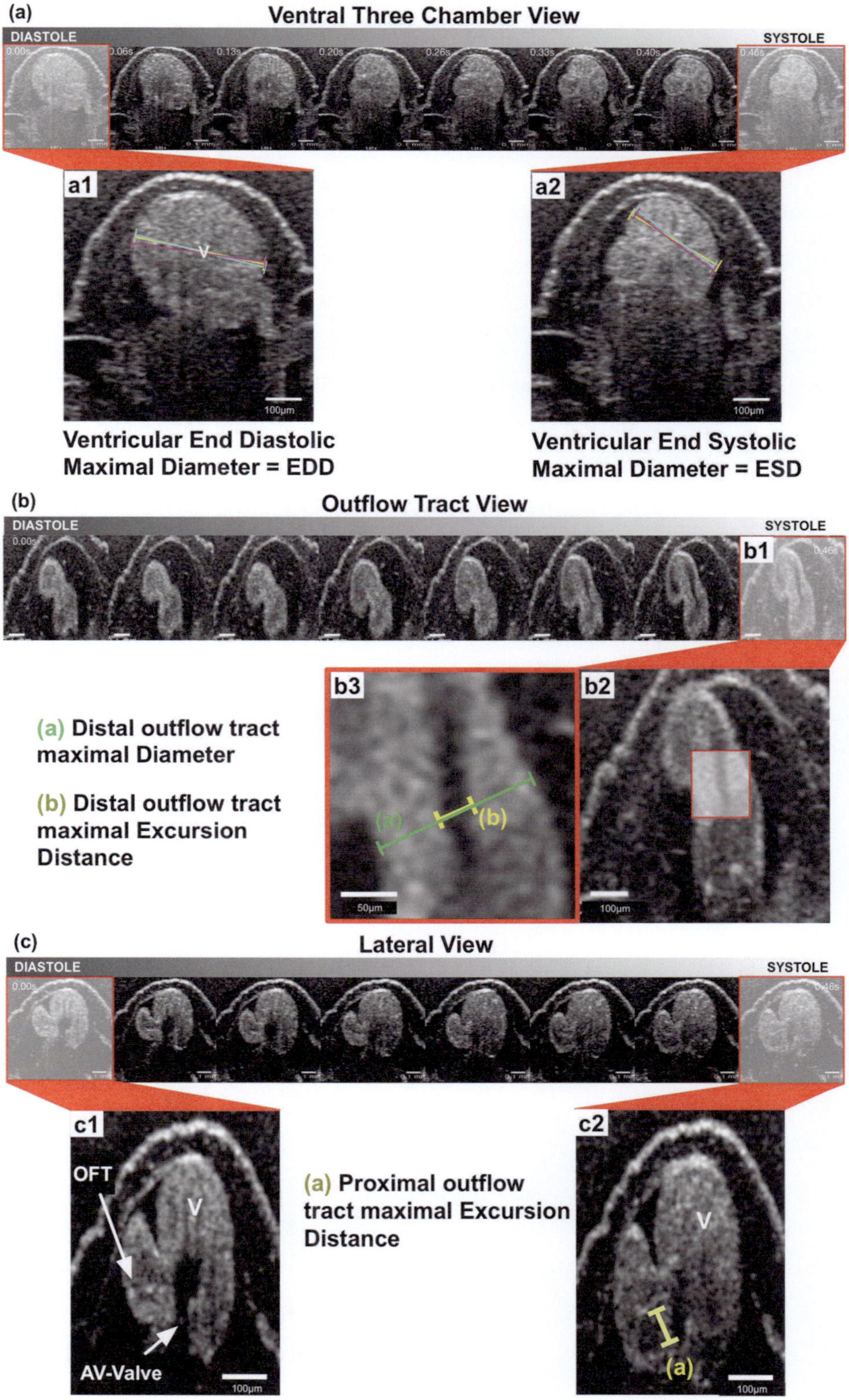

Fig. 7 Imaging Planes and Measurements of *Xenopus tropicalis* heart with OCT. (**a**) Stage 45 tadpole heart live imaging from diastole to systole. (**a1**) Shows the maximal ventricle diastole with three separate measurements recorded from wall-to-wall. (**a2**) Shows the maximal systole with three separate measurements recorded from

Table 2

Suggested measurements to analyze cardiac structures by OCT imaging

Anatomical structure	Parameters measured	Target frame during cardiac cycle	Best imaging plane
Atrium (left and right)	Atrium diameter/surface area	Maximal atrial diastole	VTCV
	Atrial wall thickness	Maximal atrial diastole	VTCV
Atrial septum	Septal length	Mid-atrial diastole	OTV
Ventricle	Apicobasal diameter	Mid-ventricular diastole	VTCV, LV
	Ventricle diameter/surface area	End of ventricular diastole	VTCV
Ventricle trabeculation	Wall thickness	End of ventricular diastole	VTCV
Atrioventricular Valve	Valve diameter	Mid-atrial systole	VTCV
Outflow tract	Outflow tract length	Mid-ventricular systole	OTV, LV
	Outflow tract wall thickness	Mid-ventricular systole	OTV, LV
	Looping	Mid-ventricular systole	OTV, LV
Truncus arteriosus (left and right)	Truncus diameter	Mid-ventricular systole	OTV

VTCV ventral two-chamber view, *OTV* outflow tract view, *LV* lateral view

Table 3

Cardiac measurements for wild type *Xenopus tropicalis*

N = 85	Mean (μm)	StdD	SEM	CoV (%)
End diastolic diameter (EDD)	323	33	5	12
End systolic diameter (ESD)	249	30	4	12
Atrioventricular valve diameter	71	14	2	19
Outflow tract distal-end diameter	131	10	2	8
Outflow tract distal-end leaflet excursion distance	29	5	1	17
Outflow tract proximal-end leaflet excursion distance	43	14	2	31
	Mean (%)	StdD	SEM	CoV (%)
Ventricular shortening fraction ((EDD − ESD)/EDD) × 100	25	5	1	19

4 Notes

1. Cas9 protein should be resuspended in nuclease-free water by gently pipetting up and down. Bubbles can form if mixed too vigorously. Pulse spin. After resuspending, make aliquots of 2.4 μl (two-use) volumes into PCR tubes in order to avoid freeze-thaw cycles. Freeze immediately at −80 °C.
2. CRISPRscan utilizes an algorithm that looks at sgRNA stability and retention within the embryo, as well as ability of the sgRNA to induce double stranded breaks at the target site. Pseudo-colored bright green targets on CRISPRscan are high stability targets with only a single target in the genome. Gray targets are either less stable or have multiple target locations within the genome. Though stability of the sgRNA is less essential when co-injecting with Cas9 protein due to the complex formation and subsequent stabilization of the sgRNA within the Cas9 protein, the higher scoring targets are more likely to be specific.
3. Best practice is to use the best-annotated version of your gene of interest and use that to BLAT the UCSC genome browser. If not annotated as a gene, the expanded CRISPR target sites on the whole genome can be used to find the sgRNA target sequence.
4. If no bright green, high scoring targets can be found in the 5′ end of the gene, extend the search throughout the gene transcript. Additionally, lower target scores can be used, but efficiency appears to be reduced in sgRNAs with scores below 50. If off target sites are predicted, blasting for the target sequence will reveal the off targets. The location of the off target can be analyzed to determine if it affects a protein-coding gene. At least two nonoverlapping sgRNAs should be selected for each gene, to control for technical and experimental difficulties.
5. To increase the number of sgRNA reactions per kit, the EnGen sgRNA synthesis kit can be used at half of the reaction volume. Yields are still more than sufficient for use in genome editing. Make sure to mix each reagent thoroughly, since small volumes and a lack of a master mix can cause a wide range of reaction yields. Standard yields for the EnGen kit are 8–50 μg.
6. Since sgRNAs are not capped or poly-adenylated, they are very unstable, especially with freeze-thaw cycles. In our hands, sgRNAs that undergo freeze-thaw cycles can suddenly cease to function. In order to maximize the number of experiments that can be completed from a single EnGen sgRNA synthesis reaction, dilute 4.8 μg of newly synthesized sgRNAs in 8 μl of nuclease-free water

to a concentration of 600 ng/μl. Aliquot 1 μl into individual PCR tubes for single use. Freeze immediately at −80 °C.

7. Injection mixes can be frozen at −80 °C after mixing. If used within a week of mixing, CRISPR/Cas9 efficiency does not appear to be affected.

References

1. Jin SC, Homsy J, Zaidi S et al (2017) Contribution of rare inherited and de novo variants in 2,871 congenital heart disease probands. Nat Genet. https://doi.org/10.1038/ng.3970
2. Homsy J, Zaidi S, Shen Y et al (2015) De novo mutations in congenital heart disease with neurodevelopmental and other congenital anomalies. Science 350:1262–1266. https://doi.org/10.1126/science.aac9396
3. Fakhro KA, Choi M, Ware SM et al (2011) Rare copy number variations in congenital heart disease patients identify unique genes in left-right patterning. Proc Natl Acad Sci U S A 108:2915–2920. https://doi.org/10.1073/pnas.1019645108
4. Choi M, Scholl UI, Ji W et al (2009) Genetic diagnosis by whole exome capture and massively parallel DNA sequencing. Proc Natl Acad Sci U S A 106:19096–19101. https://doi.org/10.1073/pnas.0910672106
5. Fraser SE, Harland RM (2000) The molecular metamorphosis of experimental embryology. Cell 100:41–55
6. Hirsch N, Zimmerman LB, Grainger RM (2002) Xenopus, the next generation: X. tropicalis genetics and genomics. Dev Dyn 225:422–433. https://doi.org/10.1002/dvdy.10178
7. Khokha MK, Chung C, Bustamante EL et al (2002) Techniques and probes for the study of Xenopus tropicalis development. Dev Dyn 225:499–510. https://doi.org/10.1002/dvdy.10184
8. Bhattacharya D, Marfo CA, Li D et al (2015) CRISPR/Cas9: an inexpensive, efficient loss of function tool to screen human disease genes in Xenopus. Dev Biol 408:196–204. https://doi.org/10.1016/j.ydbio.2015.11.003
9. Hwang WY, Fu Y, Reyon D et al (2013) Efficient genome editing in zebrafish using a CRISPR-Cas system. Nat Biotechnol 31:227–229. https://doi.org/10.1038/nbt.2501
10. Nakayama T, Fish MB, Fisher M et al (2013) Simple and efficient CRISPR/Cas9-mediated targeted mutagenesis in Xenopus tropicalis. Genesis 51:835–843. https://doi.org/10.1002/dvg.22720
11. Deniz E, Jonas S, Hooper M et al (2017) Analysis of craniocardiac malformations in Xenopus using optical coherence tomography. Sci Rep 7:42506. https://doi.org/10.1038/srep42506
12. Zetsche B, Gootenberg JS, Abudayyeh OO et al (2015) Cpf1 is a single RNA-guided endonuclease of a class 2 CRISPR-Cas system. Cell 163:759–771. https://doi.org/10.1016/j.cell.2015.09.038
13. Moreno-Mateos MA, Vejnar CE, Beaudoin J-D et al (2015) CRISPRscan: designing highly efficient sgRNAs for CRISPR-Cas9 targeting in vivo. Nat Methods 12:982–988. https://doi.org/10.1038/nmeth.3543
14. Kent WJ, Sugnet CW, Furey TS et al (2002) The human genome browser at UCSC. Genome Res 12:996–1006. https://doi.org/10.1101/gr.229102

Chapter 13

Quantitative Proteomics of *Xenopus* Embryos I, Sample Preparation

Meera Gupta, Matthew Sonnett, Lillia Ryazanova, Marc Presler, and Martin Wühr

Abstract

Xenopus oocytes and embryos are model systems optimally suited for quantitative proteomics. This is due to the availability of large amount of protein material and the ease of physical manipulation. Furthermore, facile in vitro fertilization provides superbly synchronized embryos for cell cycle and developmental stages. Here, we detail protocols developed over the last few years for sample preparation of multiplexed proteomics with TMT-tags followed by quantitative mass spectrometry analysis using the MultiNotch MS3 approach. In this approach, each condition is barcoded with an isobaric tag at the peptide level. After barcoding, samples are combined and the relative abundance of ~100,000 peptides is quantified on a mass spectrometer. High reproducibility of the sample preparation process prior to peptides being tagged and combined is of upmost importance for obtaining unbiased data. Otherwise, differences in sample handling can inadvertently appear as biological changes. We detail and exemplify the application of our sample workflow on an embryonic time-series of ten developmental stages of *Xenopus laevis* embryos ranging from the egg to stage 35 (just before hatching). Our accompanying paper (Chapter 14) details a bioinformatics pipeline to analyze the quality of the given sample preparation and strategies to convert spectra of *X. laevis* peptides into biologically interpretable data.

Key words Proteomics, *Xenopus laevis*, Development, Sample preparation, Multiplexing, Mass spectrometry, Yolk, TMT, Protein dynamics

1 Introduction

Amphibian embryos and oocytes are classic models to study cellular and developmental processes [1–4]. Due to inexpensive maintenance, resistance to diseases, and year-round availability of eggs and embryos, *Xenopus laevis* has become the predominant amphibian model in the laboratories. Recently, the genome of this quasi-tetraploid frog has been sequenced [5], which will greatly benefit systems-level analysis on the genome, RNA, and protein levels. *Xenopus* is particularly attractive for proteomics studies. This is due to the very large amount of protein that is easily obtainable.

Kris Vleminckx (ed.), *Xenopus: Methods and Protocols*, Methods in Molecular Biology, vol. 1865,
https://doi.org/10.1007/978-1-4939-8784-9_13, © Springer Science+Business Media, LLC, part of Springer Nature 2018

Depending on the method of choice, proteomic experiments are typically performed with ~1 μg to ~1 mg of proteins. One egg/embryo contains ~25 μg of non-yolk protein [6], therefore, a collection of at most ~40 embryos would suffice for experiments that require the most material. For comparison, this is equivalent to ~2000 fly embryos [7], or ~60,000 mouse oocytes [8]. Single cell proteomics with the *Xenopus* eggs and oocytes is compatible with standard proteomics protocols [9]. More recently, the Nemes group has pushed the limits of sample limited analysis and is able to quantify ~400 proteins of individual cells in the 16-cell stage [10]. Besides providing large amounts of material, *Xenopus* oocytes and embryos can be easily physically manipulated. For example, we recently published a paper on nucleocytoplasmic partitioning where we manually isolated the nuclei of oocytes [11]. Similarly, we have manually isolated and defined the composition of the Balbiani body [12]. Another major advantage of the *Xenopus* model is the superb synchrony of developmental stages achievable with in vitro fertilization. This benefit has been used in multiple papers to study the changes in protein abundances during development [13–15]. With egg activation via electric shock, synchrony can be improved to within a few seconds, which allowed us to follow the change of protein abundance and phospho dynamics in the metaphase-anaphase transition [16]. Further, the ability to make essentially undiluted cytoplasm from *Xenopus* eggs and embryos has made it a classical biochemical system. These lysates are still "alive" in many aspects as is evident in their ability to form mitotic spindles and nuclei in the test tube [17–19]. We suspect that the combination of classical biochemical approaches like isolation of organelles or fractionation with proteomics is an emerging and powerful tool to study the organization of cytoplasm [20, 21].

Since the mid-1990s, mass spectrometry-based proteomics has undergone an impressive evolution starting with the ability to identify a few proteins to experiments that regularly identify ~10,000 proteins [22–24]. Diverse options for experimental design on quantification of protein abundances are available. Among them are SILAC, label-free, multiplexed proteomics, and DIA [25–27, 39]. Due to a beneficial combination of sensitivity, measurement quality, and the ability to compare many samples at once, our method of choice is multiplexed proteomics. For multiplexed proteomics, peptides of the different conditions are labeled with a barcode, i.e., an isobaric tag (Fig. 1). The conditions (currently up to 11) are combined and ionized. In the MS1 spectrum the peptides from the different conditions are indistinguishable. Upon fragmentation, the reporter ions are released and encode the different conditions. The relative signal can be used to quantify relative protein abundance. However, multiplexed proteomics' data quality was initially limited by the artifacts that arise from co-isolation and co-fragmentation of unintended peptides.

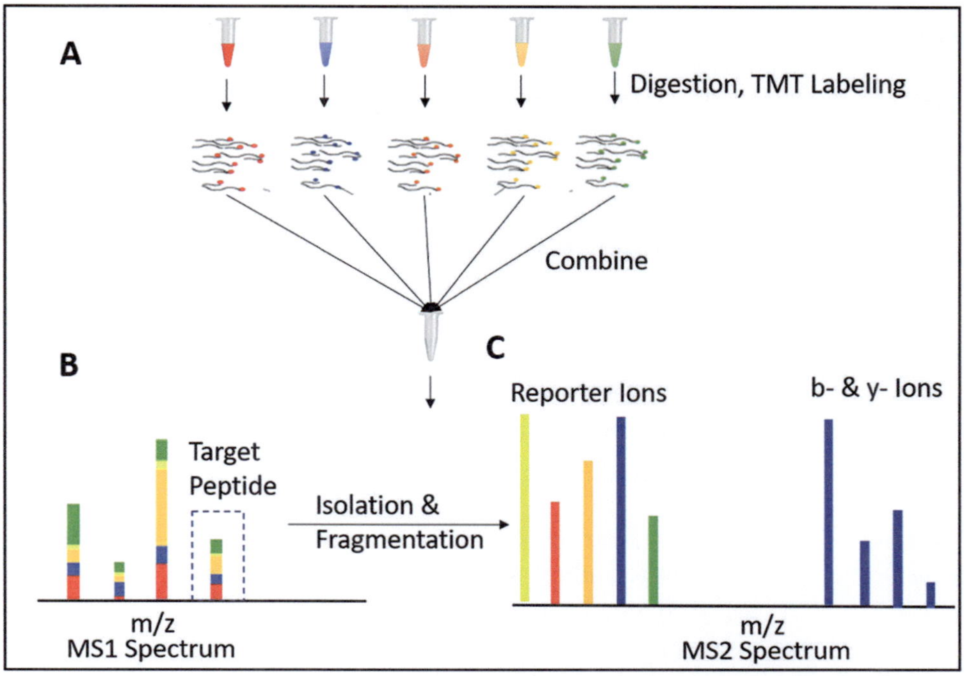

Fig. 1 Principle of multiplexed proteomics. (**A**) Proteins from multiple conditions (replicates, time points, etc.) are digested and labeled with isobaric tags (e.g., TMT). The distinct colors represent different tags, which serve as barcodes for the sample origin. After tagging, different conditions are mixed and together ionized onto a mass spectrometer. (**B**) In the MS1 spectrum the isobaric tags have the identical mass, thus peptide peaks from different conditions are indistinguishable. (**C**) The isolated peptides are fragmented and produce condition-specific reporter ions. The intensities of these reporter ions can be used for relative quantification of the associated protein. The b- and y-ions, which result from breakage of the peptide's backbone, are used for identification

Nonetheless, MultiNotch MS3, QuantMode, and TMTc can overcome this problem [28–31, 40]. MultiNotch MS3 (TMT-MS3) is currently the most widely used method and has allowed detection of protein abundance changes of less than 10% as significant [11].

Here, we detail the sample preparation and data acquisition for a TMT-MS3 experiment exemplified with the quantification of protein dynamics of an embryonic time-series. We collected ten developmental stages spanning from egg to stage 35 which is the stage just before hatching. Importantly, for multiplexed proteomics labeling with the isobaric tag occurs at the peptide level. Upstream preparation is performed in parallel. Hence, it is extremely important for any of these sample manipulations to be as reproducible as possible. Here, we detail how we achieve reproducible sample preparation to the best of our knowledge. The accompanying paper (Chapter 14) discusses how to analyze the acquired spectra and use the information in these to perform sample preparation quality analysis and how to convert collections of spectra into biologically interpretable data.

2 Materials

2.1 Embryo Lysis and Yolk Removal

1. 10 mM Combretastatin: stock solution in acetonitrile (store at −80 °C).
2. 10 mM Cytochalasin D: stock solution in acetonitrile (store at −80 °C).
3. Lysis Buffer: 250 mM Sucrose,1% Nonidet P-40 (NP 40), 10 mM Ethylenediaminetetraacetic acid (EDTA), 25 mM HEPES pH = 7.2, 10 μM Cytochalasin D, Complete Roche Mini (1 tablet per 10 ml). Lysis buffer should be freshly prepared.
4. Sodium Dodecyl Sulfate (SDS): Prepare 10% stock solution.
5. Liquid nitrogen.

2.2 Cysteine Protection

1. Dithiothreitol (DTT): 500 mM stock solution in HPLC grade water, adjust pH to ~7 (store at −80 °C).
2. N-Ethylmaleimide (NEM): 1 M stock solution in acetonitrile (HPLC grade) (store at −80 °C).

2.3 Methanol–Chloroform Precipitation [32]

1. Methanol (HPLC Grade).
2. Chloroform (HPLC grade).
3. Water (HPLC grade).
4. 10 mM 3-[4-(2-Hydroxyethyl)-1-piperazinyl]propanesulfonic acid (EPPS), pH 8.5: Prepare in HPLC grade water.
5. 6 M Guanidine Hydrochloride (GuaCl) (e.g., Chem-Implex International, Pure GuaCl >99.5%), in 10 mM EPPS pH 8.5 (use HPLC grade water).
6. Ultrasonic cleaner (e.g. Branson M2800).

2.4 Estimate Protein Amount (BCA Assay)

1. Pierce BCA Protein Assay Kit (Thermo Scientific).
2. Multi-Mode Microplate Reader (e.g., Synergy HT, Biotek).

2.5 Digestion

1. Lysyl Endopeptidase (LysC) (Wako chemicals, 10 AU, store at −80 °C): Prepare aliquots of 2 μg/μl in HPLC grade water and flash freeze them in liquid nitrogen. Check lot specifications, typically 10 AU is ~2 mg of LysC.
2. Trypsin (Promega, sequencing grade, 0.5 μg/μl, store at −80 °C).
3. 10 mM EPPS, pH 8.5: Prepare in HPLC grade water.
4. Vacuum concentrator, consisting of centrifugal evaporator SC100 (Savant), refrigerated vapor trap RVT4104 (Savant), vacuum pump RV8 (Edwards).

2.6 TMT Labeling

1. 200 mM EPPS, pH 8.0.
2. TMT10-plex reagent set (Thermo Fisher Scientific): Prepare aliquots of 20 μg/μl in dry acetonitrile (Sigma), flash freeze, and store at −80 °C until use. Recently, an 11-plex TMT kit has become available.
3. Hydroxylamine solution, 50% wt (Sigma, store at 4 °C). *We replace hydroxylamine yearly.*
4. Vacuum concentrator, consisting of centrifugal evaporator SC100 (Savant), refrigerated vapor trap RVT4104 (Savant), vacuum pump RV8 (Edwards).

2.7 Reverse Phase Solid Phase Extraction/Sep-Pak

1. Phosphoric acid.
2. Methanol.
3. 70% acetonitrile + 1% formic acid solution.
4. 35% acetonitrile + 1% formic acid solution.
5. 1% formic acid.
   All the reagents should be prepared in HPLC grade water and stored in glass scintillation vials (e.g., Sci Spec, 20 ml). Solutions are stable at room temperature but the vials should be tightly closed to prevent evaporation of acetonitrile. For corrosive solutions like formic acid, we place a parafilm sheet between cap and vial before closing.
6. C-18 Sep-Pak cartridges (Waters Corporation, 50 mg, 1.3 ml column volume).
7. Solid Phase Extraction Vacuum Manifold (e.g. Sigma 57250-U).
8. Vacuum concentrator, consisting of centrifugal evaporator SC100 (Savant), refrigerated vapor trap RVT4104 (Savant), vacuum pump RV8 (Edwards).

2.8 Medium pH Reverse Phase Prefractionation

1. HPLC buffers.

   Buffer A: 10 mM ammonium bicarbonate, pH 8.0, 0.5% acetonitrile.

   Buffer B: 10 mM ammonium bicarbonate, pH 8.0, 95% acetonitrile:

   Sonicate the buffers for ~ 30 min to degas them. Solutions are stable at room temperature but the vials should be tightly closed to prevent evaporation of acetonitrile. All buffers should be prepared with HPLC grade water.
2. 2,2,2-Trifluoroethylene (TFE) (Acros Organics, 99.8%).
3. C18 column (Agilent, ZORBAX Extend-C18, 4.6 × 150 MM, 3.5-μm, 80 A, Part # 763953-902).
4. Injection Loop (Agilent Technologies, 500 μl, Part # 0101-1246).

5. Syringe (HAMILTON, 500 μl, Part # 81216).
6. 96-well Deep well plates (e.g., Axygen Scientific, 1.1 ml, Part # 391-01-101).
7. HPLC (Agilent Technologies,1260 Infinity II LC System).
8. Vacuum concentrator, consisting of centrifugal evaporator SC100 (Savant), refrigerated vapor trap RVT4104 (Savant), vacuum pump RV8 (Edwards).

2.9 Stage Tip

1. Methanol.
2. 1% formic acid.
3. 35% acetonitrile + 1% formic acid solution.
4. 70% acetonitrile + 1% formic acid solution.
   All the reagents should be prepared with HPLC grade solvents. Solutions are stable at room temperature but the vials should be tightly closed to prevent evaporation.
5. Empore C-18 Solid Phase Extraction Disc (Fischer Scientific, 47 mm, Cat # 2216-C18).
6. Blunt needle (Monoject, 16 × 1–1/2 BLUNT, CE0123).
7. Paper Clips (e.g., Acco., Stock # 72380).
8. Vacuum concentrator, consisting of centrifugal evaporator SC100 (Savant), refrigerated vapor trap RVT4104 (Savant), vacuum pump RV8 (Edwards).

2.10 Liquid Chromatography-Mass Spectrometry Analysis

1. 1% formic acid: Prepare in HPLC grade water.
2. Vial inserts (Agilent Technologies, 250 μl, Part # 5181-8872).
3. Glass vials (Agilent Technologies, 2 ml, Part# 5181-3375).
4. Seal (SUN-Sri, 11 mm, Part # 200100).
5. Crimper (Agilent Technologies, 11 mm, Part # 5040-4667).
6. Decapper (Agilent Technologies, 11 mm, Part # 5040-4668).
7. Easy-nLC 1200 (Thermo Fisher Scientific).
8. Orbitrap Fusion Lumos (Thermo Fisher Scientific).

3 Methods

3.1 Embryo Lysis and Yolk Removal

Efficient lysis, extraction, and denaturing of proteins from the cells is necessary for accessibility of proteins in the sample. However, yolk proteins constitute approximately 90% of the *Xenopus* proteins in eggs and early embryos. Having yolk in the sample can hinder the identification of other proteins which are less abundant. For this reason, we remove most yolk via a soft spin (Fig. 2A, B). We optimized the lysis buffer to be as destructive as possible and solubilize proteins to the maximum extent possible, while keeping

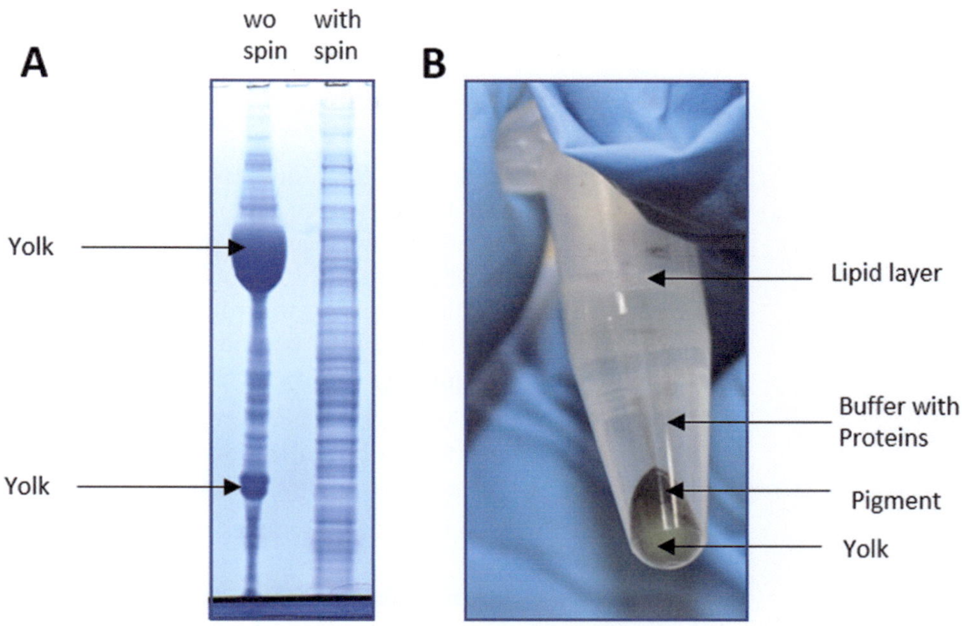

Fig. 2 Yolk spin out. (**A**) ~90% of the protein mass in Xenopus embryos are yolk. Shown is a Coomassie-stained gel of a lysed Xenopus egg (left) and egg proteins after removing yolk via centrifugation (right). (**B**) Eppendorf tube after centrifugation step as detailed in the paper. Yolk and pigments sediment. Solubilized proteins can be removed from the supernatant

yolk platelets mostly intact, so that we can remove them via centrifugation.

The buffer's pH at 7.2 is compatible with subsequent cysteine protection with NEM, higher pH would lead to solubilization of yolk and side-reactions of NEM. The nonionic NP-40 is the harshest detergent we could find that does not significantly solubilize yolk. We add the chelator EDTA, and protease inhibitors to prevent protein degradation. Combretastatin and Cytochalasin depolymerize the microtubules and the actin cytoskeleton respectively and are added to prevent the spin out of the cytoskeletal macro assemblies. After the yolk is spun out, addition of SDS denatures proteins and keeps them inactivated during cysteine protection.

1. Collect 20 embryos (approximately 20 × 25 μg) per stage, and remove the buffer. Be careful not to lyse the embryos while collecting them. Lysing them at this point might destroy consistency across different conditions.
2. Flash freeze the embryos in liquid nitrogen. This allows downstream handling of all collected samples in parallel, which improves reproducibility.
3. After the collections are complete, lyse every condition in 1 ml of lysis buffer by pipetting up and down with a 1 ml pipette at least 15 times.

4. Incubate on ice for approximately 10 min to help disassemble large structures such as the cytoskeleton.
5. Vortex for 10 s. Do not vortex longer as it might dissolve the yolk proteins.
6. Spin down the yolk at 2500 rcf for 4 min at 4 °C. This step will spin out yolk and a thin layer of pigment granules (Fig. 2B). Flick the tubes gently to resuspend the proteins that are associated with lipids collected on the surface. Extract most of the supernatant without disturbing the yolk layer (~950 μl). It is acceptable to leave some buffer behind.
7. Add HEPES to 100 mM (pH 7.2). High salt concentration tends to solubilize yolk. Therefore, we keep the salt concentration in the lysis buffer low and add additional buffering capacity before cysteine protection.
8. Denature the proteins in lysed samples by adding SDS to 2%.

3.2 Cysteine Protection

The thiol side chain of cysteines is very reactive. The unspecific modifications resulting from reactions with, e.g., oxygen make it hard or impossible to identify cysteine containing peptides. Therefore, the cysteines are protected with NEM, which makes them inert (*see* **Note 1**). DTT is first added to break disulfide bonds. Please note that one DTT molecule will react with two NEM molecules.

1. Reduce the sample by adding DTT to 5 mM. Vortex followed by a quick spin. Incubate at 60 °C for 20 min to help solubilize membrane proteins. Make sure that the pH is below 7.5 (*see* **Note 2**).
2. Cool down to room temperature and add NEM to 20 mM. Vortex followed by a quick spin.
3. Incubate it at room temperature for 20 min.
4. Add additional 10 mM of DTT to consume the unreacted NEM in the sample. Vortex followed by a quick spin.
5. Incubate at room temperature for 10 min.

3.3 Methanol-Chloroform Precipitation

We include the Methanol-Chloroform precipitation to remove detergents and salts that could interfere with the digestion, prefractionation, or the mass spectrometer [32]. MeOH precipitation is our method of choice due to the high reproducibility between multiple samples. However, it requires at least 100 μg of proteins. For samples with less material we use trichloroacetic acid precipitation [33].

1. Take 200 μl of the sample per condition in a 2 ml Eppendorf tube.
2. Add 800 μl of methanol. Vortex well.

3. Add 400 μl of chloroform. Vortex well.
4. Add 600 μl of water. Vortex well.
5. Centrifuge at 20,000 rcf for 2 min at room temperature.
6. Carefully remove the supernatant and discard the upper layer without disturbing the interphase.
7. Add 600 μl of methanol. Vortex well.
8. Centrifuge again at 20,000 rcf for 2 min at room temperature to pellet the protein.
9. Remove the supernatant.
10. Repeat the methanol wash with 500 μl of methanol. Vortex well.
11. Centrifuge again at 20,000 rcf for 2 min at room temperature.
12. Remove the supernatant and air-dry the pellet to evaporate the left over organic solvents.
13. Take up the pellet in 6 M GuaCl, 10 mM EPPS, pH 8.5 to approximately 5 mg/ml protein concentration. The pH of 8.5 seems to be optimal for efficient and reproducible digestion. To help redissolve the protein pellet, heat to 60 °C and sonicate indirectly in water bath.

3.4 Estimate Protein Amount (BCA Assay)

To determine how much protease and isobaric tags to add, we need to know the approximate protein amount. We often skip this step and approximate that one embryo/egg will produce ~25 μg of non-yolk protein.

1. Prepare a working reagent stock solution by mixing 50 parts of reagent A with 1 part of reagent B.
2. Prepare BSA standard for a range of concentrations 2, 1, 0.5, 0.25, 0 mg/ml (approximately 20 μl each).
3. Add 5 μl of sample (well within the working range of 2–0.1 mg/ml) and the standard BSA solution to 150 μl of the working reagent in the wells of the plate. Measure BSA results in triplicates.
4. Incubate for 30 min at 37 °C.
5. Cool plate to RT. Measure the absorbance at or near 562 nm on a plate reader.
6. Subtract the average 562 nm absorbance measurement of the blank standard replicates from the 562 nm measurements of all other individual standard and unknown sample replicates.
7. Prepare a standard curve by plotting the average blank-corrected 562 nm measurement for each BSA standard vs. its concentration. Use the standard curve to determine the protein concentration of each unknown sample.

3.5 Digestion

Digestion occurs in parallel for different conditions. Our main goal for this step is reproducibility. Therefore, we perform a two-step digestion. The first step of digestion is carried out with 2 M GuaCl and LysC only. High concentration of chaotrope will (partly) unfold most proteins and allow LysC to digest them into smaller pieces. While LysC is active at 2 M GuaCl, it will miss many potential cleavage sites. The GuaCl is diluted to 0.5 M and fresh LysC and Trypsin are added. In comparison with Trypsin, LysC is superior in digesting peptides ending with lysine. Trypsin is added to digest arginine. For the highest quality MultiNotch MS3 data, one can perform a LysC digest only. This will result in all b- and y-ions to be TMT-tagged and will suppress interference. However, a LysC only digest will come at the cost of quantifying fewer proteins. For the embryonic time-series example, we omitted the trypsin digestion.

1. Dilute sample with 10 mM EPPS (pH 8.5) to 2 M GuaCl (*see* **Note 3**), vortex well and spin.
2. Add LysC (Based on whatever is larger: 1/100 per protein weight or 20 ng/μl LysC concentration).
3. Gently flick the tubes to mix the LysC well in the sample. Do not vortex to avoid denaturation of protease.
4. Incubate at room temperature overnight.
5. Dilute with 10 mM EPPS (pH 8.5) to 0.5 M GuaCl (*see* **Note 3**). Vortex well and spin.
6. Add LysC to 20 ng/μl. Optionally, if besides cleavage after lysine, also cleavage after arginine is desired, add Trypsin to 10 ng/μl.
7. Gently flick the tubes to mix the proteases well in the sample. *Do not vortex.*
8. Incubate at 37 °C overnight. To prevent condensation on lid, heat the entire tube.
9. Vacuum concentrate the samples. This reduces the amount of hydroxyl ions, which interfere with TMT-labeling.

3.6 TMT Labeling

The reaction of isobaric TMT tags with the peptide N-terminus or alpha amino group of lysine to form an amide linkage is pH sensitive. At too low of a pH, the free amines get protonated and will no longer react. At too high a pH, the TMT tags react with hydroxyl ions in solution and are degraded. We found that a pH of 8.0 is optimal for this reaction to reach completion (*see* **Note 4**). In the previous step we removed water to minimize competing hydroxyl ions. Acetonitrile concentrations significantly higher than 30% will decrease reaction efficiency. We use at least 4 μg of TMT to label 1 μg of peptide and at least 2 μg of TMT per 1 μl of added 200 mM EPPS.

1. Resuspend 100 μg with 50 μl of 200 mM EPPS, pH 8.0.
2. Add 20 μl of TMT-stock solution (20 μg/μl in acetonitrile). Flick to mix well.
3. Incubate at room temperature for 2 h.
4. Quench with hydroxylamine. Hydroxylamine degrades the excess TMT and removes TMT that reacted with tyrosine. Hydroxylamine is pre-diluted to 5%. Add 5 μl of 5% to the reaction. Incubate for 30 min at room temperature.
5. Combine all conditions into one tube. Mix well.
6. Acidify with 5% phosphoric acid. DO NOT use formic acid, as we have previously observed formaldehyde adducts on peptides. We suspect that formaldehyde formed by formic acid reacting with hydroxylamine.
7. Vacuum concentrate the sample to remove acetonitrile from TMT labeling step. Acetonitrile would interfere with the following reverse phase purification step.

3.7 Reverse Phase Solid Phase Extraction/Sep-Pak

At this stage, sample contains lots of salt (GuaCl) and might contain some undigested proteins that can interfere with the downstream prefractionation step, e.g., by clogging the column. The disposable Sep-Pak will desalt the sample and retain (most) undigested proteins (*see* **Note 5**).

1. Prepare sample: The sample is resuspended in 500 μl of HPLC water. Check the pH (it should be below 2) (*see* **Note 6**). If the pH is not approximately 1, add additional phosphoric acid.
2. Condition/Equilibrate the column:
   (a) Wet the column with 100% methanol, run through at a flow rate of 5–10 ml/min for ~15 s.
   (b) Equilibrate the column with 1% formic acid, run through quickly at a flow rate of 5–10 ml/min for ~15 s.
3. Load sample: Load the sample under gravity or low vacuum at a flow rate of 0.2–1 ml/min for ~5 min.
4. Wash:
   (a) Wash the tube with 500 μl of 1% formic acid for ~1 min.
   (b) Wash with 500 μl 1% formic acid for ~1 min.
   (c) Wash with 500 μl 1% formic acid for ~1 min once again.
5. Elute: Remember to elute into a fresh tube.
   (a) Elute with 1 ml of 35% acetonitrile, 1% formic acid under gravity or low vacuum for ~5 min.
   (b) Elute with 1 ml of 70% acetonitrile, 1% formic acid under gravity or low vacuum for ~5 min.
6. Dry: Vacuum concentrate the sample to remove acetonitrile.

3.8 Quality Control (Detailed in Accompanying Paper)

The multiple sample preparation steps allow many possibilities for errors. Mass spectrometry data is rich in information and can, to some extent, inform us about problems. At this point, we evaluate the quality of the prepared sample before committing significant additional time for pre-fractionation and mass spectrometry analysis of pre-fractionated samples.

The analysis of the data to identify problems with labeling or digestion efficiency is detailed out in the accompanying paper (Chapter 14).

- Take 1 µg of the sample after Sep-Pak or perform Stage tip (*see* Subheading 3.10) and analyze on the mass spectrometer (*see* Subheading 3.11 on how to do the liquid chromatography-mass spectrometry analysis).

3.9 Medium pH Reverse Phase Prefractionation

Fractionating the sample prior to mass spectrometry analysis has the capability of increasing the depth of proteome coverage by reducing the sample complexity of the eluate. The performance of this two-dimensional chromatographic technique depends on separation efficiency in both dimensions and orthogonality of separation. A reverse phase system with significantly different pH in both dimensions performs well and is highly robust. The change in the separation selectivity on modifying the pH is attributed to the alteration in charge distribution of the peptides. Pooling multiple early, middle, and late fractions further provides an advantage in terms of time efficiency, better proteome coverage, and orthogonality. Therefore, an offline medium pH reverse phase (RP) HPLC separation with a concatenation strategy is employed prior to a low pH RP in front of the mass spectrometer. This procedure is designed for total protein content of ~1000 µg in the sample.

1. Run blank sample: Load 450 µl TFE and run through the column [34]. Elute with the two-gradient buffer system. TFE helps elute potential contaminations from the previous experiments. Refer to Table 2 for the gradients of the buffers, pressure limits and the flow rates used.
2. Prepare sample: Dissolve sample in ~450 µl of buffer A. The volume of the sample is limited by the injection loop volume.
3. Fractionate sample: Load the sample and fractionated (for gradient *see* Table 1). Collect 96 fractions starting at 17 min every 42 s, until 84.5 min (Fig. 3).
4. Combine fractions: Combine the 96 fractions into 24 fractions (Fig. 4) [35].
5. Dry: Vacuum concentrate the 24 combined fractions to remove acetonitrile.

Table 1
Gradients for elution of peptide fractions

Time [min]	*A* [%]	*B* [%]	Flow [ml/min]
0.00	100.0	0.0	0.500
17.00	100.0	0.0	0.500
18.00	93.0	7.0	0.500
75.00	65.0	35.0	0.500
76.00	0.0	100.0	0.500
80.00	0.0	100.0	0.500
86.00	100.0	0.0	0.500
90.00	100.0	0.0	0.500

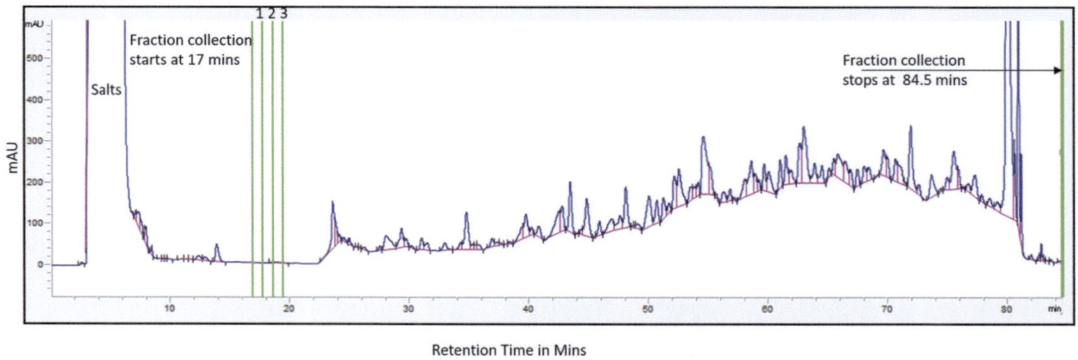

Fig. 3 Elution profile of the peptides during prefractionation. The green lines mark the fractions collected at various times starting at 17 min with 42 s increments until the 96th fraction at 84.5 mins (The first three fractions are marked as 1, 2, and 3). The large peak at the beginning corresponds to small molecules

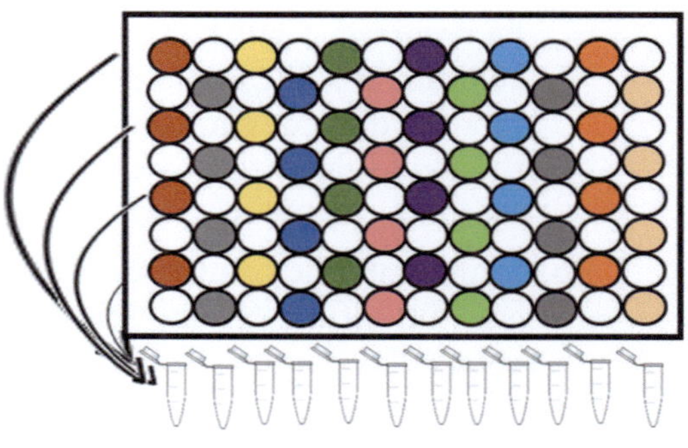

Fig. 4 Schematic representation of concatenation strategy. Each color in the vertical column represents fractions that are combined into a sample, which is analyzed via LC-MS. After the first 12 collections, collect another 12 by following the same scheme using the uncolored wells in the vertical rows

3.10 Stage Tip [36]

The sample contains ammonium bicarbonate from the pre-fractionation step. While ammonium bicarbonate is volatile and should mostly decompose on the mass spectrometer inlet tube. However, some will accumulate inside the mass spectrometer, which can severely hamper instrument performance.

1. Column production.
   - (a) Column holder: Take a 0.5 ml Eppendorf tube, cut off bottom half. Also cut bottom half of a 200 μl. Insert cut pipette tip into cut Eppendorf tube.
   - (b) Column: Press the blunt needle into the C18 disc to cut out the desired column material. Bend open the paper clip to obtain a straight wire. Insert the needle, with C18 material inside, into a new 200 μl tip. With the wire inserted into the blunt needle, push out the C18 material into the 200 μl tip. Repeat this procedure with a second disc. We use one disc per ~25 μg of peptides. Press the wire down on the C18 disc, to generate a tight seal between the C18 material and the 200 μl tip. Make sure that there are no gaps, which would allow the sample to pass by the C18 material.
   - (c) Column assembly: Insert the column into the column holder. We discard columns after use, but preserve the column holders for future use.
2. Prepare sample: Acidify the samples by taking them up in 30 μl of 5% phosphoric acid. Make sure the pH is close to 1.
3. Condition/equilibrate the column:
   - (a) Wet the column with 30 μl of 100% methanol. Centrifuge at ~420 rcf for 15 s. Visually inspect the column to observe if methanol is still present above the C18. If all the methanol has already flowed through, the C18 disc will not effectively bind the peptides and a new stage-tip needs to be prepared. Centrifuge at ~420 rcf for 2 min. Check the column. If some liquid is still present spin down for 30 s more and check the column. Repeat the process until all the liquid flows through. The disc should not dry out, it should look pale like wax (*see* **Note 7**).
   - (b) Equilibrate the column with 30 μl of 1% formic acid. Spin at 1700 rcf for 1 min. Spin more if needed as described above. Do not dry the disc.
4. Load sample: Load the sample. Spin at 1700 rcf for 1 min. Spin more if needed as described above. Do not dry the disc.
5. Wash:

(a) Wash with 30 μl of 1% formic acid. Spin at 1700 rcf for 1 min. Spin more if needed as described above. Do not dry the disc.

(b) Wash again with 30 μl of 1% formic acid. Spin at 1700 rcf for 1 min. Spin more if needed as described above. Do not dry the disc.

6. Elute:

(a) Insert the vial insert into a 2 ml Eppendorf. Take the entire column assembly and transfer it to a torpedo tube for elution.

(b) Elute with 30 μl of 35% acetonitrile + 1% formic acid. Spin at 3800 rcf for 1 min. Spin more if needed.

(c) Elute one more time with 30 μl of 70% acetonitrile + 1% formic acid. Spin at 3800 rcf for 1 min. Spin more if needed. Drying the disc at this point is acceptable.

7. Dry: vacuum concentrate the samples to remove acetonitrile.

3.11 Liquid Chromatography–Mass Spectrometry Analysis

1. Resuspend the dry sample in 1% formic acid to approximately 1 μg/μl (*see* **Note 8**).
2. Transfer the vial inserts into the mass spectrometer vials.
3. Create an instrument method with the settings described in Fig. 5.
4. Aim to shoot about 3 μg of the total protein per prefractionated sample.

4 Notes

1. We recently learnt that maleimide ring hydrolyzes over time [37]. It therefore might be advantageous to replace the NEM with other alkylating reagents like iodoacetamide.
2. A higher pH can deprotonate the lysines. NEM may react with deprotonated amines or undergo hydrolysis at a more alkaline pH.
3. Higher chaotrope concentration decreases protease activity.
4. Though we found pH 8.0 to be optimal for this reaction, the alpha amino groups of N-terminus cysteines frequently do not react.
5. We have recently started to replace this step with ultracentrifugation of the acidified sample at 200,000 rcf at 4 °C for an hour. The undigested proteins aggregate with the low pH. These aggregates, together with other large assemblies like glycogen, spin out [38]. The supernatant is collected and subjected to medium pH reverse phase prefractionation.

MS1	
Detector Type	Orbitrap
Orbitrap Resolution	120000
Mass Range	Normal
Use Quadrupole Isolation	True
Scan Range	350-1350 m/z
RF Lens	60 %
AGC Target	1.0 e6
Maximum Injection Time	100 ms
Micro Scans	1
Data Type	Centroid
Polarity	Positive
Source Fragmentation	Disabled
Use Easy-IC	False
Monoisotopic Peak Determination	Peptide
Intensity	
Filter Type	Intensity Threshold
Intensity Threshold	5.0 e3
Charge State	
Include Charge States	2-6
Include Undetermined Charge States	False
Include Charge States 25 and higher	False
Dynamic Exclusion	
Exclude after n times	1
Exclusion Duration	90 s
Mass Tolerance	Ppm
Low	10
High	10
Exclude Isotopes	True
Exclude within cycle	True
Perform dependent scan on single charge state per precursor only	True
Data Dependent	
Data Dependent Mode	Number of Scans
Number of Dependent Scans	10

Fig. 5 Parameter setting for TMT-MS3 method

MS2	
Isolation Mode	Quadrupole
Scan Polarity	1
Use Isolation m/z offset	False
Activation Type	CID
CID Collision Energy	35%
Activation Q	0.25
Multistage Activation	False
Detector Type	Ion Trap
Scan Range Mode	Auto m/z Normal
Ion Trap Scan Rate	Rapid
AGC Target	1.5 e4
Injection Ions for All Available Parallelizable Times	False
Maximum Injection Time	75 ms
Micro Scans	1
Data Type	Centroid
Precursor Selection Range	
Mass Tolerance	Mass Range
Mass Range	400-2000 m/z
Isobaric Tag Loss Exclusion	
Reagent	TMT
Precursor Ion Exclusion	
Exclusion Mass Width	m/z
Low	70
High	5
Data Dependent	
Number of SPS Precursors	10

MS3	
Synchronous Precursor Selection	True
Number of SPS Precursors	10
Use Isolation m/z Offset	False
Activation Type	HCD
HCD Collision Energy	55%
Detector Type	Orbitrap
Scan Range Mode	Auto m/z Normal
Orbitrap Resolution	50000
First Mass	110
AGC Target	1.5 e5
Injection Ions for All Available Parallelizable Times	False
Maximum Injection Time	100 ms
Micro Scans	1
Data Type	Centroid
Use Easy-IC	False
MS Isolation Window	0.5
MS2 Isolation Window	2

Fig. 5 (continued)

The prefractionation step simultaneously acts as a desalting step. We therefore load the sample with continuous flow of buffer A and only after 17 min start to increase the acetonitrile concentration and fractionation collection (Fig. 3). However, we still don't know if desalting with the medium pH reverse phase column is a sustainable option. Even occasional clogging of these very expensive columns would not be acceptable.

6. pH of the sample should be lower than pKa of the carboxyl groups. This makes the peptides more hydrophobic and helps them to stick better to the columns.

7. Repeat the process in ~30 s time increments for additional spins to avoid the disc drying out. Different stage-tips will vary by how much centrifugation force/time is required for all liquid to flow through. If the flow rate is too slow, we often have to increase centrifugation speeds at various steps from ~420 rcf to e.g., ~1000 rcf or higher. However, we try to load the sample in at least 2 min so that peptides have enough time to bind to the C18 material.

8. A fraction of the sample is lost in the multistep purification process. Therefore, the protein content at this stage is just an estimate.

Acknowledgments

We thank Thao Nguyen for help collecting the *Xenopus* embryonic time series and Felix Keber for comments and suggestions on the manuscript. M.P. was supported by NIH grant R01GM103785. M.S. was supported by NIH F31 predoctoral fellowship 5F31GM116451. This work was supported by NIH grant 1R35GM128813 and Princeton University startup funding.

References

1. Swammerdam J (1737) Bibilia Naturae; Sive historia insectorum, in classes certas redact 2
2. Prevost JL, Dumas J-B (1824) Nouvelle théorie de la génération. Ann Sci Nat 2
3. Gurdon JB, Elsdale TR, Fischberg M (1958) Sexually mature individuals of Xenopus laevis from the transplantation of single somatic nuclei. Nature 182(4627):64–65
4. Paine PL, Moore LC, Horowitz SB (1975) Nuclear envelope permeability. Nature 254(5496):109–114
5. Session AM, Uno Y, Kwon T, Chapman JA, Toyoda A, Takahashi S, Fukui A, Hikosaka A, Suzuki A, Kondo M, van Heeringen SJ, Quigley I, Heinz S, Ogino H, Ochi H, Hellsten U, Lyons JB, Simakov O, Putnam N, Stites J, Kuroki Y, Tanaka T, Michiue T, Watanabe M, Bogdanovic O, Lister R, Georgiou G, Paranjpe SS, van Kruijsbergen I, Shu S, Carlson J, Kinoshita T, Ohta Y, Mawaribuchi S, Jenkins J, Grimwood J, Schmutz J, Mitros T, Mozaffari SV, Suzuki Y, Haramoto Y, Yamamoto TS, Takagi C, Heald R, Miller K, Haudenschild C, Kitzman J, Nakayama T, Izutsu Y, Robert J, Fortriede J, Burns K, Lotay V, Karimi K, Yasuoka Y, Dichmann DS, Flajnik MF, Houston DW, Shendure J, DuPasquier L, Vize PD, Zorn AM, Ito M, Marcotte EM, Wallingford JB, Ito Y, Asashima M, Ueno N, Matsuda Y, Veenstra GJ, Fujiyama A, Harland RM, Taira M, Rokhsar DS (2016) Genome evolution in the allotetraploid frog Xenopus laevis. Nature 538(7625):336–343. https://doi.org/10.1038/nature19840
6. Gurdon JB, Wakefield, L (1986) Microinjection of amphibian oocytes and eggs for the analysis of transcription. Microinjection and Organelle Transplantation Techniques 269–299.
7. Markow TA, Beall S, Matzkin LM (2009) Egg size, embryonic development time and ovoviviparity in Drosophila species. J Evol Biol 22(2):430–434. https://doi.org/10.1111/j.1420-9101.2008.01649.x
8. Tartia AP, Rudraraju N, Richards T, Hammer MA, Talbot P, Baltz JM (2009) Cell volume regulation is initiated in mouse oocytes after ovulation. Development 136(13):2247–2254. https://doi.org/10.1242/dev.036756
9. Smits AH, Lindeboom RG, Perino M, van Heeringen SJ, Veenstra GJ, Vermeulen M (2014) Global absolute quantification reveals tight regulation of protein expression in single Xenopus eggs. Nucleic Acids Res 42(15):9880–9891. https://doi.org/10.1093/nar/gku661

10. Lombard-Banek C, Moody SA, Nemes P (2016) High-sensitivity mass spectrometry for probing gene translation in single embryonic cells in the early frog (Xenopus) embryo. Front Cell Dev Biol 4:100. https://doi.org/10.3389/fcell.2016.00100
11. Wühr M, Güttler T, Peshkin L, McAlister GC, Sonnett M, Ishihara K, Groen AC, Presler M, Erickson BK, Mitchison TJ, Kirschner MW, Gygi SP (2015) The nuclear proteome of a vertebrate. Curr Biol 25(20):2663–2671. https://doi.org/10.1016/j.cub.2015.08.047
12. Boke E, Ruer M, Wühr M, Coughlin M, Lemaitre R, Gygi SP, Alberti S, Drechsel D, Hyman AA, Mitchison TJ (2016) Amyloid-like self-assembly of a cellular compartment. Cell 166(3):637–650. https://doi.org/10.1016/j.cell.2016.06.051
13. Peshkin L, Wühr M, Pearl E, Haas W, Freeman RM Jr, Gerhart JC, Klein AM, Horb M, Gygi SP, Kirschner MW (2015) On the relationship of protein and mRNA dynamics in vertebrate embryonic development. Dev Cell 35(3):383–394. https://doi.org/10.1016/j.devcel.2015.10.010
14. Lombard-Banek C, Moody SA, Nemes P (2016) Single-cell mass spectrometry for discovery proteomics: quantifying translational cell heterogeneity in the 16-cell frog (Xenopus) embryo. Angew Chem Int Ed Engl 55(7):2454–2458. https://doi.org/10.1002/anie.201510411
15. Sun L, Bertke MM, Champion MM, Zhu G, Huber PW, Dovichi NJ (2014) Quantitative proteomics of Xenopus laevis embryos: expression kinetics of nearly 4000 proteins during early development. Sci Rep 4:4365. https://doi.org/10.1038/srep04365
16. Presler MS, Van Itallie E, Klein AM, Kunz R, Coughlin P, Peshkin L, Gygi S, Wühr M, Kirschner M (2017) Proteomics of phosphorylation and protein dynamics during fertilization and meiotic exit in the Xenopus egg. Proc Natl Acad Sci U S A. 2017 Dec 12;114(50):E10838-E10847. doi: 10.1073/pnas.1709207114
17. Sawin KE, Mitchison TJ (1991) Mitotic spindle assembly by two different pathways in vitro. J Cell Biol 112(5):925–940
18. Reinsch S, Karsenti E (1997) Movement of nuclei along microtubules in Xenopus egg extracts. Curr Biol 7(3):211–214
19. Wühr M, Chen Y, Dumont S, Groen AC, Needleman DJ, Salic A, Mitchison TJ (2008) Evidence for an upper limit to mitotic spindle length. Curr Biol 18(16):1256–1261. https://doi.org/10.1016/j.cub.2008.07.092
20. Gache V, Waridel P, Winter C, Juhem A, Schroeder M, Shevchenko A, Popov AV (2010) Xenopus meiotic microtubule-associated interactome. PLoS One 5(2):e9248. https://doi.org/10.1371/journal.pone.0009248
21. Liska AJ, Popov AV, Sunyaev S, Coughlin P, Habermann B, Shevchenko A, Bork P, Karsenti E, Shevchenko A (2004) Homology-based functional proteomics by mass spectrometry: application to the Xenopus microtubule-associated proteome. Proteomics 4(9):2707–2721. https://doi.org/10.1002/pmic.200300813
22. Wühr M, Freeman RM Jr, Presler M, Horb ME, Peshkin L, Gygi S, Kirschner MW (2014) Deep proteomics of the Xenopus laevis egg using an mRNA-derived reference database. Curr Biol 24(13):1467–1475. https://doi.org/10.1016/j.cub.2014.05.044
23. Deshmukh AS, Murgia M, Nagaraj N, Treebak JT, Cox J, Mann M (2015) Deep proteomics of mouse skeletal muscle enables quantitation of protein isoforms, metabolic pathways, and transcription factors. Mol Cell Proteomics 14(4):841–853. https://doi.org/10.1074/mcp.M114.044222
24. Eng JK, McCormack AL, Yates JR (1994) An approach to correlate tandem mass spectral data of peptides with amino acid sequences in a protein database. J Am Soc Mass Spectrom 5(11):976–989. https://doi.org/10.1016/1044-0305(94)80016-2
25. Ong SE, Blagoev B, Kratchmarova I, Kristensen DB, Steen H, Pandey A, Mann M (2002) Stable isotope labeling by amino acids in cell culture, SILAC, as a simple and accurate approach to expression proteomics. Mol Cell Proteomics 1(5):376–386
26. Cox J, Hein MY, Luber CA, Paron I, Nagaraj N, Mann M (2014) Accurate proteome-wide label-free quantification by delayed normalization and maximal peptide ratio extraction, termed MaxLFQ. Mol Cell Proteomics 13(9):2513–2526. https://doi.org/10.1074/mcp.M113.031591
27. Rosenberger G, Bludau I, Schmitt U, Heusel M, Hunter CL, Liu Y, MacCoss MJ, MacLean BX, Nesvizhskii AI, Pedrioli PGA, Reiter L, Rost HL, Tate S, Ting YS, Collins BC, Aebersold R (2017) Statistical control of peptide and protein error rates in large-scale targeted data-independent acquisition analyses. Nat Methods 14(9):921–927. https://doi.org/10.1038/nmeth.4398
28. McAlister GC, Nusinow DP, Jedrychowski MP, Wühr M, Huttlin EL, Erickson BK, Rad R, Haas W, Gygi SP (2014) Multinotch MS3 enables accurate, sensitive, and multiplexed detection of differential expression across

cancer cell line proteomes. Anal Chem 86(14):7150–7158. https://doi.org/10.1021/ac502040v

29. Wenger CD, Lee MV, Hebert AS, McAlister GC, Phanstiel DH, Westphall MS, Coon JJ (2011) Gas-phase purification enables accurate, multiplexed proteome quantification with isobaric tagging. Nat Methods 8(11):933–935. https://doi.org/10.1038/nmeth.1716

30. Ting L, Rad R, Gygi SP, Haas W (2011) MS3 eliminates ratio distortion in isobaric multiplexed quantitative proteomics. Nat Methods 8(11):937–940. https://doi.org/10.1038/nmeth.1714

31. Wühr M, Haas W, McAlister GC, Peshkin L, Rad R, Kirschner MW, Gygi SP (2012) Accurate multiplexed proteomics at the MS2 level using the complement reporter ion cluster. Anal Chem 6;84(21):9214–9221

32. Wessel D, Flugge UI (1984) A method for the quantitative recovery of protein in dilute solution in the presence of detergents and lipids. Anal Biochem 138 (1):141-143. doi:0003-2697(84)90782-6 [pii]

33. Chevallet M, Diemer H, Van Dorssealer A, Villiers C, Rabilloud T (2007) Toward a better analysis of secreted proteins: the example of the myeloid cells secretome. Proteomics 7(11):1757–1770. https://doi.org/10.1002/pmic.200601024

34. Mitulovic G, Stingl C, Steinmacher I, Hudecz O, Hutchins JR, Peters JM, Mechtler K (2009) Preventing carryover of peptides and proteins in nano LC-MS separations. Anal Chem 81(14):5955–5960. https://doi.org/10.1021/ac900696m

35. Edwards A, Haas W (2016) Multiplexed quantitative proteomics for high-throughput comprehensive proteome comparisons of human cell lines. Methods Mol Biol 1394:1–13. https://doi.org/10.1007/978-1-4939-3341-9_1

36. Rappsilber J, Mann M, Ishihama Y (2007) Protocol for micro-purification, enrichment, pre-fractionation and storage of peptides for proteomics using StageTips. Nat Protoc 2(8):1896–1906. https://doi.org/10.1038/nprot.2007.261

37. Lyon RP, Setter JR, Bovee TD, Doronina SO, Hunter JH, Anderson ME, Balasubramanian CL, Duniho SM, Leiske CI, Li F, Senter PD (2014) Self-hydrolyzing maleimides improve the stability and pharmacological properties of antibody-drug conjugates. Nat Biotechnol 32(10):1059–1062. https://doi.org/10.1038/nbt.2968

38. Dworkin MB, Dworkin-Rastl E (1989) Metabolic regulation during early frog development: glycogenic flux in Xenopus oocytes, eggs, and embryos. Dev Biol 132(2):512–523

39. Thompson A, Schafer J, Kuhn K, Kienle S, Schwarz J, Schmidt G, Neumann T, Johnstone R, Mohammed AK, Hamon C (2003) Tandem mass tags: a novel quantification strategy for comparative analysis of complex protein mixtures by MS/MS. Anal Chem 75(8):1895–1904

40. Sonnett M, Yeung E, Wühr M (2018) Accurate, Sensitive, and Precise Multiplexed Proteomics Using the Complement Reporter Ion Cluster. Analytical Chemistry 90(8): 5032–5039

Chapter 14

Quantitative Proteomics for *Xenopus* Embryos II, Data Analysis

Matthew Sonnett, Meera Gupta, Thao Nguyen, and Martin Wühr

Abstract

The oocytes, embryos, and cell-free lysates of the frog *Xenopus laevis* have emerged as powerful models for quantitative proteomic experiments. In the accompanying paper (Chapter 13) we describe how to prepare samples and acquire multiplexed proteomics spectra from those. As an illustrative example we use a 10-stage developmental time series from the egg to stage 35 (just before hatching). Here, we outline how to convert the ~700,000 acquired mass spectra from this time series into protein expression dynamics for ~9000 proteins. We first outline a preliminary quality-control analysis to discover any errors that occurred during sample preparation. We discuss how peptide and protein identification error rates are controlled, and how peptide and protein species are quantified. Our analysis relies on the freely available MaxQuant proteomics pipeline. Finally, we demonstrate how to start interpreting this large dataset by clustering and gene-set enrichment analysis.

Key words Quantitative multiplexed proteomics, TMT, Mass spectrum, MaxQuant, *Xenopus laevis*, Development, False discovery rate, Gene symbols, k-means clustering, Gene-set enrichment

1 Introduction

Xenopus laevis is a powerful amphibian model organism routinely used in numerous laboratories to investigate a wide range of problems in biology. The high synchrony of in vitro fertilization coupled with the large amounts of protein that are available (~25 μg/egg) [1] make the system ideal for proteomic analysis where a full experiment requires up to ~1 mg of protein. In a parallel publication we detail our sample preparation protocol for this workflow (Chapter 13) and how to acquire multiplexed proteomics spectra. Here, we describe how to identify proteins and extract their respective quantitative information from this data.

All of our protein identification and quantification is done with a mass spectrometer. Mass spectrometers measure the mass to charge ratio (m/z) of analytes. The goal of a proteomics experiment is to identify and quantify peptides via characteristic masses and

Kris Vleminckx (ed.), *Xenopus: Methods and Protocols*, Methods in Molecular Biology, vol. 1865,
https://doi.org/10.1007/978-1-4939-8784-9_14,

their peak heights. However, for complex mixtures such as *X. laevis* lysates, simply measuring the *m/z* of a peptide is not sufficient for identification. Rather, a peptide is isolated inside the mass spectrometer, fragmented, and a second mass spectrum (MS/MS or MS2) is acquired. By comparing the obtained fragment ions with theoretical spectra from a reference database, peptides can be identified [2]. Matching peptide fragments to a sequence requires a protein reference database. An important step forward in performing sensitive *X. laevis* proteomics was the generation of a high-quality protein reference [3]. This matching process is imperfect and false positives (both on the peptide and protein level) occur. Below we will outline how to control these errors with a false discovery rate (FDR) so that only a user-specified percentage of peptides have been incorrectly assigned a sequence. Errors also arise when mapping peptide sequences to protein sequences which can also be controlled at a user-specified FDR [4]. Typically, 1% FDR is used on both the peptide and the protein level.

To understand many biological processes, identification of proteins present in a sample is often inadequate. Rather, quantification of the respective protein amounts in each sample is preferable. Many different forms of quantitative proteomics like label-free, SILAC, or DIA are available [5–7]. Our method of choice is multiplexed proteomics. For this approach up to 11 different conditions can be barcoded with tandem mass tags (TMT), mixed together, and ionized simultaneously onto a mass spectrometer. This approach allows the simultaneous quantification of peptides from all 11 conditions and the routine quantification of >8000 proteins in 2 days of instrument time (*see* Fig. 1). In its simplest implementation, TMT barcodes from each condition can be distinguished and quantified on the MS2 level [8, 9]. While the TMT-MS2 approach is advantageous in that it is implementable on a wide range of mass spectrometers, it has significant measurement distortion [10–12]. A recently introduced improvement is the TMT-MS3 approach where an additional gas purification step is performed in the mass spectrometer which drastically reduces measurement distortion [13]. Using this enhanced multiplexing approach, we have successfully extracted biological insight regarding nuclear composition in *X. laevis* oocytes [14], protein dynamics during early *X. laevis* development [15], and the proteome and phoshpho-protoeme changes during fertilization [16]. Below we detail how to quantify peptides and proteins from a TMT-MS3 experiment. In our experience, the most difficult part of quantitative proteomics experiments is to extract meaningful biological insight from the large datasets that are generated. This remains a massive challenge, and each study typically requires novel approaches. Nevertheless, we have found several common steps can be performed when a dataset is first acquired that can help to shed a preliminary understanding. Below we detail how we use

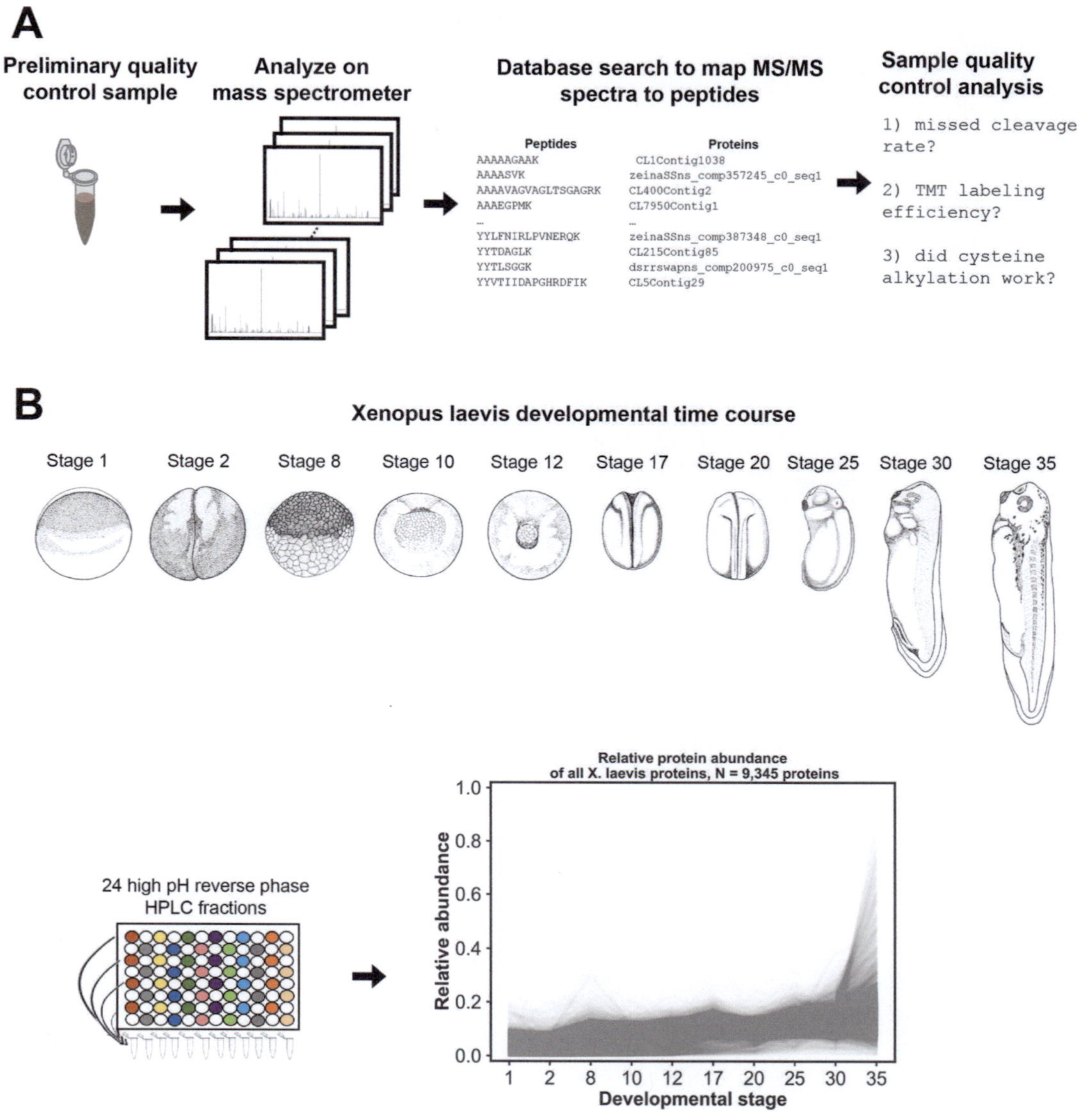

Fig. 1 Overview of quantitative multiplexed proteomics data analysis pipeline. (**A**) A preliminary quality control sample is made and analyzed on a mass spectrometer with a TMT-MS3 method. Peptide sequences are identified through a database search and matched to protein sequences. The protease digestion efficiency, tandem mass tagging efficiency, and cysteine alkylation efficiency are all estimated to determine if the sample preparation was satisfactory. (**B**) Full developmental time course *X. laevis* TMT-MS3 experiment. Samples at ten different stages in development were collected and processed as in Gupta et al. Analysis of these samples leads to the quantification of 9345 *X. laevis* proteins across ten stages in development. Illustrations of developmental stages courtesy of Garland Science [[21]]

k-means clustering to group similarly behaving proteins during early *X. laevis* development together and then use gene ontology term membership enrichment to determine which biological processes (if any) are over- or under-represented in a cluster.

2 Materials

2.1 MaxQuant

Version 1.6.0.16 at the time of publication. *See* below for download and installation instructions.

2.2 Reference X. laevis Proteome

On our lab website (https://scholar.princeton.edu/wuehr/sample_prep) we provide a *X. laevis* proteome that can be used as a reference when matching mass spectra to peptide sequences. PHROG_09_26_2017. FASTA is a compilation of 79,215 *X. laevis* protein sequences that were generated by mRNA sequencing [3]. As of this publication PHROG_09_26_2017.FASTA is slightly superior in terms of number of peptides identified compared to JGI 9.1, the latest gene models release of the frog proteome available on Xenbase. However, we suspect that as newer versions are released newer versions of JGI will become superior. We encourage the reader to check for updated versions of JGI at http://www.xenbase.org/other/static/ftpDatafiles.jsp. Newer releases can be mapped to human gene symbols with the conveniently available web tool in Subheading 2.5.

2.3 X. laevis Developmental Time Series Mass Spectrometry Raw Data

The mass spectrometry proteomics data have been deposited to the ProteomeExchange Consortium via the PRIDE partner repository with the dataset identifier PXD007915 [20].

TGR_02788.raw is our quality control sample. The remaining raw files (TGR_02793.raw to TGR_02823.raw) correspond to the 24 reverse phase fractionated samples.

2.4 X. laevis Developmental Time Series Processed Data (Post MaxQuant)

On our website (https://scholar.princeton.edu/wuehr/sample_prep) we provide the MaxQuant output that should be obtained for both the sample quality control outlined below as well as for the entire dataset analysis using PHROG_09_26_2017.FASTA as a *X. laevis* reference proteome.

2.5 Mapping X. laevis Protein Sequences to Human Gene Symbols

We previously developed a convenient online resource for matching reference proteomes from *X. laevis* (or any other organism) to human gene symbols [3]. The tool is available at http://kirschner.med.harvard.edu/tools/genesym_assignment.html and can be used with future JGI releases. On our website (https://scholar.princeton.edu/wuehr/sample_prep) we provide spreadsheets with the human gene symbol mappings for PHROG_09_26_2017.FASTA.

2.6 Anaconda Python Distribution Package

Anaconda is a freely available compilation of python and many commonly used python libraries in scientific computing. The python scripts associated with this paper make use of several of these libraries: Pandas, NumPy, Matplotlib, and SciKitLearn. We encourage the reader to download the entire package, which should be able to run all scripts associated with this manuscript without any additional installation.

2.7 Python Scripts for Data Analysis of Data After Being Processed in MaxQuant

All of our Python scripts are available for download on our website in the folder maxquant_bioinformatics.zip.

tmt_labeling_efficiency.py estimates what fraction of peptides were labeled with TMT during the sample quality control search.

missed_cleavage_rate.py estimates what fraction of peptides were missed cleavages during the sample quality control search.

percent_cysteines.py estimates what fraction of peptides that were identified contain a cysteine during the sample quality control search.

map_gene_symbol.py takes a MaxQuant output file and maps the *X. laevis* proteins from the PHROG database to a gene symbol (if a good mapping exists) and generates a new .csv file containing this information.

k_means_clustering.py will perform k-means clustering on processed data that has been mapped to human gene symbols, generating a new .csv file containing this information. The script generates plots of the individual protein dynamics of each cluster in .png and .svg formats. Finally, separate .csv files of all of the gene symbols in each cluster are generated so that gene ontology term biological enrichment can be easily performed.

2.8 Reference MaxQuant Parameters and Screenshots of All Steps

Detailed screenshots of all operations within MaxQuant both for the sample quality control as well as the fully fractionated *X. laevis* developmental time series have been uploaded to our website (https://scholar.princeton.edu/wuehr/sample_prep). All MaxQuant parameters used in both searches are available in the parameters.txt file that is present in each output folder.

3 Methods

3.1 Downloading and Installing MaxQuant

3.1.1 Rationale

MaxQuant is a high-quality freely available proteomics software that allows users to analyze TMT-MS3 quantitative multiplexed proteomics experiments [17]. MaxQuant is fast and contains all features necessary to match mass spectra to peptide and protein sequences at controlled false discovery rates and faithfully extract their quantitative information.

3.1.2 Procedure

- Go to http://www.coxdocs.org/doku.php?id=maxquant:common:download_and_installation
- Register and accept licensing agreement. After receiving an email you can download the software.
- Extract MaxQuant and place the MaxQuant_1.6.0.16 folder on your Desktop.

- Open MaxQuant by opening the MaxQuant_1.6.0.16 folder, opening the MaxQuant folder and then running MaxQuant.exe.

3.2 Specifying Peptide Modifications in MaxQuant

3.2.1 Rationale

Our intention during the sample preparation procedure is to digest all proteins into peptides, modify all cysteines with N-ethyl maleimide (NEM) (*see* **Note 1**), and label all free n-termini and lysines with a tandem mass tag (TMT) reagent. Both of these reagents alter the mass of each peptide to which they are attached. MaxQuant must be told to add these mass modifications to all peptide sequences it is searching in the uploaded *X. laevis* protein database. Below we outline how to specify these modifications.

3.2.2 Procedure

1. *Adding the cysteine NEM modification*:
   - Navigate to the Configuration tab in MaxQuant.
   - Select Modifications in the Data tab.
   - Select Add in the Table tab.
   - Name the modification NEM.
   - Leave description as New modification.
   - On the Composition row select Change and enter H(7) C(6) N O(2) (*see* **Note 2**). The mass should be 125.0476784741.
   - Position should be Anywhere.
   - Type is Standard.
   - New terminus is None.
   - Under Specificities select the K and click (*see* **Note 3**).
   - Select + under Specificities and select C.
   - Under the Actions tab in Configuration section select Modify table.
   - Under the Table actions tab in Configuration section select Save changes. The NEM modification is successfully added and is displayed at the bottom of the table on left.
2. *Adding the n-terminus TMT modification* (*see* **Note 4**):
   - Navigate to the Configuration tab in MaxQuant.
   - Select Modifications.
   - Select Add.
   - Name the modification dynamicTMTnTerm.
   - Leave description as New modification.
   - On the Composition row select Change and enter H(20) C(8) Cx(4) N O(2) Nx (*see* **Note 5**). The mass should be 229.162932141.
   - Position should be Any N-term.

- Type is Standard.
- New terminus is None.
- Under Specificities select the K and click –.
- Under Specificities select the + and add –.
- Under the Actions tab in Configuration section select Modify table.
- Under the Table actions tab in Configuration section select Save changes. The dynamicTMTnTerm modification is successfully added and is displayed at the bottom of the table on left.

3. *Adding the lysine TMT modification* (*see* **Note 4**):
   - Navigate to the Configuration tab in MaxQuant.
   - Select Modifications.
   - Select Add.
   - Name the modification dynamicTMTonK.
   - Leave description as New modification.
   - On the Composition row select Change and enter H(20) C(8) Cx(4) N O(2) Nx. The mass should be 229.162932141.
   - Position should be Anywhere.
   - Type is Standard.
   - New terminus is None.
   - Under Specificities K should be on the only AA present (*should be this way by default without doing anything*).
   - Under the Actions tab in Configuration section select Modify table.
   - Under the Table actions tab in Configuration section select Save changes. The dynamicTMTonK modification is successfully added and is displayed at the bottom of the table on left.
4. *Specifying the Isotopic impurities of the TMT reagents used* (*see* **Note 6**):
   - Navigate to the Configuration tab in MaxQuant.
   - Select Modifications.
   - If you scroll down through the list you will find TMT10plex-Nter126C and further below it TMT10 plex-Lys126C. These both correspond to the TMT 126 tag from Thermo.
   - Select one of them and scroll to the bottom on the right side. There will be rows denoted % – 2, % – 1, % + 1, % + 2. Here you can enter the isotopic impurity corrections from the manufacturer. Enter the same isotopic impurity corrections for both the "Nter" and Lys" version

of each reagent (E.g., reagent 1 is 126C, reagent 2 is 127N, and so on).

- After each isotopic impurity has been entered, select Modify table and then Save changes to update the table.
- After all isotopic impurities have been entered MaxQuant must be restarted for this information to be correctly loaded.

5. *After adding all desired modifications restart MaxQuant for them to be selectable in the drop down menus* (*see* **Note 7**).

3.3 Providing a Protein Reference Database to MaxQuant

3.3.1 Rationale

Performing sensitive proteomics requires a reference database of protein sequences for MaxQuant to match acquired mass spectra to. We have provided a *X. laevis* protein reference database in the appropriate FASTA format that can be loaded directly into MaxQuant for use.

3.3.2 Procedure

- Download PHROG.FASTA from (https://scholar.princeton.edu/wuehr/sample_prep), unzip and place PHROG_09_26_2017.FASTA in the MaxQuant directory (*see* **Note 8**).
- Navigate to the Configuration tab in MaxQuant.
- Select Sequence databases.
- Select Add.
- To the right of Fasta File name click Select and choose PHROG_09_26_2017.FASTA.
- To the right of Identifier parse rule press Select and choose identifier parse rule: >(.*).
- Under the Actions tab in Configuration section select Modify table.
- Under the Table actions tab in Configuration section select Save changes. PHROG_09_26_2017.FASTA is successfully added and is displayed at the bottom of the table on left (*see* **Note 9**).
- Close and restart MaxQuant for the changes to take effect (*see* **Note 10**).
- Under the Global Parameters tab select Sequences.
- To the right of the Fasta files header select Add File and then select PHROG_09_26_2017.FASTA and can be saved in any directory on your PC.
- Ensure Include Contaminants is checked.

3.4 Specification of Various Mass Spectrometer Instrument Parameters in MaxQuant

3.4.1 Rationale

MaxQuant must be instructed how to generate all possible peptide sequences in silico that could be present in the sample given the protein sequences from the database. The type(s) of peptide sequences present are a consequence of the specificity of the enzyme(s) that were used during the digestion step in the sample preparation (Chapter 13). Typically, the specificity is either all peptides must have a c-terminal K (LysC only digest) or, all peptides must have a c-terminal K or R (LysC + Trypsin digest). The *X. laevis* time series was digested with only LysC. If the digestion was perfect every identified peptide sequence (excluding c-termini) would end in K.

To perform an initial sample quality control step (described in depth below) we also have to add TMT as a dynamic modification so that the database considers all peptide sequences with the TMT modification either present or absent. This will allow us to determine the fraction of peptides that were modified with TMT. We also need to specify any other mass modifications that were made. In this case, we need to add NEM as a fixed modification on cysteine.

3.4.2 Procedure

- Navigate to the Global parameters tab.
- Under Fixed Modifications select the Carbamidomethyl (C) fixed modification on the right and select < to remove it.
- On the left hand side find NEM (sorted alphabetically) and select > to add it as a fixed modification.
- Min. peptide length should be 7.
- Max peptide mass [Da] should be 4600.
- Min peptide length for unspecific search should be 8.
- Max peptide length for unspecific search should be 25.
- Navigate to the Group-specific parameters tab select Digestion.
- Digestion mode should be Specific.
- Under Enzyme select Trypsin/P and select < to remove it.
- Under Enzyme select LysC and select > to add it (*see* **Note 11**).
- Max. missed cleavages should be 2.
- Select the Modifications panel.
- Under Variable Modifications find dynamicTMTnTerm and select >.
- Under Variable Modifications find dynamicTMTonK and select >.
- On the right, remove Acetyl(Protein N-term) by selecting it and then pressing < (*see* **Note 12**).
- Ensure that the max. Number of modifications per peptide is 5.
- Navigate to the Global parameters tab. Select the Protein quantification panel.
- To the right of Modifications used in protein quantification, select Acetyl (N-term) on the right and press <.
- Select NEM and press >.
- Select dynamicTMTnTerm and press >.
- Select dynamicTMTonK and press >.

- Select the Tables panel and deselect all boxes.
- All other parameters can be left as the default.

3.5 Control Peptide and Protein Sequence Matching Error Rates

3.5.1 Rationale

Matching acquired mass spectra of peptides to the correct sequence is an imperfect process that has an unknown error rate. To estimate this error rate, the gold-standard in the field is to take the true protein sequences from the reference database and construct a second set of "decoy" sequences—sequences that do not exist in nature but have the same amino acid properties as the corresponding actual sequences [18]. This is accomplished by reversing all protein sequences (e.g., MYPEPTIDE becomes EDITPEPYM) and then generating a second set of in silico peptides which are also considered when each mass spectrum is matched to a peptide sequence from the database. Previous work has shown that peptides from these reverse sequences that are at least 7 amino acids in length almost never exist in the actual proteome [18]. Thus, one can assume that for every peptide that maps to a "decoy" peptide there is an equally likely chance of a peptide mapping to another peptide sequence that is from the actual proteome. Therefore, one can count the total number of "decoy" peptides that were matched and then double this number to estimate the number of incorrect peptide sequences from the actual proteome. Typically, a large fraction of spectra are incorrectly matched with the wrong sequence if no post-filtering is done. To overcome this, MaxQuant uses a machine learning approach to determine how various parameters from each mass spectrum (e.g., mass error of the match, number of matched peptide fragments observed, etc.) relates to the probability of whether a peptide is from the correct sequence or not. After predicting these probabilities, a user-defined false discovery rate (FDR) is specified to filter the matched peptide sequences down to say where only 1% of all peptide sequences are identified incorrectly. For large datasets, even if only 1% of all peptide sequences have been incorrectly assigned, large false discovery rates on the protein level (e.g., >20%) can be observed [19]. Thus, additionally a 1% FDR is also set when matching peptide sequences to their protein sequences [17] . Below we outline the parameters in MaxQuant that are used to set the identified peptide and protein false discovery rates.

3.5.2 Procedure

- Under Global parameters select the Identification panel.
- The PSM FDR stands for Peptide Spectral Match False Discovery Rate. We recommend using the default of 0.01 which corresponds to a 1% False Discovery Rate. This means 1% of all peptide sequences in the final peptide dataset from MaxQuant will be incorrect.

- Similarly, we recommend using a Protein FDR of 0.01 (the default setting). This means that approximately 1% of all protein groups in the final protein dataset from MaxQuant will be incorrect.

3.6 Determining Sample Quality

3.6.1 Rationale

During the sample preparation protocol (Chapter 13), a prepared sample is subjected to an initial quality control step before it is fully processed and analyzed. This initial quality control step analyzes the sample with a TMT-MS3 method on a mass spectrometer for two hours. We will analyze a single unfractionated TMT-MS3 sample of a *X. laevis* developmental time series in MaxQuant to determine whether the sample preparation is of high quality. Specifically, we wish to determine (1) What was the labeling efficiency of n-terminal and lysine groups of peptides with TMT? Ideally >98% of all peptides detected are labeled with TMT. (2) How efficient was our protease digestion of each sample? Ideally, the number of missed cleavages (number of peptides that have a non-c-terminal Lysine is <10% for LysC only digests and a non-c-terminal Lysine or Arginine <20% for Trypsin/LysC digests). (3) Did the cysteine alkylation work? We expect to see least 5% of peptides containing a cysteine (*see* **Note 1**). First we need to load our raw data of a single unfractionated TMT-MS3 experiment into MaxQuant and adjust various parameters.

3.6.2 Procedure

1. Loading the data.
   - Download TGR_02788.raw from our ProteomeXchange deposit for this paper (*see* Subheadings 2 and 3 for link). Place this file in the same directory as the MaxQuant.exe.
   - Open MaxQuant.
   - In MaxQuant select the Raw data tab, select Load, and select the TGR_02788.raw file and click Open.
   - If not already done, configure and upload the PHROG_09_26_2017.FASTA reference database as described above and select the correct protease specificity.
   - All modifications should be as specified above.
   - In the bottom left hand corner change the Number of processors to the # of CPUs on your PC ignoring hyper-threading—1 (*see* **Note 13**).
   - Select Start in the bottom left hand corner to start the search (*see* **Note 14**).
   - Progress of the search can be monitored under the Performance table.
2. Determining TMT labeling efficiency:
   - Once the search is complete by default the files will be stored in the same directory that MaxQuant is located in. Within this directory should be a folder named combined. Open it and then open the txt folder.

- Open modificationSpecificPeptides.txt in a spreadsheet viewing program such as Microsoft Excel. modificationSpecificPeptides.txt is a list of all peptides identified, including those with a variable modification present (Oxidation (M), dynamicTMTnTerm and dynamicTMTonK).
- To obtain a quantitative estimate of TMT labeling efficiency either use a custom function in your spreadsheet program to count the number of peptides in modificationSpecificPeptides.txt that have the string dynamicTMTnTerm in the Modifications column or use the tmt_labeling_efficiency.py python script that we provide on our website.
- To use the python script, place the tmt_labeling_efficiency.py script in the txt folder.
- Open the terminal by simultaneously pressing the windows button + R and typing cmd.
- In the terminal type cd Desktop\MaxQuant_1.6.0.16\MaxQuant\combined\txt and press enter.
- Type python tmt_labeling_efficiency.py and press enter. Your labeling efficiency results will be printed (*see* **Note 15**).

3. *Determining missed cleavage rate*: The goal is to determine how many peptides present in the sample contain an internal amino acid (K in the case of LysC only digest and K or R in the case of a trypsin/LysC digest). These peptides are missed cleavages, the internal amino acid should have been digested but was not. If this number is too high (>10% for LysC only or >20% for LysC/Trypsin) then the digestion was not efficient and will introduce error in the measurements.
   - Open the modificationSpecificPeptides.txt file.
   - The Sequence column contains the sequences of all peptides that were identified.
   - If it is convenient, you can specify a custom function either in your spreadsheet program or with your programming language of choice to count how many peptides have an internal K (lysC only) or K/R (trypsin + LysC).
   - Alternatively, we provide a python script that will do this for you.
   - Download missed_cleavage_rate.py from our website and place this in the txt folder.
   - Open the terminal by pressing Windows button + R, typing cmd and pressing enter.
   - In the terminal type cd Desktop\MaxQuant_1.6.0.16\MaxQuant\combined\txt and press enter.

- In the terminal type python missed_cleavage_rate.py --protease <protease_used> where <protease_used> is either lysc or trypsin (e.g., python missed_cleavage_rate.py --protease lysc).
- Your missed cleavage rate will be printed.

4. *Determining cysteine alkylation efficiency*: During sample preparation (Chapter 13) cysteines are reduced to free thiols and alkylated with NEM. While we cannot quantitatively evaluate how well this worked, we can get a semi-quantitative evaluation by asking what fraction of all peptide sequences contain a cysteine. Typically for *X. laevis* samples coming from a complex mixture (egg extract, developing embryos, etc.) this number should be ~5–10%. If less than 3% of all peptides contain cysteines this suggests that alkylation efficiency was poor and cysteine containing peptides should be discarded prior to quantification (*see* **Note 16**). If cysteine alkylation is poor but TMT labeling efficiency was high and missed cleavage rates are acceptable, a full experiment can quantify proteins accurately with ~5% less proteins being quantified if all cysteine peptides are ignored.
   - If convenient, the number of peptides containing a cysteine can be determined with a custom script of your choice by counting how many peptides contain a C in the Sequence column of modificationSpecificPeptides.txt. Alternatively, we provide a python script to calculate this.
   - Download percent_cysteines.py from our website and place this in the txt folder.
   - In the terminal type cd Desktop\MaxQuant_1.6.0.16\MaxQuant\combined\txt and press enter.
   - In the terminal type python percent_cysteines.py.
   - The percent of peptides containing cysteine will be printed.

3.7 Searching a Full TMT–MS3 *X. laevis* Developmental Time Course Experiment with MaxQuant

3.7.1 Rationale

Once a high-quality sample has been produced and then fractionated by medium pH reverse phase HPLC into 24 fractions and each fraction ionized onto a mass spectrometer (Chapter 13) the full experiment can be analyzed.

3.7.2 Procedure

- Download the 24 fractionated samples from our ProteomeXchange deposit (*see* Subheadings 2 and 3).
- Store all of the samples in a folder titled 24fracs in the MaxQuant directory.
- Select the Raw data tab in MaxQuant and then select Load folder and select 24fracs.
- Select the Group-specific parameters tab and then select the Type panel. Under Type click the drop down menu and select Reporter ion MS3.

- Select the 10plex TMT button below all of the label options. A total of 20 different TMT10plex options (e.g. Lys-126C, Lys-127N etc.) will appear on the right hand side.
- Select the Digestion panel and select Trypsin/P from the right and select <. Then select LysC (*see* **Note 17**) on the left and select >.
- Select the Global parameters tab and then select the Sequences panel. Under Fasta files click Add file and then select PHROG_09_26_2017.FASTA. Under Fixed modifications select carbamidomethyl on the right and select <. Then select NEM on the left and select >. Select the Protein quantification panel and under Modifications used in protein quantification select NEM and select > (*see* **Note 18**). Under tables deselect all options.
- In the bottom left hand corner change the number of processors to # of CPUs on your computer (ignoring threads)—1.
- Select Start at the bottom and the progress of the searches can be monitored under the Performance tab. This will take at least several hours.

3.8 Obtaining *X. laevis* Protein Relative Abundance Quantitation and Mapping *X. laevis* Sequences to Human Gene Symbols

3.8.1 Rationale

MaxQuant will write a .tsv file that can be opened in excel to find the measured protein abundances for each *X. laevis* protein. We find it very convenient to map the *X. laevis* protein sequences to human gene symbols which have a lot of metadata associated with them from the literature [3] (*see* Fig. 2). Most, but not all *X. laevis* proteins map to a human gene symbol with the PHROG reference database. A total of 15,672 unique human gene symbols are present in the entire database.

3.8.2 Procedure

- After the search is finished, open the 24fracs folder, open the combined folder, and then open the txt folder.
- The proteinGroups.txt file can be opened in the spreadsheet viewing program of your choice.
- The Protein IDs column lists all of the *X. laevis* protein identifiers for a given row (*see* **Note 19**). The number of proteins in a given row is indicated by the Number of proteins column. The Peptides column indicates how many different peptides were identified and quantified from the protein(s) in that row.
- The TMT-MS3 quantitative relative abundance metrics are stored in the columns titled Reporter intensity correct *x* where *x* ranges from 0 to 9. Reporter intensity correct 0 corresponds to the TMT 126C TMT tag, Reporter intensity correct 1 corresponds to the TMT 127N TMT tag, and so on. Reporter intensity correct 9 corresponds to the TMT 131N tag.
- These values can be used for any analysis the user is interested in doing on their own.

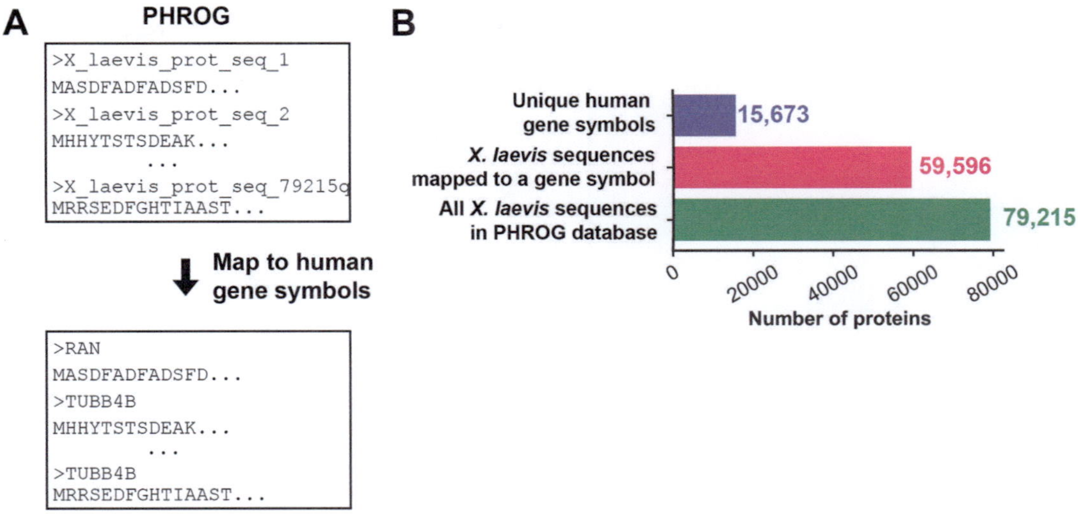

Fig. 2 Mapping *X. laevis* protein sequences to human gene symbols. (**A**) *X. laevis* protein sequences are mapped to human gene symbols by using a bi-directional blast approach [3]. (**B**) Of the 79,215 *X. laevis* sequences present in our database, 59,596 of these have enough homology with the human sequences to map to the human gene symbol. These 59,596 mapped sequences collapse to a total of 15,673 unique human gene symbols. Part of this redundancy is from protein isoforms (*X. laevis* is pseudotetraploid) and part of it may come from splice-isoforms or errors during the protein reference database construction, e.g., fragmented proteins

- To create a new spreadsheet that maps the *X. laevis* protein(s) in each row to a gene symbol, copy the map_gene_symbol.py script from our website into the txt folder.
- Download PHROG_annotation.csv from the phrog_annotation.zip file on our website (Materials and Equipment section). Unzip the folder and place PHROG_annotation.csv into the txt folder.
- In the terminal type
- cd Desktop\MaxQuant_1.6.0.16\MaxQuant\24fracs\combined\txt and press enter.
- Type python map_gene_symbol.py and press enter. A new file proteins_mapped_to_gene_symbols.csv will be created.

3.9 k-Means Clustering of Proteins During X. laevis Development

3.9.1 Rationale

One useful way of analyzing large *X. laevis* proteomic datasets is to cluster proteins into distinct groups based on their relative abundance within the set of collected conditions. Similarly behaving proteins will be grouped within the same cluster (*see* Fig. 3a–c).

3.9.2 Procedure

- Copy the k_means_clustering.py script from our website into your txt folder after you have made the proteins_mapped_to_gene_symbols.csv file as outlined in Subheading 3.8.
- In the terminal type

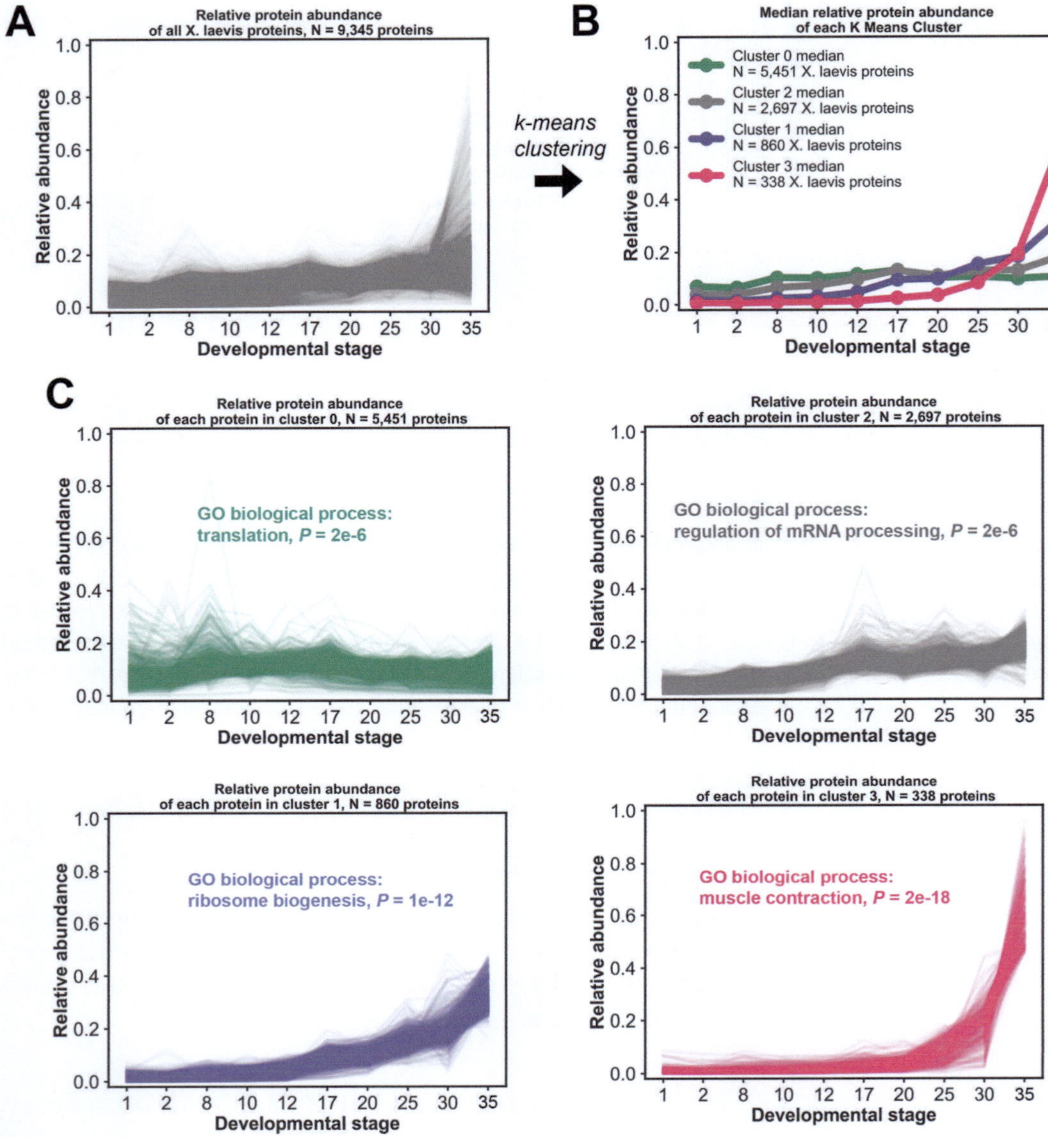

Fig. 3 k-Means clustering and gene ontology biological enrichment of developmental time series. (**a**) Time series profiles throughout developmental stages for all 9345 *X. laevis* proteins that were quantified. (**b**) Median relative abundances and their membership number using a k-means clustering approach based on Euclidean distance with the number of clusters set to 4. (**c**) Relative abundances of proteins in individual clusters and the top gene ontology (GO) biological enrichment term associated with each cluster

- cd Desktop\MaxQuant_1.6.0.16\MaxQuant\24fracs\combined\txt and press enter.
- Type python k_means_clustering.py --num_clusters <number> --line_opacity <opacity> --folder_name <name> where num_clusters is the number of clusters you wish to make. line_opacity is a tuning parameter that controls the opacity of each line in the cluster, decrease the number (a good starting point

is 0.01) to make lines more opaque. folder_name is the name of the folder all of the generated files will be stored (e.g., python k_means_clustering.py --num_clusters 4 --line_opacity 0.01 --folder_name clustering_results). We recommend playing around with this number and inspecting the generated plots of each cluster to determine if increasing the number of clusters produces additional clusters with distinct behavior. In the case of the *X. laevis* developmental time series dataset the most informative number appears to be 4.

- A new file proteins_mapped_to_gene_symbols_w_kmeans_clusters.csv is generated. This contains all previous information as well as column cluster_num which indicates the cluster number each protein belongs to.
- For each cluster, a cluster_num_x.svg and cluster_num_x.png file (where *x* is a cluster number) will be generated that can be used to visually inspect the individual behaviors in each cluster.
- A cluster_num_x.csv (where x is a cluster number) will be generated which contains a list of human gene symbols that can be used for gene ontology biological enrichment analysis (*see* Subheading 3.10).
- A protein_background.csv file is generated which contains a list of the unique gene symbols identified in the entire dataset and will be used in the gene enrichment analysis below.

3.10 Gene Ontology Biological Function Enrichment Analysis of k-Means Clusters

3.10.1 Rationale

The human gene symbols of the proteins in each protein cluster can be analyzed with the bioinformatics tool Gene Ontology which determines if certain gene sets belonging to specific biological processes are either over- or under-enriched (see Fig. 3c). This can often give a helpful course grained picture of the underlying biological response. An important aspect of enrichment analysis is the notion of a background: the set of all proteins from which subsets will be analyzed for enrichment. By default the background is set to be the entire genome. However, while the genome contains ~20,000 gene symbols, typically we only map all *X. laevis* proteins quantified in a TMT-MS3 experiment to ~7000 gene symbols. Therefore, we typically only use the ~7000 gene symbols that were detected in the experiment as background.

3.10.2 Procedure

Open a web browser and go to http://pantherdb.org/webservices/go/overrep.jsp

- Under Step 1. Press choose file and select one of the cluster_num_x.csv files located in the txt file from the MaxQuant searches.
- Under Step 2. Select Homo sapiens.
- Under Step 3. Select Statistical overrepresentation test and use the default settings.

- Select submit.
- To the right of Reference list (this is the background) select Change. Select the protein_background.csv file that was generated in Subheading 3.9.
- To the right of Annotation Data Set you can select different types of analyses. We recommend starting with GO biological process complete.
- The category, fold enrichment, and multiple hypothesis testing corrected P values will be displayed.

4 Notes

1. If cysteines are not modified with NEM (as in Chapter 13), free thiols are extremely reactive and will almost certainly be oxidized. There are several oxidation states that can be produced, which are difficult to reliably detect with current mass spectrometry technology and will not be identified. Thus, if alkylation was not carried out successfully, essentially no peptides identified will contain a cysteine.
2. To successfully enter a chemical formula the drop down menu for an element must be selected and then the correct number entered. The drop down menu for the next element must then be selected for the previous selection to be registered. On the last element select ok.
3. By default MaxQuant puts a Lysine (K) here. Remove it in this case because with the correct pH NEM will only modify cysteines.
4. MaxQuant already has the TMT modification incorporated if one wants to assume it is a static modification, which we will make use of later. However, we first use it as a dynamic modification, allowing MaxQuant to consider the full database for all possible peptides with or without TMT present so that we can determine how many peptides were modified with TMT.
5. Cx is C^{13} and Nx is N^{15}.
6. Each condition in a multiplexed experiment is barcoded by being labeled with a different TMT reagent, which are distinguishable after fragmentation and have different numbers of heavy isotopes. However, these TMT reagents are commercially synthesized and there are isotopic impurities in the tags that will distort the resulting ratios unless MaxQuant normalizes them. The exact impurity ratios depend on the LOT# you have purchased from Thermo and can be obtained from them.
7. These modifications will be saved in the future and only need to be configured once until you install a new version of MaxQuant.

8. E.g., if the MaxQuant folder is on your desktop the directory would be C:\Users\YourUserName\Desktop\MaxQuant_1.6.0.16\MaxQuant where you replace YourUserName with the name of your PC.
9. If this step is not done correctly when you initiate the search you will obtain an error saying "WARNING: fasta file <fasta_directory_here> is not configured." If MaxQuant does not know how to correctly parse the header of the FASTA file it will error out when writing out the spreadsheets after the search is over and you will not be able to obtain your searched data.
10. The fasta file, isotopic impurity modifications as well as the NEM, dynamicTMTonK, and dynamicTMTnTerm will be saved in the future and only need to be configured once unless a new version of MaxQuant is installed.
11. Trypsin/P is appropriate when a trypsin digest alone is carried out, as Trypsin cannot cut lysines that are followed by a proline. However, when a tandem LysC/Trypsin digest is used as we outlined in our previous paper, LysC is able to digest lysines that are followed by proline and Trypsin should be used. If only LysC was used then select LysC.
12. Including acetylated n-termini is useful when analyzing a full dataset but during sample quality control analysis we want to estimate what fraction of n-termini on peptides were labeled with TMT so we will ignore any peptides with endogenous acetyl groups (this is typically a small fraction: <1% of all peptides).
13. E.g., if you have an intel i7-6700K with four cores and eight threads this number would be 3.
14. This should take anywhere from 30 min to several hours depending on how many cores your CPU has.
15. Ideally this number should be >98%. Lower labeling efficiency will typically result in noisier data.
16. Cysteine peptides can be ignored for quantification by not adding NEM under the protein quantification panel.
17. If samples were digested with Trypsin and LysC then choose Trypsin instead. Trypsin/P should be used if only Trypsin (and not LysC) was used during the digestion.
18. Only do this if cysteine alkylation worked.
19. More than one *X. laevis* protein is often assigned to a given row because a peptide mapped to both sequences and it cannot be determined which sequence the peptide came from (e.g., protein isoforms).

Acknowledgments

We thank Lillia Ryazanova for help with the sample preparation, and Felix Keber for comments and suggestions on the manuscript. MS was supported by a NIH F31 pre-doctoral fellowship 5F31GM116451. This work was supported by NIH grant 1R35GM128813 and by Princeton University startup funding.

References

1. Gurdon, J. B., & Wakefield, L. (1986). Microinjection of amphibian oocytes and eggs for the analysis of transcription. Microinjection and Organelle Transplantation Techniques, 269-299.
2. Eng JK, McCormack AL, Yates JR (1994) An approach to correlate tandem mass spectral data of peptides with amino acid sequences in a protein database. J Am Soc Mass Spectrom 5(11):976–989. https://doi.org/10.1016/1044-0305(94)80016-2
3. Wühr M, Freeman RM Jr, Presler M, Horb ME, Peshkin L, Gygi S, Kirschner MW (2014) Deep proteomics of the *Xenopus laevis* egg using an mRNA-derived reference database. Curr Biol 24(13):1467–1475. https://doi.org/10.1016/j.cub.2014.05.044
4. Savitski MM, Wilhelm M, Hahne H, Kuster B, Bantscheff M (2015) A scalable approach for protein false discovery rate estimation in large proteomic data sets. Mol Cell Proteomics 14(9):2394–2404. https://doi.org/10.1074/mcp.M114.046995
5. Cox J, Hein MY, Luber CA, Paron I, Nagaraj N, Mann M (2014) Accurate proteome-wide label-free quantification by delayed normalization and maximal peptide ratio extraction, termed MaxLFQ. Mol Cell Proteomics 13(9):2513–2526. https://doi.org/10.1074/mcp.M113.031591
6. Ong SE, Blagoev B, Kratchmarova I, Kristensen DB, Steen H, Pandey A, Mann M (2002) Stable isotope labeling by amino acids in cell culture, SILAC, as a simple and accurate approach to expression proteomics. Mol Cell Proteomics 1(5):376–386
7. Selevsek N, Chang CY, Gillet LC, Navarro P, Bernhardt OM, Reiter L, Cheng LY, Vitek O, Aebersold R (2015) Reproducible and consistent quantification of the Saccharomyces cerevisiae proteome by SWATH-mass spectrometry. Mol Cell Proteomics 14(3):739–749. https://doi.org/10.1074/mcp.M113.035550
8. Thompson A, Schafer J, Kuhn K, Kienle S, Schwarz J, Schmidt G, Neumann T, Johnstone R, Mohammed AK, Hamon C (2003) Tandem mass tags: a novel quantification strategy for comparative analysis of complex protein mixtures by MS/MS. Anal Chem 75(8):1895–1904
9. Ross PL, Huang YN, Marchese JN, Williamson B, Parker K, Hattan S, Khainovski N, Pillai S, Dey S, Daniels S, Purkayastha S, Juhasz P, Martin S, Bartlet-Jones M, He F, Jacobson A, Pappin DJ (2004) Multiplexed protein quantitation in Saccharomyces cerevisiae using amine-reactive isobaric tagging reagents. Mol Cell Proteomics 3(12):1154–1169. https://doi.org/10.1074/mcp.M400129-MCP200
10. Ting L, Rad R, Gygi SP, Haas W (2011) MS3 eliminates ratio distortion in isobaric multiplexed quantitative proteomics. Nat Methods 8(11):937–940. https://doi.org/10.1038/nmeth.1714 pii
11. Wühr M, Haas W, McAlister GC, Peshkin L, Rad R, Kirschner MW, Gygi SP (2012) Accurate multiplexed proteomics at the MS2 level using the complement reporter ion cluster. Anal Chem 84(21):9214–9221. https://doi.org/10.1021/ac301962s
12. Hebert AS, Merrill AE, Bailey DJ, Still AJ, Westphall MS, Strieter ER, Pagliarini DJ, Coon JJ (2013) Neutron-encoded mass signatures for multiplexed proteome quantification. Nat Methods 10(4):332–334. https://doi.org/10.1038/nmeth.2378
13. McAlister GC, Nusinow DP, Jedrychowski MP, Wuhr M, Huttlin EL, Erickson BK, Rad R, Haas W, Gygi SP (2014) MultiNotch MS3 enables accurate, sensitive, and multiplexed detection of differential expression across cancer cell line proteomes. Anal Chem 86(14):7150–7158. https://doi.org/10.1021/ac502040v
14. Wühr M, Guttler T, Peshkin L, McAlister GC, Sonnett M, Ishihara K, Groen AC, Presler M, Erickson BK, Mitchison TJ, Kirschner MW, Gygi SP (2015) The nuclear proteome of a vertebrate. Curr Biol 25(20):2663–2671. https://doi.org/10.1016/j.cub.2015.08.047
15. Peshkin L, Wühr M, Pearl E, Haas W, Freeman RM Jr, Gerhart JC, Klein AM, Horb

M, Gygi SP, Kirschner MW (2015) On the relationship of protein and mRNA dynamics in vertebrate embryonic development. Dev Cell 35(3):383–394. https://doi.org/10.1016/j.devcel.2015.10.010

16. Presler MS, Van Itallie E, Klein AM, Kunz R, Coughlin P, Peshkin L, Gygi S, Wühr M, Kirschner M (2017) Proteomics of phosphorylation and protein dynamics during fertilization and meiotic exit in the Xenopus egg. bioRxiv 2017:145086
17. Tyanova S, Temu T, Cox J (2016) The MaxQuant computational platform for mass spectrometry-based shotgun proteomics. Nat Protoc 11(12):2301–2319. https://doi.org/10.1038/nprot.2016.136
18. Elias JE, Gygi SP (2007) Target-decoy search strategy for increased confidence in large-scale protein identifications by mass spectrometry. Nat Methods 4(3):207–214. https://doi.org/10.1038/nmeth1019 nmeth1019 [pii]
19. Huttlin EL, Jedrychowski MP, Elias JE, Goswami T, Rad R, Beausoleil SA, Villen J, Haas W, Sowa ME, Gygi SP (2010) A tissue-specific atlas of mouse protein phosphorylation and expression. Cell 143(7):1174–1189. https://doi.org/10.1016/j.cell.2010.12.001
20. Vizcaíno JL, Csordas A, del-Toro N, Dianes JA, Griss J, Lavidas I, Mayer G, Perez-Riverol Y, Reisinger F, Ternent T, Xu Q-W, Wang R, Hermjakob H, (2016) 2016 update of the PRIDE database and its related tools. Nucleic Acids Research 44 (D1):D447-D456
21. Hubrecht-Laboratorium (Embryologisch Instituut), Nieuwkoop PD, Faber J (1967). Normal Tables of Xenopus Laevis:(Daudin) a Systematical and Chronological Survey of the Development from the Fertilized Egg Till the End of the Metamorphosis. North-Holland.

Chapter 15

Dye Electroporation and Imaging of Calcium Signaling in *Xenopus* Nervous System

Lukas Weiss, Thomas Offner, Thomas Hassenklöver, and Ivan Manzini

Abstract

Electroporation is an efficient method of transferring charged macromolecules into living cells in order to study their morphology, function, and connectivity within neuronal networks. Labeling cells with fluorophore-coupled macromolecules can be used to trace projections of whole neuronal ensembles, as well as the fine morphology of single cells. Here, we present a protocol to visualize pre- and postsynaptic components of a sensory relay synapse in the brain, using the olfactory system of *Xenopus laevis* tadpoles as a model. We apply bulk electroporation to trace projections of receptor neurons from the nose to the brain, and single cell electroporation to visualize the morphology of their synaptic target cells, the mitral-tufted cells. Labeling the receptor neurons with a calcium-sensitive dye allows us to record stimulus-induced presynaptic input to the dendrites of the postsynaptic cells via functional calcium imaging.

Key words Electroporation, Neuronal tracing, Single cell morphology, Fluorescent dyes, Calcium imaging, In vivo imaging, *Xenopus laevis*, Olfactory system, Olfactory receptor neurons, Mitral-tufted cells

1 Introduction

To understand the formation and wiring principles of neuronal networks, it is extremely important to have a robust mode of tracing cells and their processes in living neuronal tissue. Electroporation is an established method that has been used in a variety of cell types and brain regions and has proven to deliver fast results, surpassing some of the drawbacks of other common transfection methods like microinjection, viral transfection, or the creation of transgenic animal lines [1]. The method relies on the principle of transiently permeabilizing the membrane of cells through the exposure to an electric field. When the trans-membrane voltage exceeds a certain threshold, membrane molecules start to rearrange and form pores that increase the membrane's permeability for different kinds of molecules [2]. Charged macromolecules like fluorophore-coupled dextrans or plasmid DNA can be transported along the electric

Kris Vleminckx (ed.), *Xenopus: Methods and Protocols*, Methods in Molecular Biology, vol. 1865,
https://doi.org/10.1007/978-1-4939-8784-9_15,

field through the membrane pores into the cytoplasm. Upon collapse of the electrical field, the pores re-seal and the cell membrane returns to its prior state. The exact mechanisms leading to pore formation and sealing still remain a topic of discussion [3–5].

The first electroporations performed in neuronal tissue had the goal of genetically transfecting rat [6] or chicken embryos [7]. Protocols were soon optimized to target neurons in a variety of in and ex utero tissues, as well as postnatal neuronal tissue ([8], for an extensive review). Due to its exposed position, the nervous system of *Xenopus* tadpoles soon proved to be a well-suited model for various approaches of electroporation. By varying the type of electrodes used (plate-like, needle-like, wires, micropipettes), electroporation can be applied to examine the projections of whole nerve fibers [9], the structure of microcircuits [10], or even fine morphological details of single cells [1, 9]. In addition to the variable cell number labeled using different protocols, the characteristics of variable dyes allow different experimental approaches. Fluorophore-coupled dextrans allow us to immediately visualize morphological structures. They are biologically inert and relatively small in size compared to plasmid DNA or fluorescent proteins [9]. Cell transfection using plasmid DNA encoding fluorescent proteins provides the opportunity to monitor the morphological changes of electroporated cells over a longer period of time, and under cell type-specific genetic promoters. For example, in vivo time lapse experiments have been successfully conducted in different parts of the brain of *Xenopus laevis* tadpoles, the optic tectum [9, 11–13], and the olfactory system [14]. Furthermore, electroporation with calcium-sensitive dyes provides the opportunity to examine cellular activity through transient changes in intracellular calcium levels, as demonstrated in *Danio rerio* [15] and *Xenopus laevis* [16].

In this chapter, we describe a combination of two electroporation methods applied to study the functional connectivity between axon terminals of olfactory receptor neurons (presynapse) and single mitral-tufted cells (postsynapse) in the olfactory bulb of *Xenopus laevis* tadpoles. Bulk electroporation using the calcium-sensitive dye Oregon Green BAPTA dextran is used to label the axonal projections originating in the olfactory epithelium of the animals, and measure their activity upon odorant stimulation via functional calcium imaging. In addition, single mitral-tufted cells and granule cells are electroporated with fluorophore-coupled dextrans, delivered via micropipette electroporation. The colocalization between presynaptic calcium transients following odorant stimulation and mitral-tufted cell dendritic morphology provides insight into the odorant information conveyed at the respective synaptic connection [16, 17].

2 Materials

2.1 Animals

Xenopus laevis tadpoles of either sex, staged 50–55 (staged after [18]).

2.2 Solutions and Dyes

1. Frog saline ringer: 98 mM NaCl, 2 mM KCl, 1 mM $CaCl_2$, 2 mM $MgCl_2$, 5 mM glucose, 5 mM Na-pyruvate, 10 mM Hepes, pH 7.8, osmolarity 230 mOsmol/l, stored at 4 °C.
2. Anesthetic: 0.02% MS-222 (ethyl-3-aminobenzoat methanesulfonate; Sigma Aldrich) in tap water, pH 7.8, stored at 4 °C.
3. Dextran coupled dyes (e.g., 3 mM Alexa 594 dextran, 10,000 MW, Life Technologies; 3 mM Oregon Green 488 BAPTA-1 dextran, 10,000 MW, Life Technologies). Both dyes were dissolved in frog saline ringer.
4. Odorant stimuli for calcium imaging diluted in frog saline ringer (e.g., 100 μM L-histidine, L-lysine, L-arginine, L-phenylalanine, L-tryptophan, L-methionine, L-leucine, L-isoleucine, L-valine).

2.3 Micropipette Fabrication

1. Micropipette puller (e.g., P 1000, Sutter Instruments).
2. Borosilicate glass capillaries containing an internal filament (outer diameter 1.5 mm, inner diameter 0.86 mm, length 100 mm).

2.4 Electroporation Devices and Microscopes

1. Dissection microscope with fluorescent illumination and appropriate filter sets. The objective needs to have a large working distance for tissue preparation.
2. Voltage pulse generator (e.g., ELP-01D, NPI Electronics) connected to two platinum wire electrodes (e.g., diameter 0.2 mm). Optimally, a capacitor is connected into the circuit in parallel.
3. Upright microscope with fluorescence detection and appropriate filter sets for single cell electroporation (with water immersion objective with numerical aperture = 1 and working distance = 2 mm).
4. Single Cell Electroporator (e.g., Axoporator 800A, Axon Instruments/Molecular Devices). The headstage of the Axoporator contains an electrode holder and is equipped with a wire electrode. The electrode is preferably coated with AgCl and connects to the dye in the micropipette. Additionally, a reference electrode is connected to close the electrical circuit.
5. Micromanipulators (e.g., Scientifica PatchStar). The micromanipulator should be motorized and remotely controllable.
6. Confocal or multiphoton microscope.

3 Methods

3.1 Bulk Electroporation of Fluorophore-Coupled Dextrans in the Tadpole Nose

1. Prior to electroporation, produce small dye crystals by dissolving the fluorophore-coupled dextrans in frog saline ringer at a concentration of 3 mM and carefully place drops of the solution on a piece of parafilm. Let the dye drops dry at room temperature for at least 30 min before you use the resulting dehydrated dye crystals for electroporation, or store them in a freezer at −20 °C (*see* **Note 1**). Keep the dye in the dark at all times during the procedure to prevent photo-bleaching.
2. Select a *Xenopus* tadpole of desired developmental stage and anesthetize it in a small glass beaker containing 0.02% MS-222 dissolved in tap water for up to 5 min. Test the tadpole's responsiveness to mechanical stimuli in order to confirm proper anesthesia. If not responsive, transfer the tadpole to a wet paper tissue placed in a Petri dish (*see* **Note 2**).
3. Center the region of interest (e.g., the nostrils of the tadpole) in the field of view of the dissection microscope (Fig. 1a) and gently dry the moisture using a paper towel to prevent the fluorescent dye from spreading (*see* **Note 3**). Apart from the region of interest, the animal should be covered in ringer solution throughout the procedure to keep it moist.
4. Use fine forceps to pick up a crystal (e.g., Oregon Green BAPTA dextran crystal) and place it in the nasal cavity, where it will dissolve in the residual moisture (white arrowheads in Fig. 1b). Use the fluorescent illumination of the dissection microscope to monitor the confined spread of the fluorophore-coupled dextran in the nasal cavity (*see* **Note 4**).
5. Connect two platinum wire electrodes (e.g., diameter 0.2 mm) to a voltage pulse generator with a capacitor that is connected in parallel. Insert each electrode into one of the dye filled nasal cavities without injuring the olfactory mucosa (*see* **Note 5**, Fig. 1c). Apply 12 pulses (25 ms duration at 2 Hz) with alternating polarity at a voltage of 20 V (*see* **Note 6**).
6. After removing the electrodes, carefully transfer the tadpole into a beaker with tap water for recovery. The water column in the recovery beaker should not exceed 2–3 cm (*see* **Note 7**).
7. The dextran dye is transported along the nerve fibers, in this case the olfactory nerve, to the axonal terminals located in the olfactory bulb (Fig. 1d).
8. Check for labeled sensory neurons in the nose and olfactory bulb using the fluorescence illumination of the dissection microscope (white arrowheads in Fig. 1d) before continuing with the labeling of single neurons in the brain, e.g., mitral-tufted cells.

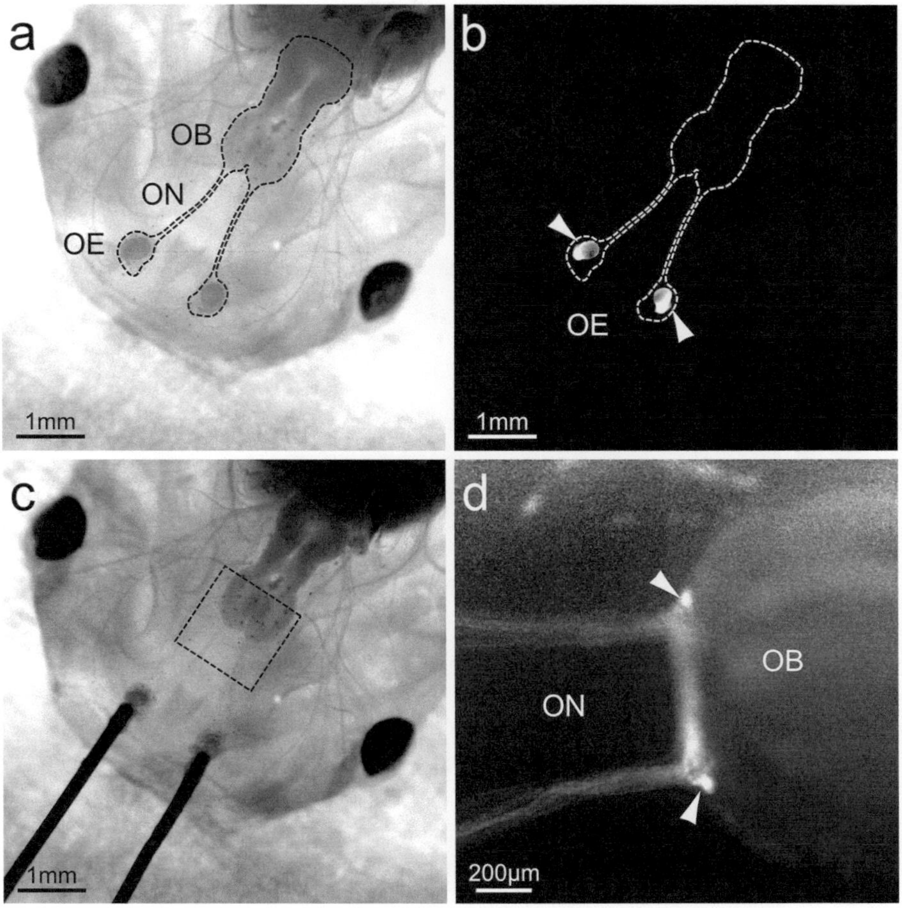

Fig. 1 Bulk electroporation in the nose of *Xenopus laevis* tadpoles using fluorophore-coupled dextran crystals. (*a*) Overview of the tadpole head under the dissection microscope. The black dotted line highlights the olfactory system and the anterior part of the brain. (*b*) After placing the dye crystals in the nostrils, they dissolve in the nasal moisture (white arrowheads). The image shows the olfactory system and the anterior part of the brain (dotted line) using fluorescent illumination. (*c*) Two wire electrodes are placed in the nasal cavities for electroporation. The dotted square indicates the section shown in d. (*d*) The electroporated dye (Oregon Green BAPTA dextran) is transported along the axons of olfactory receptor neurons via the olfactory nerve to the olfactory bulb 24 h after electroporation. Image was taken using fluorescent illumination. The white arrowheads highlight the axon terminals that are filled with the dye and arranged in characteristic clusters. *OE* olfactory epithelium, *ON* olfactory nerve, *OB* olfactory bulb

3.2 Pipette Fabrication for Sparse Cell Labeling in the Brain

1. Fabricate electroporation micropipettes from borosilicate glass capillaries using a horizontal micropipette puller.
2. Modify the protocol to obtain micropipettes with a small tip opening, long taper, and a high pipette resistance of around 15–20 MΩ (*see* **Note 8**). The pipette tip opening should be around 1–2 μm [11] in diameter and the pipette should be rigid enough not to break when penetrating the tissue.
3. Slowly fill the pipettes with 3–5 μl of the solution (e.g., Alexa 594 dextran) using a microloader pipette tip and make sure to

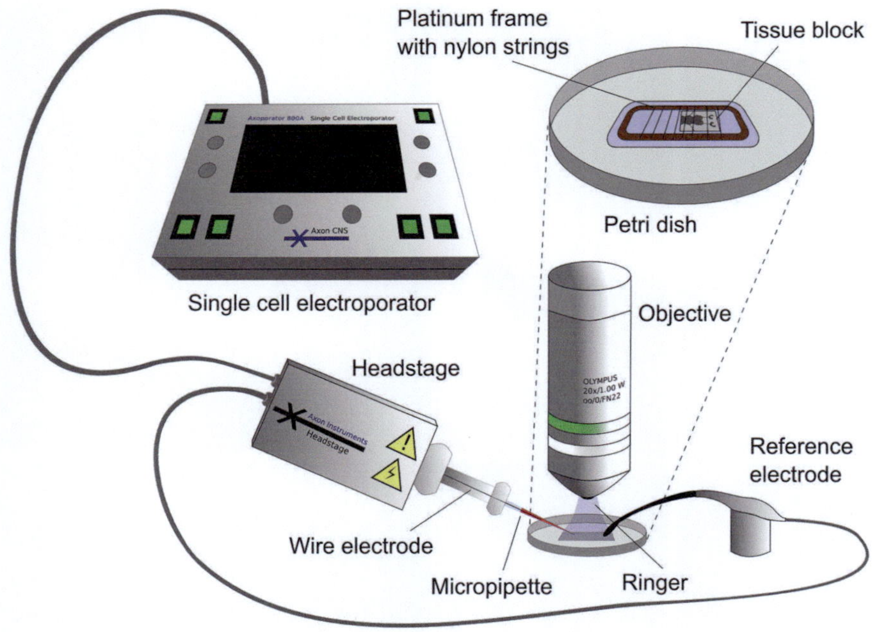

Fig. 2 Schematic of single cell electroporation setup. The single cell electroporator is linked to the headstage and in contact to the fluorophore solution in the micropipette via a wire electrode. The reference electrode is connected to the headstage and immersed in the ringer solution surrounding the tissue block to close the circuit. The tissue block is stabilized on a Petri dish with a platinum frame stringed with nylon threads

avoid the formation of bubbles within the fluorophore solution, especially in the taper of the micropipette (*see* **Note 9**).

4. Mount the micropipette on the holder of the micromanipulator, ensuring that the AgCl-coated silver wire of the headstage of the Axoporator is in contact with the fluorophore solution (*see* **Note 10**, Fig. 2).

3.3 Whole Mount Preparation

1. Anesthetize the tadpole in a small glass beaker containing 0.02% MS-222 as described above (Subheading 3.1, **step 2**).
2. Sacrifice the tadpole by severing the region between brainstem and spinal cord with a scalpel. Subsequently cut a rectangular tissue block out of the upper jaw and palate of the animal containing both nostrils, the olfactory nerves and the olfactory bulb.
3. Make sure to always keep the tissue block covered with frog saline ringer to keep the tissue from drying out.
4. Carefully pin the tissue block to a silicone filled preparation dish using fine needles with the ventral surface of the block facing up. Remove the soft palatial tissue covering the olfactory bulb with fine forceps. Cut alongside the olfactory bulb on both sides using fine scissors until the point where the olfactory nerve connects to the bulb. Be sure to not damage the

olfactory nerves or the brain tissue. Lift the cartilaginous palatial lid covering the olfactory bulb and remove it from the tissue block to fully expose the olfactory bulb.

3.4 Electroporation of Mitral-Tufted Cells Using Fluorophore-Coupled Dextrans

1. Transfer the tissue explant to a plastic Petri dish with a small rectangular recess in the middle and mechanically fix it with a grid made of a small platinum frame stringed with nylon threads (*see* **Note 11**, Fig. 2). Cover the tissue block with ringer solution.
2. Carefully place the reference electrode connected to the headstage of the Axoporator near the tissue block, and make sure it is in contact with the solution to close the electrical circuit (Fig. 2).
3. Immerse the objective of the upright microscope (*see* **Note 12**) into the solution, and position the tissue explant in a way that your region of interest (in this case the mitral cell layer of the olfactory bulb) is centered in the field of view (*see* **Note 13**, Fig. 3a).
4. Using the fluorescent illumination of the microscope, you can use the fluorescent signal of the previously electroporated sensory neuron axons (Subheading 3.1, Fig. 1d) as orientation to help locate your region of interest (*see* **Note 14**, green signal in Fig. 3b).
5. Remotely control the micromanipulator to position the micropipette in the solution column between objective and tissue (Fig. 2), and carefully approach the pipette tip toward the tissue block along the *z*-axis (*see* **Note 15**, black arrowhead in Fig. 3a).
6. As soon as the pipette touches the surface of the olfactory bulb, activate the virtual approach mode of the micromanipulator to be able to penetrate the tissue along the axis of the micropipette (*see* **Notes 10** and **16**).
7. By adjusting and changing the brightness of the light source illuminating the sample you should be able to see some cellular outlines within the tissue that facilitate targeting of a single cell of interest (*see* **Note 13**, Fig. 3a, c).
8. Once you have approached the micropipette to the proximity of a cell of interest (black arrows in Fig. 3c highlight single cellular outlines), trigger positive square voltage pulses using the Axoporator to transfer the dye into the cell (e.g., mitral-tufted cell, Fig. 3d; *see* **Note 17**). Use a train of multiple pulses at a voltage of 50 V, pulse length of 300 μs and a frequency of 300 Hz for 500 ms. By applying higher voltage pulses or longer pulse trains of higher frequency, the extent of neuronal labeling (i.e., the number of labeled cells or the intensity of labeling) can be increased (or decreased, if the parameters are lowered).

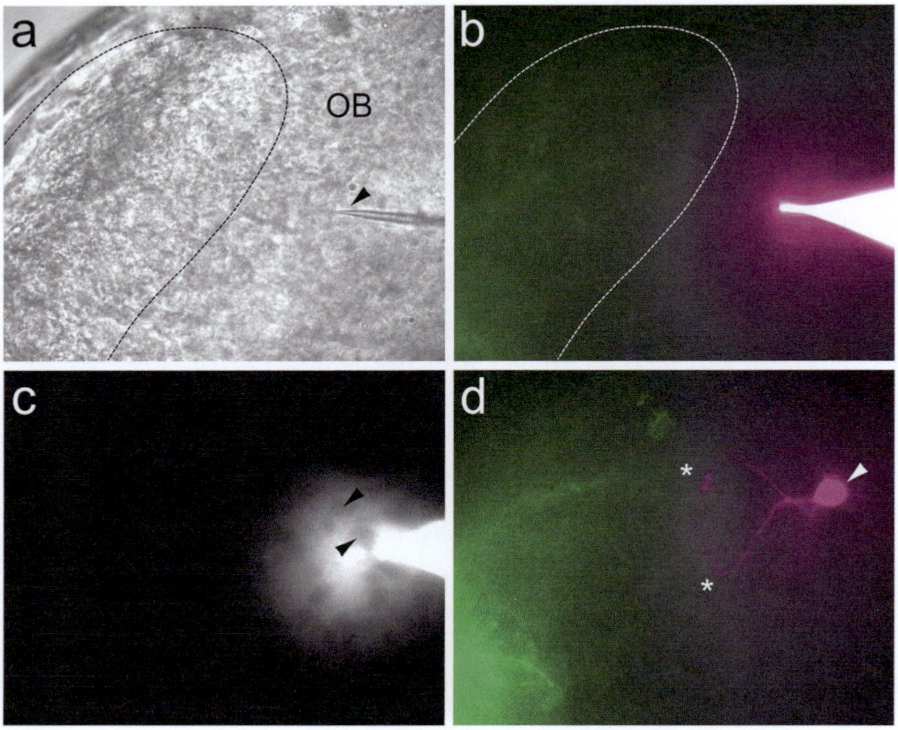

Fig. 3 Single cell electroporation in the olfactory bulb. (*a*) The micropipette (black arrowhead) is approaching the region of interest, in this case the mitral cell layer of the olfactory bulb. The black dotted line indicates the location of sensory neuron axon terminals. (*b*) The electroporated axon terminals of receptor neurons (Oregon Green BAPTA dextran, green) and the dye filled micropipette (Alexa 594 dextran, magenta) are shown using fluorescent illumination of a stereomicroscope. (*c*) After the extrusion of dye into the intercellular space, cells become visible as dark outlines (black arrowhead), which facilitates targeting of a single cell for electroporation. (*d*) Successful electroporation of a single mitral-tufted cell (magenta) projecting its apical dendrites toward the axon terminals of the receptor neurons (green) via its dendritic tufts (white asterisks). The cell soma is indicated by a white arrowhead. *OB* olfactory bulb

9. While applying the pulses, monitor the fluorescent dye extrusion and cell staining under appropriate fluorescent illumination (*see* **Note 18**, Fig. 3b). If the pipette tip was close enough to a single cell, the fluorescent dye fills the soma (white arrowhead in Fig. 3d) and neurites (white asterisks in Fig. 3d) within seconds after pulse application (*see* **Note 19**).
10. Carefully move the micropipette away from the cell and withdraw it out of the tissue using the virtual approach mode of the micromanipulator. The pipette can be used to label further cells at different regions of interest by repeating Subheading 3.4, **steps 6–9**.
11. Imaging of the cell morphology and functional calcium imaging can be conducted using a confocal or multiphoton microscope.

3.5 Functional Calcium Imaging of Odorant-Induced Activity

1. Functional calcium imaging is conducted on axon terminals of sensory neurons that have previously been electroporated with the calcium-sensitive dye Oregon Green BAPTA dextran (Subheading 3.1). We monitor presynaptic activity at the relay synapse to mitral-tufted cells (labeled in Subheading 3.4) upon stimulation with odorants. Single amino acids have been found to be potent olfactory stimuli for aquatic animals [19, 20] and are applied to the nasal cavities.
2. Mount the imaging chamber with the tissue explant containing the olfactory system of the animal on the stage of a confocal or multiphoton microscope.
3. Use a laser wavelength that properly excites the dyes used in the experiments (e.g., 800 nm wavelength for Oregon Green BAPTA dextran and Alexa 594 dextran) and appropriate emission filters.
4. To visualize the morphology of the electroporated cells, acquire a virtual image stack including the cell and all its processes (Fig. 4a presents a maximum intensity projection of a virtual image stack of mitral-tufted cells).
5. For odorant perfusion, use a gravity feed perfusion system made of plastic tubing.
6. Install the funnel for odorant applications and direct it toward the olfactory epithelium. In order to create continuous and stable flow directed to the olfactory epithelium, position a suction syringe driven by a peristaltic pump caudally to the tissue block into the recording chamber. Suction speed of the pump needs to be adjusted until continuous flow of bath solution is achieved.
7. Direct the focus of the microscope to your region of interest in the bulb via bright field illumination. Then use multiphoton excitation to screen for the fluorescent signal emitted by the electroporated axons and postsynaptic cells. Focus on the substructures of interest (e.g., the dendritic tufts of the electroporated mitral-tufted cell, Fig. 4b).
8. Acquire recordings of single or multiple imaging planes of the desired region of interest over time.
9. Increases in intracellular calcium concentration upon odorant stimulation, e.g. single L-amino acids or mixtures at 100 μM concentration can be recorded by monitoring temporal changes of fluorescence of the calcium-sensitive dye. Multiple measurements of odorant applications ensure reproducibility of the signal. Leave at least 2 min intervals between measurements to avoid olfactory receptor neuron desensitization.
10. Responsive axon terminals in the image can be further analyzed by normalizing the fluorescence changes over time

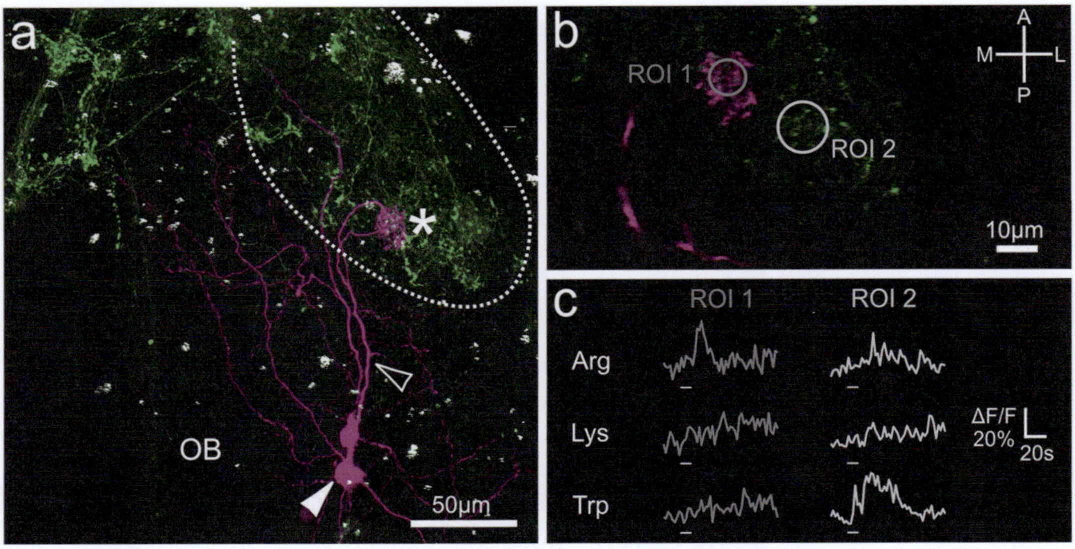

Fig. 4 Functional measurements of presynaptic input on mitral-tufted cells imaged with a multiphoton microscope. (*a*) Sparse cell labeling of postsynaptic cells (magenta; filled arrowhead: mitral-tufted cell soma, empty arrowhead: dendrites, asterisk: dendritic tuft) in the ventral olfactory bulb of *Xenopus laevis* tadpoles. Olfactory receptor neurons and their axonal projections in the olfactory bulb (green) were previously labeled with dextran-coupled calcium indicator by bulk electroporation into the nostrils. (*b*) Single imaging plane including the apical tuft of the mitral-tufted cell (magenta) and axonal fibers loaded with Oregon Green BAPTA dextran (green). Odorant-induced calcium-dependent changes in fluorescence of the two different regions of interest (ROI 1 in colocalization with the dendritic tuft and ROI 2 in a not connected structure) are shown in *c*. (*c*) The normalized fluorescence time courses ($\Delta F/F$; for 90 s) of the calcium indicator signal upon 10 s of odorant application (lower white bars) are depicted for the respective ROIs. Individual L-amino acids were used as odor stimuli and induce different response profiles in the respective ROIs. ROI1 shows a transient increase in fluorescence upon stimulation with arginine, but not to lysine or tryptophan. ROI 2 is responsive to tryptophan, but not to the stimulation with arginine or lysine. *A* anterior, *L* lateral, *P* posterior, *M* medial, *OB* olfactory bulb, *Arg* L-arginine, *Lys* L-lysine, *Trp* L-tryptophan

(Fig. 4c). Colocalization between these responsive terminals and mitral-tufted cell dendrites gives information about odor transmission across this synapse.

11. Further insight into the neuronal network can be obtained by applying electroporation with multiple dyes (Fig. 5a) or targeting different cell types (Fig. 5b mitral-tufted cell, Fig. 5c granule cell).

4 Notes

1. Drops of the fluorophore-coupled dextrans should be about 1 μl. The amount of one dye crystal should suffice to fill up the entire nasal cavity once it is dissolved in the nasal moisture (Fig. 1b).

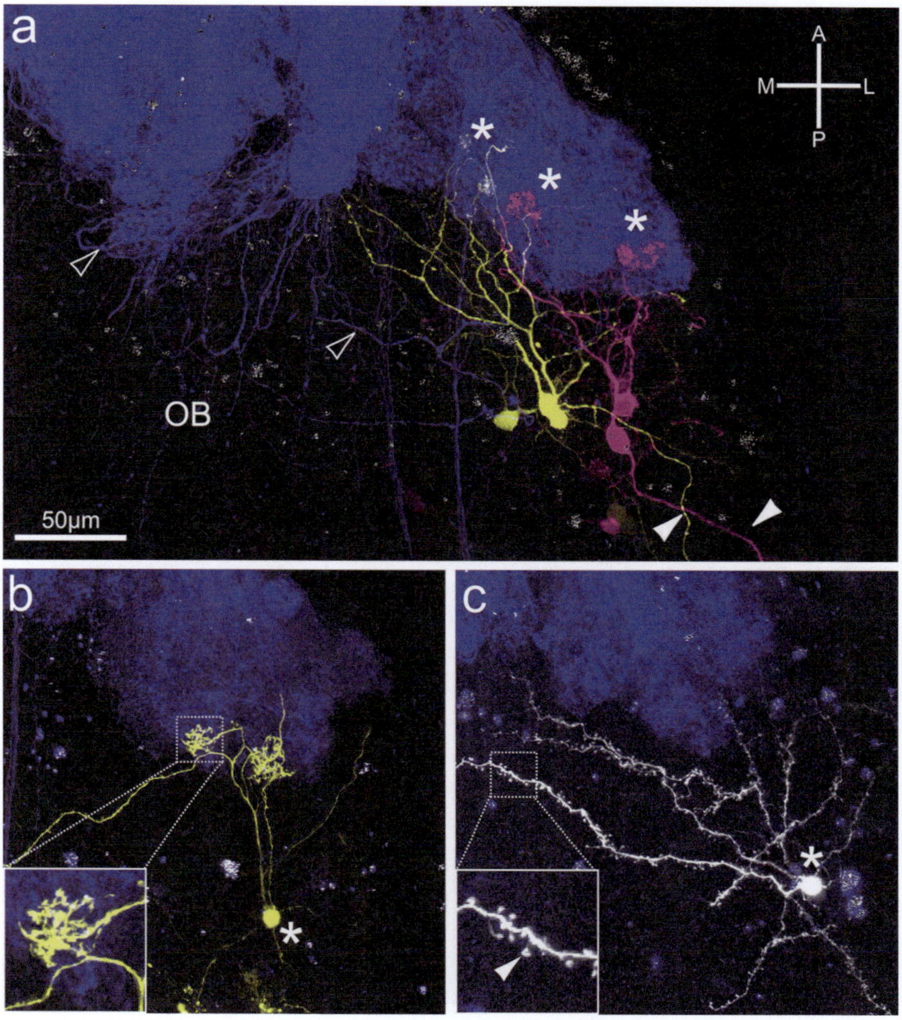

Fig. 5 Examples of successfully labeled cells via electroporation imaged with a multiphoton microscope. The olfactory receptor neuron terminals are electroporated using Cascade blue dextran (blue). (*a*) Single receptor neuron axons are indicated by empty arrowheads. Multiple mitral-tufted cells were electroporated using two different fluorescent dyes (yellow: Alexa 488 dextran, magenta: Alexa 594 dextran). Their dendritic tufts connect to the receptor neuron terminals (white asterisks). Mitral-tufted cells project their axons toward higher brain centers (filled arrowheads). (*b*) Example of a single electroporated mitral-tufted cell. One dendritic tuft is displayed in a higher magnification in the inset. The white asterisk indicates the cell soma. (*c*) A single granule cell is labeled. Granule cells are characterized by long branches that extend over the entire olfactory bulb bearing fine spines on their dendrites (arrowhead in the inset). The cell soma is indicated by an asterisk. All pictures are shown as maximum intensity projections of virtual image stacks acquired with a multiphoton microscope. *A* anterior, *L* lateral, *P* posterior, *M* medial, *OB* olfactory bulb

2. The time required to fully anesthetize the animal varies between animals of different stages. All following references to animal stages are referenced to Nieuwkoop and Faber [18].
3. When applying bulk electroporation in the nose of postmetamorphotic froglets (stage 66+), it is important to make sure

that the dye enters all nasal cavities. Because of the more segregated, bony structure and increased size of the nose in higher staged animals, the dye may not diffuse into the deeper parts of the nasal cavities.

4. If desired, the two nostrils can be electroporated with different fluorophores to label axonal projections from both sides separately. In that case, it is crucial to monitor the dye spread to ensure that the fluorophores do not mix in the moisture of the nostrils.
5. Make sure that the two electrodes are in contact with the fluid in the nasal cavities, otherwise you will not obtain a proper staining of the neurons. If the electrodes are in direct contact with the epithelial tissue in the nose, neurons might be damaged.
6. By increasing the voltage or the number of pulses applied, you can increase the amount of receptor neurons electroporated. Nevertheless, the voltage range is limited, with too low voltages failing to electroporate cells, and too high voltages resulting in cellular damage and even cell death. The ideal voltage should be tested for the electrodes used for the electroporation. The voltage proven to be most effective using our setup ranged from 20 to 30 V for 12 pulses at a frequency of 2 Hz. The use of a capacitor connected to the electroporation circuit increases successful labeling of cells as voltage pulses show an exponential decay due to capacitance discharge [1].
7. To improve recovery of the animals, lower the water level and/or pump oxygen enriched water through their gills with a Pasteur pipette. The bulk electroporation procedure and handling of anesthetized animals should not take longer than 5 min.
8. The fabrication of the micropipette might vary according to the experimental question and available equipment. Generally, a small tip opening (1–2 μm) and a long taper are desirable to confine labeling to a small number of cells or a single cell. If the pipette taper is too long and fragile, it might break when penetrating the tissue. In case a larger number of cells needs to be labeled (Fig. 5a), the tip opening can also be bigger. The right size of the tip opening and taper length can be modified by changing the heat and pulling velocity parameters of the puller [11].
9. If air bubbles form, remove them by flicking the finger against the pipette shaft. In case this does not work, use a new pipette as the bubble can affect the electroporation protocol.
10. Optimally, the micromanipulator should be controllable along a "virtual approach" axis in addition to the *x*-, *y*-, and *z*-axes. This enables the diagonal approach to the sample, along the axis of the pipette and thus facilitates precise placement of

the pipette and cellular targeting. Move the pipette along this fourth axis to approach single cells as well as when retracting the micropipette from the tissue.

11. It is essential to fix the tissue block to prevent it from moving while penetrating with the micropipette. The region of interest should optimally be placed between two strings of the grid.
12. Use an upright fluorescent microscope with a large working distance objective to have enough space for the micropipette.
13. With increasing thickness of the tissue (due to higher staged animals for example), the bright field illumination needs to be increased in order to see cellular outlines in the region of interest (Fig. 3a).
14. Switching between bright field (Fig. 3a) and fluorescent illumination (Fig. 3b), using the previously labeled receptor neurons as a guide (green, Fig. 3b), is the best mode of orientation to find mitral-tufted cells in the region of interest.
15. Use the fluorescence filter of the microscope to monitor the pipette tip closely. Always keeping the pipette tip in focus while approaching the tissue and the cells of interest helps to prevent breaking the pipette at the objective or inserting it too far into the tissue block.
16. Before approaching the target cell layer, make sure your bath electrode is in contact to the solution and that there is no outflow of solution from the pipette tip. To fully eliminate the outflow, change the offset voltage on the Axoporator stepwise. If the pipette tip opening is too big, bath solution might be streaming into the tip due to the capillary effect.
17. Monitor the resistance or current flow on the Axoporator closely to estimate whether the pipette is approaching cells, or whether it is clogged. Resistances between 20 and 80 MΩ are generally a good range for electroporating; if resistances increase drastically above 100 MΩ, the pipette might be clogged and electroporation might not work properly. If very little dye is extruded from the pipette tip because of an elevated resistance, it usually helps to either increase the voltage to 60–70 V or increase the offset voltage to 0.5–2 V. If that does not solve the problem, retract the pipette from the tissue and penetrate a different region while monitoring changes in the pipette resistance. If the pipette remains clogged, it is advisable to exchange the pipette.
18. There are multiple reasons for the lack of dye extrusion from the pipette tip. Double check the reference electrode and the offset voltage that can be adjusted on the Axoporator as well as the resistance (compare **Notes 16** and **17**). If dye is leaving the pipette tip, but no cell is labeled, you should try to move

the pipette tip to a different position within the tissue. If dye is extruded into the intercellular space, you might be able to see black outlines of cells nearby (black arrowheads in Fig. 3c). These outlines make it much easier to orientate and point the pipette tip to a single cell soma. The dye in the interstitial area usually leaves no to very little background staining when imaged in a confocal or multiphoton microscope.

19. Electroporated and fluorescently labeled cells often show signs of cell death after the electroporation, e.g., decomposition of their appendages. This cell death can be seen in the form of little fluorescent droplets in the tissue. If a cell is electroporated correctly, but the fluorescent signal is weak, the cell can also be electroporated again by repositioning the pipette.

References

1. Haas K, Sin WC, Javaherian A, Li Z, Cline HT (2001) Single-cell electroporation for gene transfer in vivo. Neuron 29(3):583–591
2. Chen C, Smye SW, Robinson MP, Evans JA (2006) Membrane electroporation theories: a review. Med Biol Eng Comput 44(1–2):5–14
3. Ho SY, Mittal GS (1996) Electroporation of cell membranes: a review. Crit Rev Biotechnol 16(4):349–362
4. Neumann E, Kakorin S, Tœnsing K (1999) Fundamentals of electroporative delivery of drugs and genes. Bioelectrochem Bioenerg 48(1):3–16
5. Falk J, Drinjakovic J, Leung KM, Dwivedy A, Regan AG, Piper M, Holt CE (2007) Electroporation of cDNA/Morpholinos to targeted areas of embryonic CNS in *Xenopus*. BMC Dev Biol 7(107):1–16
6. Miyasaka N, Arimatsu Y, Takiguchihayashi K (1999) Foreign gene expression in an organotypic culture of cortical anlage after in vivo electroporation. Neuroreport 10(11):2319–2323
7. Araki I, Nakamura H (1999) Engrailed defines the position of dorsal di-mesencephalic boundary by repressing diencephalic fate. Development 126(22):5127–5135
8. De Vry J, Martínez-Martínez P, Losen M, Temela Y, Steckler T, Steinbusch HWM, De Baets MH, Prickaerts J (2010) In vivo electroporation of the central nervous system: a non-viral approach for targeted gene delivery. Prog Neurobiol 92(3):227–244
9. Haas K, Jensen K, Sin WC, Foa L, Cline HT (2002) Targeted electroporation in *Xenopus* tadpoles in vivo – from single cells to the entire brain. Differentiation 70(4–5):148–154
10. Hovis KR, Padmanabhan K, Urban NN (2010) A simple method of in vitro electroporation allows visualization, recording, and calcium imaging of local neuronal circuits. J Neurosci Methods 191(1):1–10
11. Bestman JE, Ewald RC, Chiu S-L, Cline HT (2006) In vivo single-cell electroporation for transfer of DNA and macromolecules. Nat Protoc 1(3):1267–1272
12. Cline H, Haas K (2008) The regulation of dendritic arbor development and plasticity by glutamatergic synaptic input: a review of the synaptotrophic hypothesis. J Physiol 586(6):1509–1517
13. Hewapathirane DS, Haas K (2008) Single cell electroporation in vivo within the intact developing brain. J Vis Exp 17:e705
14. Hassenklöver T, Manzini I (2014) The olfactory system as a model to study axonal growth patterns and morphology in vivo. J Vis Exp 92:e52143
15. Kassing V, Engelmann J, Kurtz R (2013) Monitoring of single-cell responses in the optic tectum of adult zebrafish with dextran-coupled calcium dyes delivered via local electroporation. PLoS One 8(5):1–10
16. Gliem S, Syed AS, Sansone A, Kludt E, Tantalaki E, Hassenklöver T, Korsching SI, Manzini I (2013) Bimodal processing of olfactory information in an amphibian nose: odor responses segregate into a medial and a lateral stream. Cell Mol Life Sci 70(11):1965–1984
17. Hassenklöver T, Manzini I (2013) Olfactory wiring logic in amphibians challenges the basic assumptions of the unbranched axon concept. J Neurosci 33(44):17247–17252

18. Nieuwkoop PD, Faber J (1994) Normal table of *Xenopus laevis* (daudin). Garland Publishing Inc, New York, NY
19. Caprio J, Byrd RP (1984) Electrophysiological evidence for acidic, basic, and neutral amino acid olfactory sites in the catfish. J Gen Physiol 84:403–422
20. Manzini I, Schild D (2004) Classes and narrowing selectivity of olfactory receptor neurons of *Xenopus laevis* tadpoles. J Gen Physiol 123:9–107

Chapter 16

X-FaCT: Xenopus-Fast Clearing Technique

Pierre Affaticati, Sébastien Le Mével, Arnim Jenett, Laurie Rivière, Elodie Machado, Bilal B. Mughal, and Jean-Baptiste Fini

Abstract

Accessibility and imaging of cell compartments in big specimens are crucial for cellular biological research but also a matter of contention. Confocal imaging and tissue clearing on whole organs allow for 3D imaging of cellular structures after being subjected to in-toto immunohistochemistry. Lately, the passive CLARITY technique (PACT) has been adapted to clear and immunolabel large specimens or individual organs of several aquatic species. We recently demonstrated tissue clearing on one-week old tadpole brain (Fini et al., Sci Rep 7:43786, 2017). We here describe a further simplified version with clearing of small tissue samples (thickness inferior to 500 μm)) carried out by immersion in a fructose-based high-refractive index solution (fbHRI). By refining steps of the protocol, we were able to reduce the overall procedure time by two thirds. This offers the advantages of reducing the time of experimentation to a week and minimizes procedure-induced tissue deformations. This protocol can be easily adapted to be performed on thick section. We present an example of immunohistochemistry performed on NF45 *Xenopus laevis* brains with anti-pH 3 (phosphorylated histone H3) antibody used to stain chromatin condensation commonly associated with proliferation.

Key words Tissue clearing, Fructose-based high-refractive index solution, 3D imaging

1 Introduction

There is an exponential increase in the number of studies on neurodevelopmental diseases, which show evidence of embryonic origin, especially during neurogenesis processes. Many molecules in the direct environment of the embryo, mammalian, or non-mammalian can interfere with neurodevelopment, especially by interfering with hormonal signaling. It is therefore essential to be able to precisely study the changes in the proliferation, migration, or maturation of neural cells in the developing brain. The external development of aquatic larvae makes the ontology of their central

Electronic Supplementary Material: The online version of this chapter (doi:10.1007/978-1-4939-8784-9_16) contains supplementary material, which is available to authorized users.

Kris Vleminckx (ed.), *Xenopus: Methods and Protocols*, Methods in Molecular Biology, vol. 1865,
https://doi.org/10.1007/978-1-4939-8784-9_16, © Springer Science+Business Media, LLC, part of Springer Nature 2018

nervous system tractable over time. Moreover, these larvae are small and have a natural almost transparent appearance. However, pigmentation and relative structure of different tissues make 3D-imaging challenging.

Recently, tissue clearing methods have been developed in parallel with confocal, biphoton, or light sheet imaging techniques. The CLARITY technique [1, 2] and the passive CLARITY technique, also called PACT [3, 4], were originally developed on rodent organs and human tissues. These clearing strategies permit deep image acquisition of thick 3D objects.

The Tefor Core Facility adapted the PACT technique to clear and immunolabel large specimens or organs of aquatic species. We modified and adapted the CLARITY method for use in aquatic species including fish, such as medaka (*Oryzias latipes*) [5], and zebrafish (*Danio rerio*) [6]. We also transposed this protocol, named Z-PaCT, to NF45 *Xenopus* larvae and brains where we were able to label neurons and oligodendrocytes in tadpole brains [7]. However, one limitation of this technique, besides the image system accessibility, is the time needed to reach sensitivity. Indeed, immuno-labeling using specific antibodies took 2.5–3 weeks for the whole procedure and slightly affected tissue integrity.

Here, we describe a simplified version of this recent protocol where we obtained tissue clearing by simple immersion in a fructose-based high-refractive index solution (fbHRI). By refining numerous steps of the protocol, we were able to reduce the overall procedure time to 1 week instead of 2.5/3 weeks previously. This offers the double advantage of reducing the time of experimentation but also minimizes tissue modification during the procedure.

We here present the use of this protocol and results of immunohistochemistry performed on the brains of NF45 *Xenopus laevis*. The tadpoles were exposed to thyroid hormone, which is known to trigger cell proliferation, for 72 h fixed and stained with anti-pH3 antibody marking chromatin condensation. We used anti-pH3 because it is commonly acknowledged as a proliferation marker. We are confident that this protocol will be useful for in toto immunohistochemistry on both heads and brains but also on thick sections.

2 Materials

2.1 Tadpole Rearing

1. Aquatic system water, chlorine-free.

2.2 Treatment with Thyroid Hormone T_3

1. T_3 stock solution: 10 mM 3,3′,5-Triiodo-L-thyronine sodium salt (T_3) in 30% NaOH) (Sigma Aldrich). The 10 mM stock solution of T^3 is made by adding 100 mg of T^3 powder to 14.86 mL of 30% NaOH (obtained by mixing 4.458 mL of 1 N NaOH with 10.402 mL Milli-Q (ultrapure) water). Vortex

the T_3 stock solution well, aliquot and store at −20 °C. Prevent multiple weighing due to the hygroscopic properties of the powder.

2. Dimethyl sulfoxide (DMSO) (Carl Roth).
3. 1 mL transfer pipettes (Samco Scientific).
4. 50 mL polypropylene conical tube (Corning).
5. 6-well plate TPP (Thermo Fisher Scientific).
6. Incubator at 23 °C protected from light.

2.3 Brain/Head Preparation

1. Dumont tweezers, #2 straight (World Precision Instruments).
2. 60 × 15 mm Petri dishes (Greiner Bio One International).
3. 1× Phosphate-buffered saline (PBS): 137 mM NaCl, 2.7 mM KCl, 10 mM Na_2HPO_4, 1.76 mM KH_2PO_4, pH 7.4.
4. 2× Phosphate-buffered saline (PBS): 274 mM NaCl, 5.4 mM KCl, 20 mM Na_2HPO_4, 3.52 mM KH_2PO_4, pH 7.4.
5. PBST: 1× PBS, 0.01% Tween® 20, pH 7.4.
6. Ethyl 3-aminobenzoate methanesulfonate salt (MS222) (Sigma-Aldrich).
7. 8% paraformaldehyde in distilled water prepared from paraformaldehyde powder (Sigma Aldrich). Prepare aliquots of 8% PFA in H_2O (dissolved at 60 °C and stored at −20 °C). For 4% PFA dilute in 2× PBST. Another option is to buy ready-to-use 16% formaldehyde (w/v), methanol-free capsules (Thermo Fisher Scientific).
8. Fixing solution: 4% paraformaldehyde (PFA) in PBST.
9. Storage solution: 0.5% PFA in PBST with 0.05% sodium azide.

2.4 Depigmentation

1. Pre-incubation solution: 0.5× SSC (150 mM NaCl, 15 mM sodium citrate, pH 7.2), 0.1% Tween® 20.
2. Depigmentation solution: 5% formamide, 0.5× SSC, 3% H_2O_2.

2.5 Immunolabeling

1. 2 mL glass vials (Carl Roth).
2. Blocking solution: 10% normal goat serum (Antibodies Incorporated), 10% DMSO, 1% Triton X-100 in PBST.
3. Primary antibody: e.g., mouse anti-pH3, 1/200 (Millipore).
4. Secondary antibody: e.g., goat anti-mouse Alexa fluor-555, 1/200 (Thermo Fisher Scientific, Invitrogen).
5. Staining solution: 2% normal goat serum, 10% DMSO, 0.1%Triton X-100, 0.05% sodium azide, in PBST.
6. DiI (DiIC18(3)) (Thermo Fisher Scientific, Invitrogen). Prepare 1 mM Dil stock solution by dissolving DiI powder in DMSO.

7. DiD (DiIC18(5)) (Thermo Fisher Scientific, Invitrogen). Prepare 1 mM DiI stock solution by dissolving DiI powder in DMSO.

2.6 Tissue Clearing in Fructose–Based High–Refractive Index Solution (fbHRI)

1. Refractometer (e.g., KRUSS Optronic DR 301-95).
2. D(−)-fructose (Carl Roth).
3. Fructose-based high-refractive index solution (fbHRI):

This clearing solution is obtained by mixing three solutions prepared separately in advance:

(a) fbHRI Diluent: 0.2× PBS, 0.002% sodium azide.

(b) 90% fructose solution (RI = 1.457 (F1457)). In a beaker, dissolve 800 g of fructose in 300 mL fbHRI diluent. Cover the solution to avoid evaporation and stir with a strong magnetic stirrer. Make up the volume to 800 mL with fbHRI diluent. Measure RI and adjust the fructose solution to an RI of 1.457 with fbHRI diluent.

(c) 85% DMSO solution (RI = 1.457 (DMSO1457)). Adjust the DMSO solution to an RI of 1.457 with fbHRI diluent.

Mix F1457 and DMSO1457 in the ratio of 4/1 (v/v). Adjust RI with 100% fructose to raise RI or diluent to lower RI. This solution can be stored for 6 months at 4 °C.

2.7 Mounting and Imaging

1. Low-melting point agarose (e.g., Sigma-Aldrich).
2. Fluorescent microscope, for example Leica TCS SP8 laser scanning microscope (Leica Microsystems, model: Leica TCS SP8) with two photomultipliers (PMT) and equipped with a Leica HC FLUOTAR L 25×/1.00 IMM motCorr objective (Leica Microsystems).

3 Methods

3.1 Tadpole Rearing

Xenopus laevis embryos and tadpoles are reared at 20 °C in chlorine-free water (photoperiod = 12 h/12 h light/dark), in groups of 70 individuals per 15 L tank. All procedures are performed in accordance with European Union Directive 2010/63/EU.

3.2 Treatment with Thyroid Hormone T_3

1. Make a 10^{-4} M intermediate solution of T^3 fresh every day by adding 5 μL of 10 mM stock solution of T^3 to 495 μL of chlorine-free water (*see* **Note 1**).
2. Add 2.5 μL of 10^{-4} M T^3 to 50 mL of chlorine-free water to create a final 5 nM T^3 solution. Vortex the solution thoroughly (*see* **Note 2**).

3. Using a 1 mL transfer pipette with open section cut, transfer fifteen st.45 tadpoles into one well of a 6-well plate containing 8 mL chlorine-free water. In this standard assay, 15 tadpoles are used per exposure group per replicate. Rinse every 6-well plate with chlorine-free aquarium water before use.
4. Empty the liquid in the 6 wells using transfer pipette. Add 8 mL of 5 nM T^3 test solution previously prepared in Subheading 3.2, **step 2** into the well. Place the 6-well plate into a dark incubator at 23 °C for the next 72 h.
5. Renew the solutions every day at the same time (±1 h). Remove any dead tadpoles (*see* **Note 3**).

3.3 Brain/Head Preparation

1. Prepare the anesthetic solution: 1 g/L MS222 solution (1 g MS222, 1 g sodium bicarbonate in 1 L chlorine-free water). Adjust the pH between 7.2 and 7.8. Protect from light, store for 3 weeks (*see* **Note 4**).
2. Under the fume hood, prepare fresh 4% PFA fixing solution from thawed 8% PFA (H_2O) in 2× PBST (*see* Subheading 2.3, **item 7**). Prepare 40 mL of fixing solution 4% PFA in PBST in a 50 mL conical tube for 5 tadpoles. Keep the tubes at 4 °C for 1 week.
3. Kill the specimens in 1 g/L MS222 solution in accordance with local regulations. Usually 2–3 min are enough for one-week-old tadpoles.
4. Fix tadpoles overnight at 4 °C in fresh fixing solution or at room temperature for 3 h.
5. If needed, store the sample in storage solution at 4 °C until further use.
6. Brain dissection: place tadpoles in 1× PBS under fume hood. Dissect brains in the liquid. Remove skin, move a small forceps under brain tissue, and cut spinal cord (see movie dissection accessible at URL www).

3.4 Depigmentation

(Optional for ventral acquisition of brain tissues)

1. Wash in 1× PBST samples at least 1 day before starting the depigmentation needed.
2. Pour 10 mL of pre-incubation solution in a 60 mm Petri dish and transfer the fixed samples to the Petri dish with 1 mL transfer pipette.
3. Incubate at room temperature for 1 h.
4. Prepare the depigmentation solution under the fume hood.
5. Under the fume hood, remove the samples with some liquid and place them in the lid of the Petri dish. Remove all of the pre-incubation solution from the dish and replace it with

depigmentation solution. Return the samples to the dish. Keep the Petri dish uncovered and under light to accelerate the depigmentation process.

6. Check the progress of the depigmentation every 10 min.
7. Once the depigmentation is complete (*see* Fig. 1), under a fume hood rinse twice in PBST. The standard depigmentation time is 90–120 min for intact *Xenopus* NF45 larvae and 45 min for dissected brains.
8. Rinse samples overnight or for a minimum of 4 h in several changes of PBST.

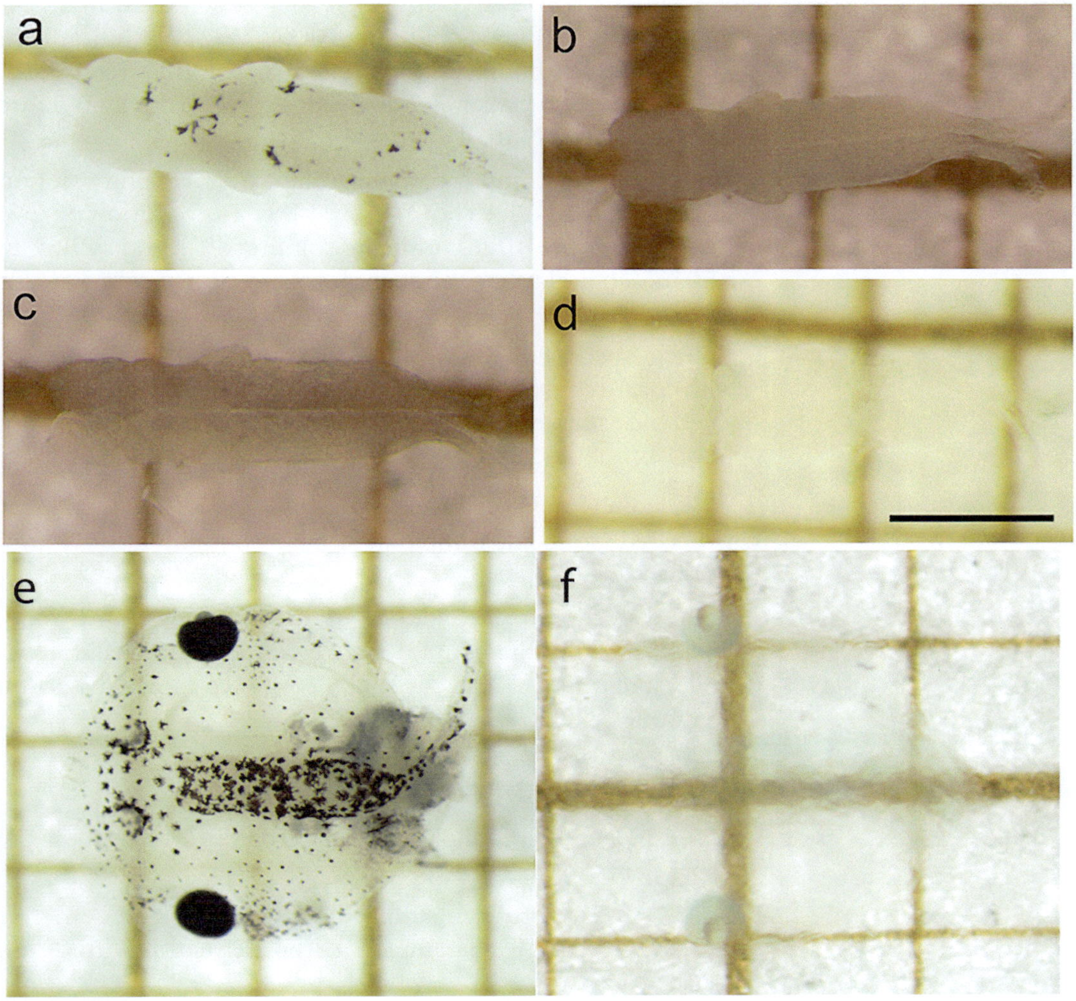

Fig. 1 Tissue clearing of Xenopus brains stage NF45. Final clearing of NF45 Xenopus laevis brains and heads after passive clearing by homogenizing the refractive index in fructose based high refractive index medium (fbHRI). (**a**) Dissected NF45 Xenopus laevis brain in PBS after PFA fixation (Subheading 3.3). (**b**) Removal of superficial pigments by H_2O_2 bleaching (Subheading 3.4). (**c**) After incubation in 50% fbHRI the opacity is significantly reduced (Subheading 3.6, **step 2**). (**d**) The tissue clearing process is finished after incubation in 100% fbHRI (Subheading 3.6, **step 4**) (**e**) NF45 *Xenopus laevis* head in PBS after PFA fixation (**f**) same head after depigmentation. Scale bar = 500 μm. All brains/heads are anterior to the left

3.5 Fluorescent Staining

1. Transfer the samples to 2 mL glass vials using a truncated transfer pipette.
2. Remove the liquid and add 1 mL of blocking solution.
3. Incubate overnight at 4 °C or for 2–4 h at room temperature.
4. Remove the blocking solution and add 600 μL of staining solution.
5. Add primary antibody to obtain the working dilution.
6. Incubate at room temperature for 2–3 days, with gentle shaking.
7. Rinse 3 times, for 20 min each, in PBST.
8. Remove all the PBST and add 600 μL of staining solution per tube.
9. Add fluorescent secondary antibody.
10. Incubate at room temperature for 2 days, with gentle shaking.
11. Rinse for 1 day in several changes of PBST.
12. Optional: Incubate the samples for 2 days in 600 μL of staining solution containing 1 μM DiI/DiD.
13. Rinse the DiI-stained sample in PBST for a minimum of 3 h.

3.6 Tissue Clearing

1. Prepare fbHRI (see 2.6 point 3), make sure to adjust the fbHRI to the desired refractive index.
2. Incubate the samples in 50% fbHRI/50% (v/v) diluent for at least 6 h (Fig. 1).
3. Remove the solution and replace by fbHRI.
4. Incubate for at least 12 h (Fig. 1).

3.7 Mounting and Imaging

Mounting strategies depend on the microscopy technique one may want to apply. Confocal microscopy, multiphoton microscopy, or light sheet microscopy are well suited techniques to image cleared samples. An example of a mounting method is given here. It is optimized for the use of an upright confocal microscope equipped with a dipping lens RI-corrected objective (Leica 15507703).

1. Place your sample in a 60 mm petri dish and remove as much liquid as possible.
2. Pipet 2–5 μL of 1% agarose (50 °C) and use it to glue the sample to the petri dish.
3. Allow the agarose to solidify at 4 °C for 2 min.
4. Fill the petri dish with fbHRI adjusted to the right RI.
5. Dip objective into the fbHRI and capture fluorescence signals by exciting the fluorophores with the appropriate laser wavelength. Tilescan imaging followed by mosaic stitching is required to capture images of the whole brain.

3.8 Visualization of 3D Data Sets (Supplementary Material)

1. Translation movies (connected to Fig. 2a, b) are accessible at the link URL under the names Movies 1, 2, and 3. These movies are showing tapdole brains, which were exposed for 72 h to T_3, processed using X-FaCT with pH3. The translation movies constitute a direct conversion of the raw data into mp4 movie file format. In this process, using a combination of fiji [8] and libav (https://libav.org/), each slice (z-step) of the raw-data stack is turned into a frame of the movie turning the z-dimension of the stack into the temporal domain of the movie. Lossy compression (×246-driven h246-compression) provides significant data reduction while keeping the image information largely intact. The original 12 bit histogram of the original data is linearly converted to the necessary 3 × 8 bit (RGB) without application of any contrast enhancement or noise reduction. This allows insight into a multi-GB-dataset while dealing with only tens of MBs.
2. Rotation movies (connected to Fig. 2c, d) are accessible at link www. Under the names Movies 4 and 5. The rotation movies can be generated using amira 6.0.1 (FEI, SAS) on a Dell T630 running Ubuntu 16.04.

Since in the original data of the pH3-channel we found a low but distinct amount of signal bleed-through from the DiD stain, we subtracted a fraction of the DiD-channel from the pH3-channel ([pH3]-([DiD]/5). The division coefficient thereby was determined empirically for selectively removing the background signal. This process allows subsequent enhancing of the contrast of the pH3-signal by global linear histogram stretching (GLHS), which results in the pH3-positive cells appearing brighter, which again

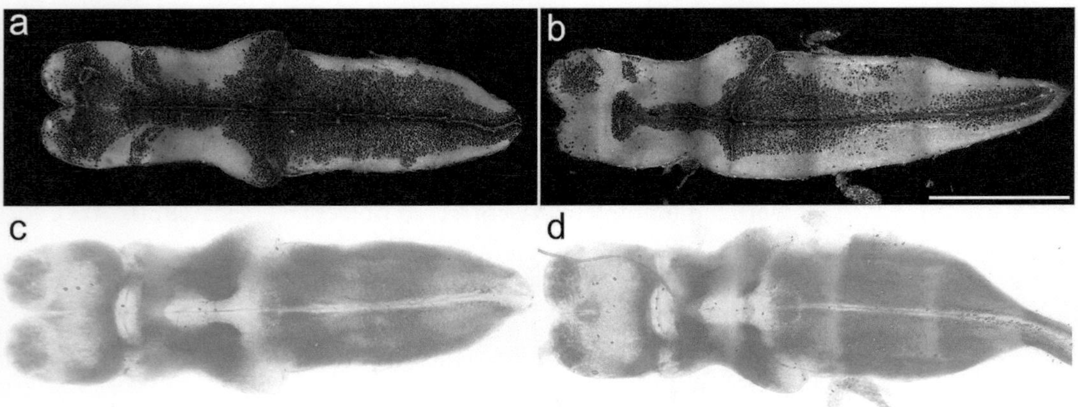

Fig. 2 NF45 Xenopus laevis whole-brain immunolabeling and imaging. Single confocal plane of a NF45 control brain (**a**) or treated with 5 nM T_3 (**b**) stained with DiD (gray), immunolabeled for pH 3 (green). The visualisation of the whole dataset is provided by two translation movies, control brain (Movie 2) or T3 treated (Movie 3) provided at the URL www. Scale bar = 500 μm. Ventral 3D views from rotation movies of control brain (**c** and Movie 4) or T3 treated (**d** and Movie 5) accessible at the URL www. All brains are anterior to the left

facilitates their detection in the 3D-visualization. To the reference channels (DiD) we applied only GLHS without any other pretreatment. Due to slightly different acquisition settings between the specimens the GLHS settings were optimized for each dataset individually.

[[untreated: pH3: 150–1300, DiD: 0–3500; treated: pH3: 150–2000, DiD: 0–2800. All values on the basis of a 12 bit histogram (0–4095)]].

4 Notes

1. Pay attention that neither precipitate nor crystals are present in the T_3 stock.
2. Be careful to rinse every 50 mL plastic tube with dechlorinated animal system water before use.
3. If the mortality exceeds 20% in a well, stop and repeat the experiment.
4. Consider making a stock solution at 4 g/L.

References

1. Chung K, Wallace J, Kim SY, Kalyanasundaram S, Andalman AS, Davidson TJ, Mirzabekov JJ, Zalocusky KA, Mattis J, Denisin AK, Pak S, Bernstein H, Ramakrishnan C, Grosenick L, Gradinaru V, Deisseroth K (2013) Structural and molecular interrogation of intact biological systems. Nature 497(7449):332–337. https://doi.org/10.1038/nature12107
2. Tomer R, Ye L, Hsueh B, Deisseroth K (2014) Advanced CLARITY for rapid and high-resolution imaging of intact tissues. Nat Protoc 9(7):1682–1697. https://doi.org/10.1038/nprot.2014.123
3. Treweek JB, Chan KY, Flytzanis NC, Yang B, Deverman BE, Greenbaum A, Lignell A, Xiao C, Cai L, Ladinsky MS, Bjorkman PJ, Fowlkes CC, Gradinaru V (2015) Whole-body tissue stabilization and selective extractions via tissue-hydrogel hybrids for high-resolution intact circuit mapping and phenotyping. Nat Protoc 10(11):1860–1896. https://doi.org/10.1038/nprot.2015.122
4. Yang B, Treweek JB, Kulkarni RP, Deverman BE, Chen CK, Lubeck E, Shah S, Cai L, Gradinaru V (2014) Single-cell phenotyping within transparent intact tissue through whole-body clearing. Cell 158(4):945–958. https://doi.org/10.1016/j.cell.2014.07.017
5. Dambroise E, Simion M, Bourquard T, Bouffard S, Rizzi B, Jaszczyszyn Y, Bourge M, Affaticati P, Heuze A, Jouralet J, Edouard J, Brown S, Thermes C, Poupon A, Reiter E, Sohm F, Bourrat F, Joly JS (2017) Postembryonic fish brain proliferation zones exhibit neuroepithelial-type gene expression profile. Stem Cells 35(6):1505–1518. https://doi.org/10.1002/stem.2588
6. Affaticati P, Simion M, De Job E, Rivière L, Hermel J, Machado E, Joly J, Jenett A (2017) zPACT: tissue clearing and immunohistochemistry on juvenile zebrafish brain. Bio-Protocol 7(23):e2636. https://doi.org/10.21769/BioProtoc.2636
7. Fini JB, Mughal BB, Le Mevel S, Leemans M, Lettmann M, Spirhanzlova P, Affaticati P, Jenett A, Demeneix BA (2017) Human amniotic fluid contaminants alter thyroid hormone signalling and early brain development in Xenopus embryos. Sci Rep 7:43786. https://doi.org/10.1038/srep43786
8. Schindelin J, Arganda-Carreras I, Frise E, Kaynig V, Longair M, Pietzsch T, Preibisch S, Rueden C, Saalfeld S, Schmid B, Tinevez JY, White DJ, Hartenstein V, Eliceiri K, Tomancak P, Cardona A (2012) Fiji: an open-source platform for biological-image analysis. Nat Methods 9(7):676–682. https://doi.org/10.1038/nmeth.2019

Chapter 17

Cell Cycle Analysis of the Embryonic Brain of Fluorescent Reporter *Xenopus tropicalis* by Flow Cytometry

Rivka Noelanders and Kris Vleminckx

Abstract

Many developmental signals are associated with changes in proliferative response. Also, growing organs and tissues can contain different cellular subpopulations with a defined status in the cell cycle, e.g., quiescent in stem cells, high proliferation in progenitors, cell cycle exit in differentiating cells. This chapter describes a method for isolation of individual cell populations from the *Xenopus* tadpole brain and determination of their cell cycle status using flow cytometry.

Key words Cell cycle, Flow cytometry, *Xenopus*, Brain

1 Introduction

Cell proliferation is tightly regulated in all organisms. The cell cycle consists of four phases: two gap or resting phases (G1 and G2), a synthesis phase, when DNA replication takes place and a mitosis phase, when physical separation of the newly formed daughter cells takes place. Upon terminal differentiation, a cell will exit the cell cycle and become quiescent (G0-phase). Transition between the different phases of the cell cycle is tightly regulated by cyclins and cyclin-dependent kinases, which in turn are regulated by a myriad of signaling events and transcription factors. Assessment of changes in cell cycle progression upon experimental manipulation is interesting in a large variety of contexts, e.g., for developmental biology studies, regeneration studies, cancer research, and drug treatments [1–5]. These changes can be obvious as hyper- or hypoplasia of the affected region/organ or can be more subtle and are only detected through the analysis of proliferation markers. Many of these markers are available (e.g., phospho-histone 3, Ki67, PCNA, halogen-conjugated deoxyuridine homologs) [6]. However, quantification of proliferation markers is labor-intensive and tedious when performed manually or has to be performed with specialized software, which is expensive

Kris Vleminckx (ed.), *Xenopus: Methods and Protocols*, Methods in Molecular Biology, vol. 1865,
https://doi.org/10.1007/978-1-4939-8784-9_17,

and might not be completely accurate. We present here a protocol for cell cycle analysis of the *Xenopus tropicalis* brain by flow cytometry. Cell cycle analysis by flow cytometry is based on the fact that during the course of the cell cycle the amount of DNA rises during the S-phase to be doubled (4n) during G2 and M-phases and returns back to normal (2n) after mitosis during G1- or G0-phase [7–9]. The amount of DNA a cell contains is thus an immediate reflection of the cell cycle phase the cell is in. DNA replication and mitosis only take place in dividing cells and the fraction of cells present in these cell cycle phases thus reflects the proliferation status of a cell population. We show that cell cycle analysis by flow cytometry is a straightforward technique that can be applied to any region or organ of interest during development or in the adult. Most importantly, high quality quantitative data are obtained that shed light on speed of progression of cells through the different phases of the cell cycle as opposed to the snapshot of often only one phase provided by proliferation markers. Moreover, we show that cell cycle analysis of a specific cell population is possible through use of transgenic reporter lines.

2 Materials

2.1 Dissection

1. MS-222 1% (w/v) stock solution.
2. 0.1× MMR: 10 mM NaCl, 200 μM KCl, 100 μM $MgSO_4$, 200 μM $CaCl_2$, 500 μM HEPES pH 7.8.
3. Black silicone dissection dish.
4. Number 5 Watchmaker's forceps.
5. 2 mL round-bottom Eppendorf tubes.
6. 70% (v/v) L15 medium.

2.2 Dissociation

1. Phosphate-Buffered Saline (1× PBS): 137 mM NaCl, 2.7 mM KCl, 10 mM Na_2HPO_4 1.8 mM KH_2HPO_4.
2. Amphibian PBS (APBS): 0.66× PBS.
3. Trypsin solution: 0.033% (w/v) Trypsin (tissue culture grade), 0.021% (w/v) EDTA in APBS.
4. Ca^{2+}-free disaggregation medium: 116.6 mM NaCl, 0.67 mM KCl, 4.62 mM Tris–HCl pH 8.0, 0.4 mM EDTA pH 7.4.
5. Bovine Serum Albumin (tissue culture grade).
6. 6% Ficoll 400 (w/v) solution.
7. Lighter.

2.3 Fixation and Permeabilization

1. 2% paraformaldehyde solution: 2% (w/v) paraformaldehyde (PFA) in 0.66× amphibian PBS).
2. Cooled benchtop centrifuge.

3. 70% (v/v) EtOH in 1× PBS (stored at −20 °C).
4. Vortex.

2.4 DNA Staining

1. Hoechst 33342 (1 mg/mL stock).

2.5 Acquisition and Analysis

1. Flow cytometer equipped with a blue (488 nm) laser (e.g., BD FACS Verse).
2. FlowJo software package.

3 Methods

The protocol outlined here has been optimized for cell cycle analysis of brains from Nieuwkoop stage 41 *Xenopus tropicalis* embryos. However, it can be applied to any tissue or cell population that can be cleanly dissected. Furthermore, the protocol was designed to assess the cell cycle specifically in cells expressing GFP. The protocol is optimized to allow detection of endogenous GFP without the need for immunofluorescent staining. This is achieved by fixation with low concentration of paraformaldehyde and a gentle permeabilization method (ice-cold ethanol) to avoid leakage of GFP out of the cells.

3.1 Dissection

1. It is important to prepare all reagents beforehand and keep them on ice so they are chilled before use. As you are working with live cells (until fixation) it is advised to work as fast as possible and under cold conditions to maintain cell viability.
2. Before dissection embryos are anesthetized in 0.01% MS-222 in 0.1× MMR.
3. Fully sedated embryos are transferred to a black silicone dish containing the anesthesia medium (*see* **Note 1**).
4. First the skin in the head region is removed by making a small incision with watchmaker's forceps posterior to the brain and peeling the skin off toward the anterior side. The brain is isolated by moving a closed forceps underneath the brain from posterior to anterior, thereby loosening the brain from the other tissues. Finally the spinal cord is severed.
5. The dissected brain is moved immediately to a round-bottom 2 mL Eppendorf on ice containing 500 μL 70% (v/v) L15 medium. Ten brains are collected for each sample (*see* **Notes 2** and **3**).

3.2 Dissociation

1. The dissected brains will sink to the bottom of the tube. Remove most of the L15 medium and carefully add 100 μL of the trypsin-EDTA solution (warmed up to RT) so that the brains remain at the bottom of the tube.
2. Incubate for 8 min at room temperature without agitation (*see* **Note 4**).

3. Carefully remove the trypsin-EDTA solution and add 100 μL Ca^{2+}-free disaggregation medium. Place tissue back on ice. Mechanically dissociate the brains by pipetting up and down using a cut P200 tip that has been passed quickly through a flame so that it becomes smooth and with which you have pipetted the Ficoll solution up and down a few times (*see* **Note 2**). Pipet up and down for 5 min while keeping the cells on ice at all times. Be careful not to introduce air bubbles.
4. Add another 100 μL of Ca^{2+}-free disaggregation medium supplemented with 2% (w/v) BSA. Pipet up and down for another 5 min. No aggregates of tissue should be present at this stage.

3.3 Fixation and Permeabilization

1. Collect cells by centrifugation for 5 min at 200 g at 4 °C using a cooled benchtop centrifuge. Remove the supernatant by decantation.
2. Add 500 μL of cold APBS and 500 μL of cold 2% paraformaldehyde solution and mix by pipetting up and down with a smoothened tip coated with Ficoll. Incubate on ice for 30 min.
3. Spin cells down by centrifugation for 5 min at 700 × *g* at 4 °C (*see* **Note 5**). Remove the supernatant by decantation.
4. Add 1 mL of cold 1× PBS. Spin cells down by centrifugation for 5 min at 700 × *g* at 4 °C (*see* **Note 5**). Remove the supernatant by decantation.
5. Add 1 mL of ice-cold 70% EtOH/1× PBS dropwise to the cell pellet while the tube is sitting on a vortex (low speed). Incubate cell suspension overnight at 4 °C (*see* **Note 6**).

3.4 DNA Staining

1. Spin cells down by centrifugation for 5 min at 700 × *g* at 4 °C. Remove the supernatant by decantation.
2. Add 1 mL of cold 1× PBS. Spin cells down by centrifugation for 5 min at 700 × *g* at 4 °C. Remove the supernatant by decantation.
3. Meanwhile prepare the Hoechst 33342 staining solution by diluting the Hoechst 33342 stock 1/500 in 1× PBS.
4. Add 200 μL of the Hoechst 33342 staining solution and incubate for 30 min at room temperature in the dark.
5. Cells are kept in the staining solution, in the dark, at 4 °C until acquisition on the flow cytometer (*see* **Note 7**).

3.5 Acquisition and Analysis

Acquisition can be performed on a flow cytometer equipped with a blue laser and appropriate filters: V450, for Hoechst 33342 detection, and FITC, for GFP detection. Running speed should be set so that not more than 500 events/second are measured. Analysis should be limited to single cells as cell doublets can be mistaken for cells in G2/M-phase. Data analysis is performed using the FlowJo software package.

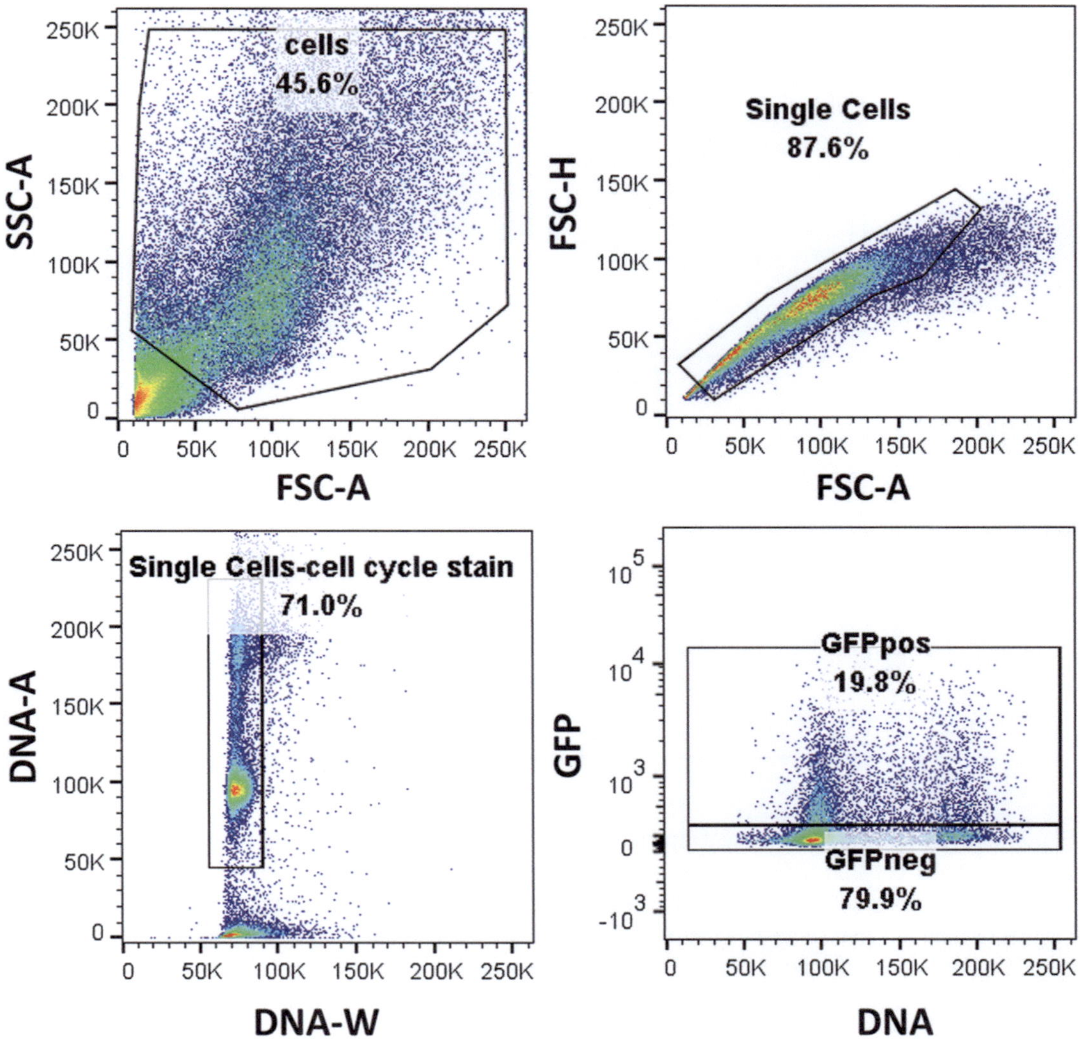

Fig. 1 Selection of single cells. Cells are first identified through forward scatter (FSC) versus side scatter (SSC). Single cells are gated in two different ways: FSC height versus area and Hoechst signal area (DNA-A) versus Hoechst signal width (DNA-W)

1. Cells are first identified on scatter characteristics (side scatter (SSC) vs. forward scatter (FSC)).
2. Single cells are identified in two ways (*see* Fig. 1).
   - Signal height and area of the forward scatter should show a linear relationship. So only cells on the diagonal of a FSC-H vs. FSC-A plot should be included.
   - For a given DNA signal area, DNA signal width should remain constant. If DNA signal width is increased to reach a same DNA signal area, this means that two 2n cells stuck together resulting in a 4n signal area. So when V450-A is

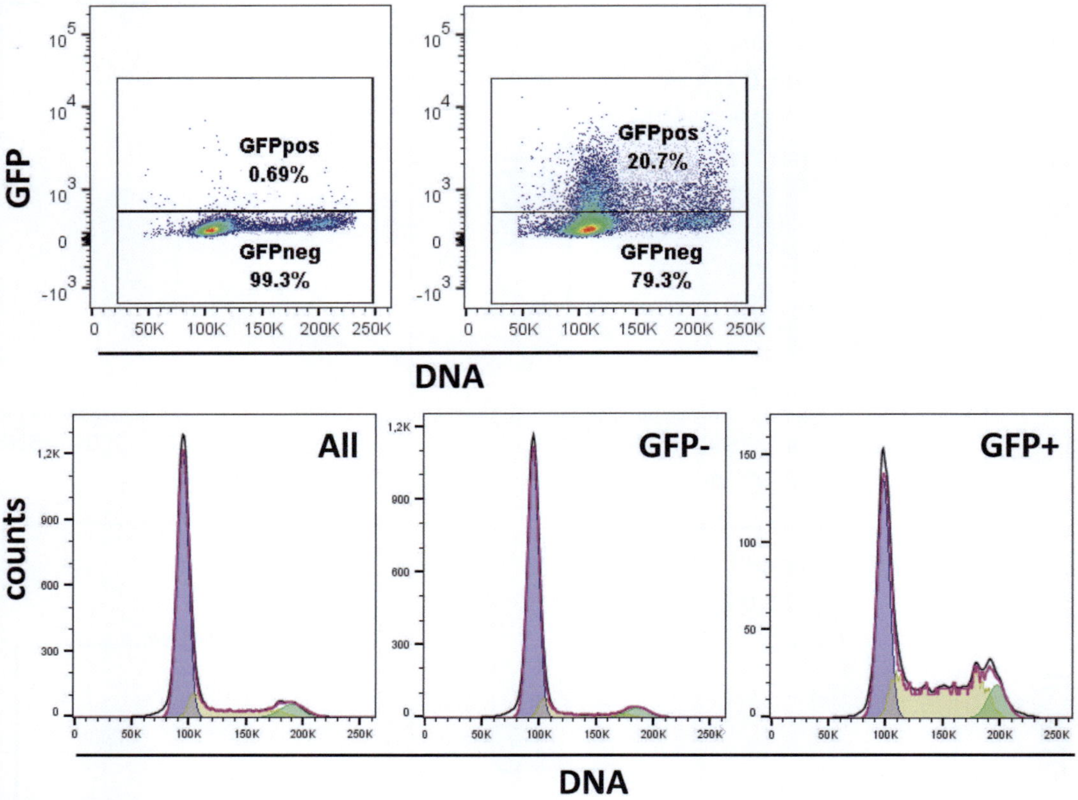

Fig. 2 Cell cycle analysis of fluorescent reporter brains (*top panels*). GFP positive cells from a transgenic reporter line were gated out from GFP negative hindbrain cells based on a GFP negative sample (*bottom panels*). Histograms showing cell cycle phase distribution of the different cell populations. Pink line represents the optimal histogram deduced by the FlowJo Cell Cycle Analysis algorithm. Blue peak and green peak represent cells in G0/G1- and G2/M-phase, respectively. Yellow/grey peak represents cells in S-phase

plotted against v450-W, all cells shifted to the right (greater DNA signal width) should be excluded. Importantly, DNA signal is plotted on a linear scale.

3. GFP positive cells are identified based on a gate set by a GFP negative sample, which should always be included. All cells with increased GFP expression compared to the GFP negative sample are considered to be GFP-positive (*see* Fig. 2).
4. DNA-histograms are obtained by plotting count vs DNA signal area and should result in a smooth outline of the histogram (*see* **Note 8**). This can be done for GFP-positive and GFP-negative cells separately (*see* Fig. 2).
5. The percentage of cells in each phase of the cell cycle is calculated with the built-in cell cycle analysis module of FlowJo, which models the ideal 2n and 4n DNA content peaks on the provided DNA histogram and then deducts the amount of cells in each phase.

4 Notes

1. A black silicone dissection dish was used for enhanced contrast with the white brain tissue. Additionally, the softness of the silicone prevents blunting of the watchmaker's forceps. For other tissues other dissection recipients may be more appropriate.
2. To avoid tissue damage, brains are transferred using a P200 with a cut tip that is smoothened by moving it through a flame. First, pipet up and down in the Ficoll solution to prevent tissue sticking inside the tip.
3. The amount of tissue to be collected is dependent on the amount of cells it contains. Take into account that you need at the very least 5000 cells to obtain a reliable DNA histogram. Dissociation of ten brains results in about 50,000 single cells.
4. Conditions for dissociation are very tissue dependent. Embryonic tissues generally dissociate quite easily. Moreover, embryonic cells are very fragile. Care must be taken to apply enzymatic treatment long enough to achieve complete dissociation of the tissue, but not longer to avoid cell death.
5. Fixation causes the cells to shrink, thereby requiring higher g-force to collect them at the bottom of the tube.
6. Cells should be in constant movement while adding the ethanol to prevent cell clumping.
7. We recommend running samples on a flow cytometer as soon as possible. However, samples can still be run the next day without obvious loss of signal strength and quality of the DNA histogram.
8. If a smooth outline of the histogram is not obtained, try reducing the running speed or increase the amount of cells analyzed.

Acknowledgments

The authors would like to thank Dr. Muriel Perron for sharing the protocol on which ours is based. Research in the authors' laboratory is supported by the Research Foundation—Flanders (FWO-Vlaanderen) (grants G0A1515N and G029413N), by the Belgian Science Policy (Interuniversity Attraction Poles—IAP7/07) and by the Concerted Research Actions from Ghent University (BOF15/GOA/011). Further support was obtained by the Hercules Foundation, Flanders (grant AUGE/11/14) and the Desmoid Tumor Research Foundation.

References

1. Alexiou GA, Vartholomatos G, Stefanaki K, Lykoudis EG, Patereli A, Tseka G, Tzoufi M, Sfakianos G, Prodromou N (2015) The role of fast cell cycle analysis in pediatric brain tumors. Pediatr Neurosurg 50(5):257–263. https://doi.org/10.1159/000439029
2. Chauhan A, Lorenzen S, Herzel H, Bernard S (2011) Regulation of mammalian cell cycle progression in the regenerating liver. J Theor Biol 283(1):103–112. https://doi.org/10.1016/j.jtbi.2011.05.026
3. Myster DL, Duronio RJ (2000) To differentiate or not to differentiate? Curr Biol 10(8):R302–R304
4. Sakaue-Sawano A, Kobayashi T, Ohtawa K, Miyawaki A (2011) Drug-induced cell cycle modulation leading to cell-cycle arrest, nuclear mis-segregation, or endoreplication. BMC Cell Biol 12:2. https://doi.org/10.1186/1471-2121-12-2
5. Williams GH, Stoeber K (2012) The cell cycle and cancer. J Pathol 226(2):352–364. https://doi.org/10.1002/path.3022
6. Romar GA, Kupper TS, Divito SJ (2016) Research techniques made simple: techniques to assess cell proliferation. J Invest Dermatol 136(1):e1–e7. https://doi.org/10.1016/j.jid.2015.11.020
7. Darzynkiewicz Z, Juan G, Bedner E (2001) Determining cell cycle stages by flow cytometry. Curr Protoc Cell Biol Chapter 8:Unit 8 4. doi:https://doi.org/10.1002/0471143030.cb0804s01
8. Nunez R (2001) DNA measurement and cell cycle analysis by flow cytometry. Curr Issues Mol Biol 3(3):67–70
9. Pozarowski P, Darzynkiewicz Z (2004) Analysis of cell cycle by flow cytometry. Methods Mol Biol 281:301–311. https://doi.org/10.1385/1-59259-811-0:301

Chapter 18

Manipulating and Analyzing Cell Type Composition of the *Xenopus* Mucociliary Epidermis

Peter Walentek

Abstract

The *Xenopus* embryonic epidermis serves as a model to investigate the development, cell biology, and regeneration of vertebrate mucociliary epithelia. Its fast development as well as the ease of manipulation and analysis in this system facilitate novel approaches and sophisticated experiments addressing the principle mechanisms of mucociliary signaling, transcriptional regulation, and morphogenesis. This protocol describes how cell type composition can be manipulated and analyzed, and how mucociliary organoids can be generated and used for "omics"-type of experiments.

Key words Xenopus, Cilia, Mucociliary, Epidermis, Multiciliated cell, Small secretory cell, Ionocyte, Goblet cell

1 Introduction

Mucociliary epithelia line various organs of animals and provide a first line protection against pathogens and harmful substances [1]. Most prominently, a mucociliary epithelium is present in the mammalian respiratory tract and accounts for clearance of the airways [2]. Pathological changes in the airway mucociliary epithelium are associated with severe diseases, but the underlying molecular mechanisms are largely unresolved [3]. The mucociliary epidermis of *Xenopus* embryos has become a powerful system to gain novel insights into the molecular, developmental, and regenerative biology of mucociliary cell types and their interactions, with the potential for translational impact on human health [4]. Over the past few years, multiple studies have revealed a high degree of evolutionary conservation between *Xenopus* and mammalian mucociliary cells and their functions, most importantly multiciliated cells [5]. Furthermore, the *Xenopus* mucociliary epidermis system offers unique advantages:

Kris Vleminckx (ed.), *Xenopus: Methods and Protocols*, Methods in Molecular Biology, vol. 1865,
https://doi.org/10.1007/978-1-4939-8784-9_18, © Springer Science+Business Media, LLC, part of Springer Nature 2018

1. Ease of manipulation using morpholino mediated knockdown, overexpression of mRNA/DNA, or F0 knock-out studies using genome editing (e.g., CRISPR/Cas9), and the application of proteins and small pharmacological compounds by simple addition to the medium.
2. Fast developmental time of a mature mucociliary epithelium within 2–3 days, as compared to 2–4 weeks in mammalian systems.
3. Easy generation of large quantities of mucociliary organoids from animal cap explants, which allow the use of almost pure mucociliary tissue in transcriptomics, genomics, and proteomics experiments.
4. Superficial localization of the mucociliary epidermis, which makes it accessible to physical and chemical manipulations as well as to live-cell imaging analysis.
5. Several studies have shown how cell type composition can be easily manipulated to obtain mucociliary epithelia in which a specific cell type is strongly enriched or virtually absent, e.g., by manipulating the Notch signaling pathway or some of the key transcription factors required for cell type specification (Multicilin, Foxa1, Foxi1).

In combination, these advantages facilitate novel experimental approaches to investigate, e.g., mucociliary signaling, cell biology, as well as transcriptional, post-transcriptional and epigenetic mechanisms.

Here, I am providing a general protocol for *Xenopus laevis*, which describes how to manipulate and analyze cell type composition and to generate mucociliary organoids for "omics"-type of experiments.

2 Materials

2.1 Microinjection into Early Xenopus Embryos

1. 3% L-Cysteine solution: dissolve 3% Cysteine-HCl monohydrate in water, adjust to pH 8.0 with 5 N NaOH stirring.
2. 1 × modified amphibian Ringers (MR): 0.1 M NaCl, 1.8 mM KCl, 2 mM $CaCl_2$, 1 mM $MgCl_2$, 5 mM HEPES-NaOH (pH 7.6).
3. 1/3 × MR with 2.5% Ficoll PM400: Add 1 part of 1 × MR, 2 parts of water and 2.5 g (per 100 mL) of Ficoll PM400 and stir until Ficoll is fully in solution.
4. 1/3 × MR with Gentamycin: Add 1 part of 1 × MR, 2 parts of water and 50 μg/mL of Gentamycin.
5. Gurdon's buffer: 88 mM NaCl, 15 mM HEPES, 1 mM KCl, 15 mM Tris–HCL, pH 7.6.

6. Kit for in vitro synthesis of 5′-capped and 3′-polyadenylated mRNAs (e.g., mMessage mMachine by Ambion, or HiScribe by NEB) (*see* **Note 1**).
7. DNA constructs for cell type manipulation:
   (a) Notch-ICD, activator of Notch signaling, described in [6].
   (b) Suppressor of Hairless DNA-binding mutant (Su(H)-DBM), inhibitor of Notch signaling, described in [6].
   (c) Multicilin-HGR (Mci, MCIDAS), master transcription factor for induction of multiciliated cells (MCCs) coupled to HGR-sequence for activation upon Dexamethasone treatment, described in [7].
   (d) Dominant-negative Multicilin (dn-Mci), inhibits MCC formation, described in [7].
   (e) Foxi1-HGR, master transcription factor for induction of ionocytes (also called ion secreting cells, ISCs) coupled to HGR-sequence for activation upon Dexamethasone treatment, described in [8].
8. Morpholino oligonucleotides for cell type manipulation:
   Morpholino-oligonucleotides (MOs) suitable for knockdown can be designed by and purchased from Gene Tools LLC (gene-tools.com) (*see* **Note 2**).
   (a) Multicilin knockdown via translational-blocking.
      (Mci-ATG MO, AAAACGCTTTCCTTCTGTTCTG CAT) or splice-blocking.
      (Mci-SPL MO, GTGCCAGTCACTCACCTGGTTGTTC) Morpholino to inhibit MCC formation, described in [7].
   (b) Foxa1 knockdown via translational-blocking.
      (Foxa1-ATGMO,ATGGCAAGTGCCCGAGAGAATCATT) Morpholino to inhibit SSC formation, our unpublished sequence.

2.2 Analyzing Cell Type Composition

2.2.1 Morphological Analysis by Immunofluorescence and Cell Staining

1. 1 × Phosphate-buffered saline (PBS): 137 mM NaCl, 2.7 mM KCl, 10 mM Na_2HPO_4, 1.8 mM KH_2PO_4.
2. PBS + 0.1% Triton X-100 (PBSt): For 1 l of 1 × PBS add 1 mL of Triton X-100.
3. CAS-Block (Invitrogen).
4. 1:10 CAS-Block/PBSt: Mix 1 part of CAS-Block with nine parts of PBSt.
5. 4% Paraformaldehyde in PBS (PFA): Add 4 g of PFA to 100 mL of PBS pH 7.2 and heat to 65 °C stirring until PFA has fully dissolved.

6. Anti acetylated α-Tubulin antibody (mouse monoclonal, 1:700; Sigma #T6793) to visualize cilia in MCCs.
7. Anti Serotonin antibody (rabbit polyclonal, 1:500, Merck #AB938) to visualize serotonin in SSCs.
8. Appropriate species-specific secondary antibodies should be used depending on microscope filter sets available at your institution/laboratory (e.g., Alexa Fluor 405 goat anti-mouse IgG antibody, 1:250, Molecular Probes #A31553; and Alexa Fluor 555 goat anti-rabbit IgG antibody, 1:250, Molecular Probes #A21428).
9. Fluorophore-conjugated Phalloidin (e.g., Alexa Fluor 488-Phalloidin, 1:50, Molecular Probes #A12379) to visualize cell borders and apical morphology.
10. Fluorophore-conjugated Lectin PNA from peanut (e.g., Alexa Fluor 594-PNA, 1:1000 from 1 mg/mL stock, Molecular Probes #L32459) to visualize mucus-like compounds in goblet cells and SSCs.
11. High vacuum silicon grease (e.g., Dow Corning high-vacuum silicone grease from Sigma-Aldrich/Merck).

2.2.2 Analysis of Gene Expression by In Situ Hybridization

Gene expression can be visualized by chromogenic staining following standard protocols for in situ hybridization in *Xenopus* [9]. Specific probes are generated from constructs specified below using in vitro RNA synthesis kits (e.g., Sp6 Promega, T7 Promega, T3 Promega) and using Digoxigenin-modified nucleotides (e.g., combination of DIG-11-UTP Roche, and rNTPs Roche). Detection of probes by chromogenic reaction is achieved best using AP-conjugated Fab-fragments antibodies (Roche) in combination with BM purple substrate (Roche).

1. MEMFA fix: Add 1 part of 10 × MEMFA salts (1 M MOPS, 20 mM EGTA, 10 mM $MgSO_4$), 1 part 37% formaldehyde solution, eight parts of water and mix.
2. To visualize MCCs: α-Tubulin (late, [6]), Foxj1 (early, [10]), Mci (early, [7]).
3. To visualize ISCs: Foxi1 (all ISCs early), ATP6V1E1 (all ISCs late), Ae1 (α-subtype ISCs), Ubp1 or Pendrin (β-subtype ISCs) (all described in [8]).
4. To visualize SSCs: Foxa1 (early), TPH1 (late) (both described in [11]).
5. To visualize goblet cells: Otogelin, Xeel (both described in [12]).
6. To visualize basal cell-like precursors: Tp63 (described in [13]) (*see* **Note 3**).

2.3 Generating Mucociliary Organoids

1. 1 × modified Bath's saline (MBS): 88 mM NaCl, 1 mM $CaCl_2$, 1 mM $MgSO_4$, 5 mM HEPES (pH 7.8 using 5 N NaOH), 2.5 mM $NaHCO_3$.
2. 0.5 × MBS with Gentamycin: Add 1 part of 1 × MBS, 1 parts of water and 50 μg/mL of Gentamycin.
3. Fine forceps, e.g., FST Dumont, size #5, Dumostar-Biology (*see* **Note 4**).

3 Methods

3.1 Injections into Early Xenopus Embryos for Mucociliary Cell Type Manipulation

In general, microinjections into early (one to four cell-stage) embryos are a standard procedure in the field and detailed information regarding this can be found in multiple publications as well as in the book:Early Development of *Xenopus laevis*, by Sive, Grainger and Harland [14] (*see* **Note 5**).

- *Xenopus* embryos are obtained by standard procedures and kept at 16 °C after in vitro fertilization. After 45 min to 1 h, fertilized oocytes are treated with 3% L-Cysteine for 10 min at room temperature (RT) (*see* **Note 6**).
- Wash fertilized oocytes 3–4 times in 1/3 × MR to remove Cysteine.
- Place fertilized oocytes at 12 °C–16 °C to slow down development (*see* **Note 7**).
- Prepare injection solutions: Dilute mRNAs, DNAs, MOs or CRISPR/Cas9 (sgRNAs and Cas9 protein cf. [15]) to desired concentration with Gurdon's buffer (add lineage tracer, *see* **Notes 5** and **8**).
- Inject *Xenopus* embryos at 4-cell stage in 1/3 × MR + 2.5% Ficoll PM400 (Fig. 1b, c):
  - For in vivo or whole embryo analysis, inject 8–10 nL into ventral-right blastomere (*see* **Note 9**).
  - For generation of mucociliary organoids, inject 6–8 nL each into all four blastomeres.
- Place injected embryos on cooling plate or 12 °C before injecting the next condition to prevent high divergence in embryonic stages between conditions during later analysis (*see* **Note 10**).
- Transfer embryos to 1/3 × MR + Gentamycin 15 min—1 h after injection.
- Raise embryos to desired stage for preparation of organoids, imaging or fixation (Fig. 1a):
  - Stage 8 for generating organoids (*see* **Note 11**).
  - Stage 10 for analysis of early cell type specification.

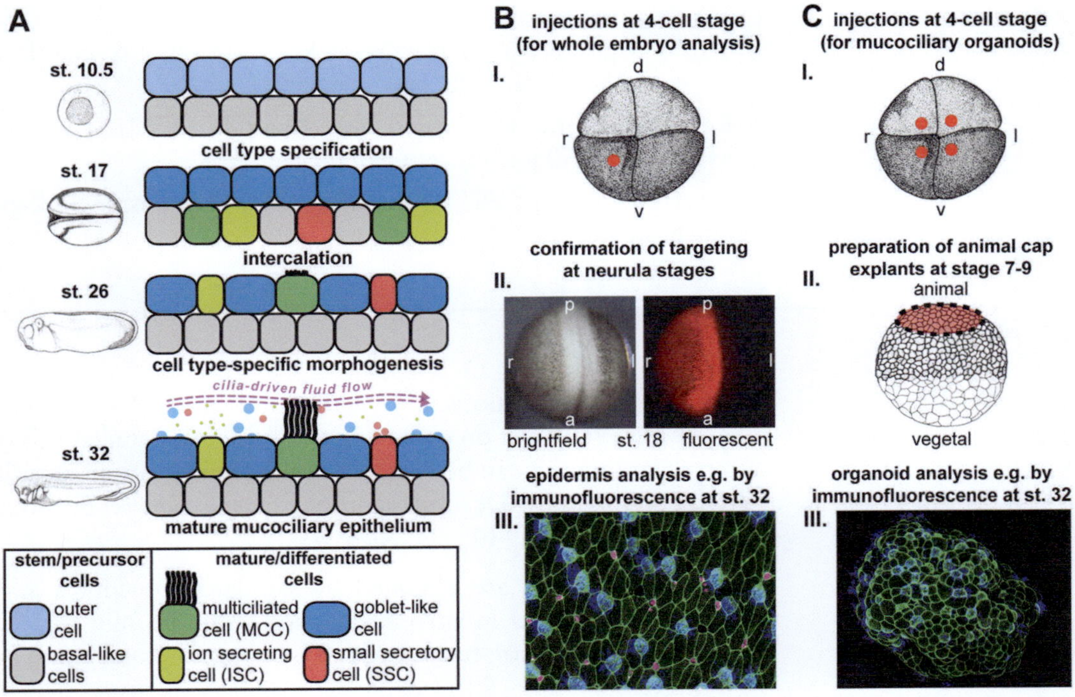

Fig. 1 (**a**) Schematic representation of mucociliary epidermis development and formation of cell types over developmental stages. (**b**) Workflow for manipulation and analysis of mucociliary cell type composition in whole embryos. (**b I.**) For unilateral targeting of the epidermis, inject into one of two ventral blastomeres (indicated by red dot). (**b II.**) Confirmation of successful targeting using Rhodamine-B dextran (red, in fluorescent channel). (**b III.**) Analysis of the epidermis using immunofluorescence and confocal microscopy (green = phalloidin-488 staining of Actin; blue = anti-acetylated-α-tubulin/Alexa-Fluor-405 antibody staining of cilia in MCCs; magenta = anti-serotonin/Alexa-Fluor-555 antibody staining of serotonin in SSCs). (**c**) Workflow for manipulation and analysis of mucociliary cell types in organoids. (**c I.**) For targeting the prospective animal ectoderm, inject in four of four blastomeres (indicated by red dots). (**c II.**) Preparation of animal cap explants by the removal of the animal ectoderm (indicated by red area). (**c III.**) Analysis of mucociliary organoids using immunofluorescence (green = phalloidin-488 staining of Actin; blue = anti-acetylated-α-tubulin/Alexa-Fluor-405 antibody staining of cilia in MCCs)

- Stage 15–17 for analysis of late cell type specification.
- Stage 22–25 for analysis of cell intercalation and early morphogenesis.
- Stage 28–30 for analysis of early cell type composition and mature MCCs/ISCs.
- Stage 32–35 for analysis of mature cell type composition and mature SSCs.

3.2 Analyzing Cell Type Composition by Immunofluorescence and Cell Staining

Fix embryos at desired stage with 4% PFA for 2 h at RT shaking or overnight at 4 °C (*see* **Note 12**).

- Wash embryos 3 times for 10–15 min in PBS at RT shaking.
- Wash embryos for 30 min in PBSt at RT shaking.

- Block embryos for 1 h using 1:10 CAS-Block/PBSt at RT shaking.
- Incubate in primary antibody over-night in 100% CAS-Block at 4 °C gently shaking (*see* **Note 13**).
- Wash embryos 2 times for 30 min in PBS at RT shaking.
- Incubate in secondary antibody in 100% CAS-Block at RT for >2 h shaking or overnight at 4 °C gently shaking (*see* **Note 14**).
- Wash embryos 2 times for 30 min in PBS at RT shaking.
- Incubate with another primary antibody following the above procedure or continue with cell stains.
- To stain the actin cytoskeleton, incubate in 1:40 dilution of phalloidin in PBSt for 15–30 min at RT shaking. Wash embryos 2 times for 15 min in PBS after staining (*see* **Note 15**).
- To stain goblet cells and SSCs, incubate in 1:1000 dilution of Lectin PNA for 60 min at RT shaking or overnight at 4 °C gently shaking. Wash embryos 2 times for 15 min in PBS after staining.
- After staining and washes are completed, mount embryos on microscope slide using high-vacuum silicone grease. For that, fill silicone grease into a 5 mL plastic syringe and place a continuous square ridge of silicon grease on microscope slide. Fill the mounting space with PBS, place embryos into PBS and cover with coverslip.
- Image epidermis using appropriate microscopy setup (e.g., confocal microscope) (Figs. 1b, c and 2).

3.3 Analyzing Cell Type Composition by In Situ Hybridization

Fix embryos at desired stage with MEMFA for 2 h at RT shaking or overnight at 4 °C.

- After fixation, transfer embryos into 100% ethanol, let them equilibrate for 15 mins at RT shaking, replace volume with fresh ethanol, and keep overnight at −20 °C (*see* **Note 16**).
- Follow standard in situ protocol for staining using anti-sense probes indicated in Subheading 2 (*see* **Note 17**).
- After staining is completed, stop staining by placing embryos into 100% methanol for 10–20 min at RT shaking (*see* **Note 18**). Replace volume with fresh methanol and keep embryos at −20 °C before imaging (Fig. 2b) (*see* **Notes 19** and **20**).

3.4 Generating Mucociliary Organoids

Animal cap explants are generated following standard procedures described in: Early Development of *Xenopus laevis*, by Sive, Grainger and Harland [14].

- Mucociliary organoids are generated from animal cap explants at stages 7–9 (*see* **Note 21**).

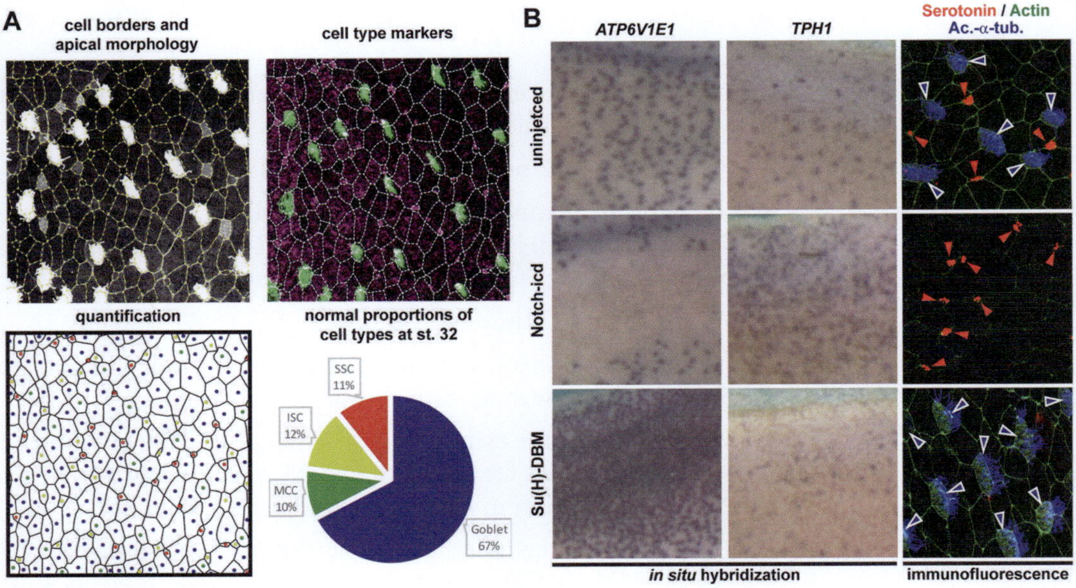

Fig. 2 (**a**) Guide for quantification of cell type composition in the mucociliary epidermis by immunofluorescence and cell staining. The st. 32 specimen was stained using a combination of phalloidin-488 (cell borders and apical morphology), anti-acetylated-α-tubulin/Alexa-Fluor-488 antibodies (cilia in MCCs), and lectin PNA-594 (mucus-like compounds in goblet cells and SSCs). Cell borders are outlined by phalloidin staining (white lines in the upper left panel). MCCs are characterized by dense cilia tuft (white in the upper left panel, green in the upper right panel). ISCs are characterized by small apical surface and lack of cilia staining (pale white in the upper left panel) as well as lack of PNA staining (magenta in the upper right panel). SSCs are characterized by small apical surface and large vesicle-like structures stained by PNA (magenta in the upper right panel). Goblet cells are characterized by large apical surface and small vesicle-like structures stained by PNA (magenta in the upper right panel). Lower left panel shows schematic representation of cell type analysis. Lower right panel shows the quantified proportions of cell types in a representative control sample at stage 32. (Image processing and outlining cell borders was performed using Image J software). (**b**) Examples of cell type markers in controls and after manipulation of the Notch signaling pathway using in situ hybridization and immunofluorescence. Uninjected = control; Notch-icd = gain of Notch signaling; Su(H)-DBM = loss of Notch signaling. ATP6V1E1 = marker for ISCs; TPH1 = marker for SSCs; Serotonin = marker for SSCs; Actin = phalloidin staining to visualize cell borders; Ac.-α-tub. = anti-acetylated-α-tubulin/Alexa-Fluor-405 antibody staining to mark cilia in MCCs

- Place embryos in agarose-coated dish filled with 1 × MBS.
- Remove fertilization membrane using sharp forceps. For that, turn the embryos vegetal side up and make an incision using the forceps. Hold embryos by gently pushing them down with the side on one pair of forceps, while grabbing the fertilization membrane with the other pair of forceps and pulling it away from the embryo.
- Turn embryos animal side up and use the forceps to cut a small piece of animal-most tissue (2 cell layers thick at this stage). Place explant tissue next to donor embryo (*see* **Note 21**).

- Let the explants "heal" for at least 30 min in 1 × MBS before transferring them (*see* **Note 22**).
- Transfer explants to 0.5 × MBS + Gentamycin using a p200 pipette and BSA-coated pipette tips (*see* **Note 18**).
- Raise explants to the desired stage, by monitoring the stage of control embryos from the same batch, which are kept at the same temperature as the explants.
- Organoids can be fixed for immunofluorescence and cell staining (cf. Subheading 3.2, Fig. 1c) or in situ hybridization (cf. Subheading 3.3) or used to extract total RNA for quantitative PCR or RNA-sequencing (10–20 organoids), to extract genomic DNA for ChIP-sequencing (100–250 organoids), or for the extraction of total protein content for proteomics experiments (300–500 organoids) (*see* **Note 23**).

4 Notes

1. All plasmids are originally based on pCS2+ and, thus, can be transcribed in vitro using commercially available SP6 polymerase-containing kits. Before transcription, plasmids should be linearized using restrictions enzymes cutting 3′ of the polyadenylation signal (often specified in the original papers).
2. MO sequences provided are specific to *Xenopus laevis*, for *Xenopus tropicalis* experiments MOs should be designed to match the species-specific sequence. A MO targeting *X. tropicalis* FoxA1 was published by Dubaissi et al. [16].
3. Sequences provided are specific to *Xenopus laevis*. Before transcription, plasmids should be linearized using restrictions enzymes cutting 5′ of the sequence (often specified in the original papers).
4. While one of the two forceps should be very sharp and used for cutting the explant, it is useful for the other forceps to be not as sharp, i.e., blunter, and to use those for grabbing the fertilization membrane and turning the embryo without harming the specimen.
5. It is advisable to co-inject either non-fixable fluorescent Dextrans (e.g., Rhodamine-B dextran, 0.5–1 mg/mL, Invitrogen) or fixable fluorescent mRNAs (e.g., membrane RFP, GFP, CFP at low dosage of 20–50 ng/μL) as lineage tracers to confirm successful injection and targeting (cf. [17–19]).
6. Mix oocytes gently every 2 min with 1 mL pipette tip or similar to allow the cysteine solution to remove all fertilization jelly.

Do not mix vigorously to prevent defects in dorso-ventral patterning.

7. At 12 °C–16 °C, it takes about 1.5 h until cleavage begins and embryos can be injected. This is a good time to start preparing injection solutions.
8. Concentrations of RNAs, DNAs, and MOs should always be determined experimentally in a preliminary experiment since synthesis quality, purity, and age of reagent can profoundly influence the efficiency. Nevertheless, the following concentrations should provide a useful starting point:
   (a) 80 ng/μL NICD mRNA—inhibition of MCCs/ISCs and enrichment of SSCs.
   (b) 120 ng/μL NICD mRNA—inhibition of MCCs/ISCs/SSCs.
   (c) 100 ng/μL SuH-DBM mRNA—increase of MCCs/ISCs and depletion of SSCs.
   (d) 100 ng/μL dn-MCI mRNA—depletion of MCCs and slight enrichment of ISCs.
   (e) 3–4 pmol FoxA1 MO—depletion of SSCs.
9. Unilateral injections have the benefit that the contralateral side can be used as internal control.
10. Until the embryonic epidermis has reached full maturity around stage 35, cell type composition and degree of cell type differentiation is quite dynamic. Thus, staging and synchronization of embryos is very important for correct interpretation of results.
11. For generating mucociliary organoids from animal cap explants, it is best to fertilize and inject in the afternoon and to keep embryos at 12 °C overnight to obtain stage 8 blastula embryos before noon the following day.
12. While 4% PFA works for all stains described in this protocol, in some cases, fixation with other reagents might be necessary, depending on the antibodies/cell stain used. An excellent summary for many MCC markers was provided by Werner and Mitchell [20].
13. The anti-acetylated Tubulin antibody is prone to attract cross-reaction of other primary antibodies (i.e., when tubulin staining is performed first, other primary antibodies applied in later steps might bind to the anti-acetylated Tubulin antibody and lead to an artifact staining of cilia leading to the impression that the protein localizes to axonemes!). Therefore, it is advised to use this antibody last when multiple antibodies are used in the experiment.

14. Always use a box or cover embryos with tin foil after the application of secondary antibody to prevent photobleaching of fluorophores.
15. Phalloidin staining is not compatible with previous washes in alcohols. The phalloidin solution can be taken off the embryos, collected and reused 2–3 times (store phalloidin solution in the dark at 4 °C between applications).
16. It is important to keep embryos in alcohol (methanol or ethanol) for at least 12 h at −20 °C as this step provides additional fixation. If this step is omitted, embryos might fall apart during hybridization at high temperatures. Both methanol and ethanol are routinely used to keep embryos before in situ hybridization, in my experience, ethanol has the additional benefit that it removes some of the lipids from the embryos, which is beneficial for preventing background staining during the staining protocol.
17. When staining embryonic stages that contain large interior cavities, e.g., blastocoel or gastrocoel, it is useful to "poke" the cavities open using sharp forces. Even small openings help with washing and prevent unspecific background staining.
18. Please note that embryos tend to be very sticky after in situ hybridization. Therefore, use agarose-coated plastic dishes or glass dishes for imaging and glass pipettes for transferring embryos. If no glass pipettes are available, make a 1% solution of Bovine serum albumin (BSA) in PBS and "coat" plastic pipettes by filling and emptying the pipette with coating solution 5–6 times. Following that, wash by filling and emptying the pipette with PBS 5–6 times before use.
19. If samples were overstained, embryos can be washed overnight in 100% methanol while shaking. This will wash out some of the excessive staining, without affecting the expression pattern.
20. It is advisable to bleach the pigment after the in situ staining to reveal full expression pattern. For bleaching of pigment protocol, see Early Development of *Xenopus laevis*, by Sive, Grainger, and Harland [14].
21. It is important to keep this time-window. If explants are generated too early, they might fall apart or become small, because of the large cell size in early embryos. If explants are generated too late, organizer-dependent induction of neural ectoderm in animal tissue might have taken place, thereby preventing development into a mucociliary epidermis (check regularly for dorsal-lip formation in embryos before and during animal cap preparation). Do not generate explants that are too big, since they might contain mesoderm (thicker than 2 cell layers) and not develop into epidermis either.

22. Animal cap explants are "healed" when they have rounded up and the area of white unpigmented inner cells has become very small. If explants are transferred too early, the explant could either fall apart in the low salt medium or break during transfer.

23. Information on a set of tested qPCR primers for cell type analysis in *Xenopus laevis* can be found in these publications [18, 19, 21].

Acknowledgment

PW and the Walentek lab are funded by the Deutsche Forschungsgemeinschaft (DFG) through the Emmy Noether Programme grant WA 3365/2-1. PW thanks the Xenopus community and resources for continuous support of his research work, in especially the National Xenopus Resource (RRID:SCR_013731) and Xenbase (RRID:SCR_003280).

References

1. Ibanez-Tallon I, Heintz N, Omran H (2003) To beat or not to beat: roles of cilia in development and disease. Hum Mol Genet 12:27R–235R. https://doi.org/10.1093/hmg/ddg061
2. Robinson M, Bye PTB (2002) Mucociliary clearance in cystic fibrosis. Pediatr Pulmonol 33:293–306. https://doi.org/10.1002/ppul.10079
3. Crystal RG, Randell SH, Engelhardt JF et al (2008) Airway epithelial cells: current concepts and challenges. Proc Am Thorac Soc 5:772–777. https://doi.org/10.1513/pats.200805-041HR
4. Walentek P, Quigley IK (2017) What we can learn from a tadpole about ciliopathies and airway diseases: using systems biology in Xenopus to study cilia and mucociliary epithelia. Genes J Genet Dev:1–13. https://doi.org/10.1002/dvg.23001
5. Brooks ER, Wallingford JB (2014) Multiciliated cells. Curr Biol 24:R973–R982. https://doi.org/10.1016/j.cub.2014.08.047
6. Deblandre GA, Wettstein DA, Koyano-nakagawa N, Kintner C (1999) A two-step mechanism generates the spacing pattern of the ciliated cells in the skin of Xenopus embryos. Development 126:4715–4728
7. Stubbs JL, Vladar EK, Axelrod JD, Kintner C (2012) Multicilin promotes centriole assembly and ciliogenesis during multiciliate cell differentiation. Nat Cell Biol 14:1–10. https://doi.org/10.1038/ncb2406
8. Quigley IK, Stubbs JL, Kintner C (2011) Specification of ion transport cells in the Xenopus larval skin. Development 714:705–714. https://doi.org/10.1242/dev.055699
9. Harland RM (1991) In situ hybridization: an improved whole-mount method for Xenopus embryos. Methods Cell Biol 36:685–695
10. Stubbs JL, Oishi I, Izpisúa Belmonte JC et al (2008) The forkhead protein Foxj1 specifies node-like cilia in Xenopus and zebrafish embryos. Nat Genet 40:1454–1460. https://doi.org/10.1038/ng.267
11. Walentek P, Bogusch S, Thumberger T et al (2014) A novel serotonin-secreting cell type regulates ciliary motility in the mucociliary epidermis of Xenopus tadpoles. Development 141:1526–1533. https://doi.org/10.1242/dev.102343
12. Hayes JM, Kim SK, Abitua PB et al (2007) Identification of novel ciliogenesis factors using a new in vivo model for mucociliary epithelial development. Dev Biol 312:115–130. https://doi.org/10.1016/j.ydbio.2007.09.031
13. Cibois M, Luxardi G, Chevalier B et al (2015) BMP signalling controls the construction of vertebrate mucociliary epithelia. Development: 1–12. https://doi.org/10.1242/dev.118679
14. Sive HL, Grainger RM, Harland RM (2000) Early development of Xenopus laevis. Cold

Spring Harbor Laboratory Press, Cold Spring Harbor

15. Campbell EP, Quigley IK, Kintner C (2017) Correction: Foxn4 promotes gene expression required for the formation of multiple motile cilia. Development 144:731–731. https://doi.org/10.1242/dev.149567
16. Dubaissi E, Rousseau K, Lea R et al (2014) A secretory cell type develops alongside multiciliated cells, ionocytes and goblet cells, and provides a protective, anti-infective function in the frog embryonic mucociliary epidermis. Development 141:1514–1525. https://doi.org/10.1242/dev.102426
17. Walentek P, Beyer T, Thumberger T et al (2012) ATP4a is required for Wnt-dependent Foxj1 expression and leftward flow in Xenopus left-right development. Cell Rep 1:516–527. https://doi.org/10.1016/j.celrep.2012.03.005
18. Walentek P, Beyer T, Hagenlocher C et al (2015) ATP4a is required for development and function of the Xenopus mucociliary epidermis—a potential model to study proton pump inhibitor-associated pneumonia. Dev Biol 408(2):292–304. https://doi.org/10.1016/j.ydbio.2015.03.013
19. Walentek P, Quigley IK, Sun DI et al (2016) Ciliary transcription factors and miRNAs precisely regulate Cp110 levels required for ciliary adhesions and ciliogenesis. eLife 5:1–24. https://doi.org/10.7554/eLife.17557
20. Werner ME, Mitchell BJ (2013) Using Xenopus skin to study cilia development and function. Methods Enzymol 525:191–217. https://doi.org/10.1016/B978-0-12-397944-5.00010-9
21. Song R, Walentek P, Sponer N et al (2014) MiR-34/449 miRNAs are required for motile ciliogenesis by repressing cp110. Nature 510(7503):115–120. https://doi.org/10.1038/nature13413

Chapter 19

Evaluating Blood Cell Populations in *Xenopus* Using Flow Cytometry and Differential Counts by Cytospin

Jacques Robert, Eva-Stina Edholm, and Francisco De Jesus Andino

Abstract

Analyses by flow cytometry and differential counts by cytospin of peripheral blood leukocytes are two reliable and convenient methods used to assess the immune status and immune responses in the amphibian *Xenopus*. Here, we describe these methods and discuss their challenges and limitations.

Key words Amphibian, Comparative immunology, Developmental immunology, Aquatic vertebrates

1 Introduction

As a representative jawed vertebrate, the amphibian *Xenopus* has an immune system remarkably conserved and similar to that of mammals (reviewed in [1]). In fact, the immune system of *X. laevis* is among the most extensively characterized in cold-blooded vertebrates. It is, as in other jawed vertebrates, composed of a thymus and spleen; T and B cells with Rag mediated rearranging T-cell receptors (TCR) and immunoglobulin (Ig) receptors, respectively; Activation-induced cytidine deaminase (AID) mediated hypermutation and Ig isotype heterogeneity; hallmark cytokines and MHC class I and class II genes, as well as most of the innate immune cells known in mammals; natural killer (NK) cells, monocyte/macrophages, neutrophils and other granulocytes [1]. In addition, *X. laevis* provides a versatile model for the study of MHC defined immune recognition, ontogeny, and phylogeny, as well as tumors, viral, bacterial and fungal immunity.

Among various ways to assess the immune status of an animal at steady state or during an immune response, collection of peripheral blood is convenient because it usually does not require euthanasia. Thus, analysis of peripheral blood leukocytes is particularly

Kris Vleminckx (ed.), *Xenopus: Methods and Protocols*, Methods in Molecular Biology, vol. 1865,
https://doi.org/10.1007/978-1-4939-8784-9_19, © Springer Science+Business Media, LLC, part of Springer Nature 2018

well suited for longitudinal studies. In *Xenopus* as in mammals, mature blood cells are categorized into two separate lineages: lymphoid and myeloid [2–4]. The lymphoid lineage consists of T, B, and natural killer (NK) cells. The myeloid lineage encompasses several morphologically, phenotypically, and functionally distinct cell types, including monocytes, macrophages, granulocytes (neutrophils, eosinophils, and basophils), megakaryocytes, and erythrocyte precursors. However, erythrocytes are by far the largest cell fraction in the blood and need to be removed for a better characterization of immune cells. In *Xenopus*, a convenient way to separate immune cells from erythrocytes (nucleated in *Xenopus*) is using a density gradient established by centrifugation with percoll (composed of colloidal silica coated with polyvinylpyrrolidone). To date, the method of preferential erythrocyte lysis in hypotonic buffer has proven unreliable in *Xenopus*. Another potential issue not encountered in mammals is the presence of nucleated thrombocytes in *Xenopus* [4]. Morphologically, these thrombocytes can easily be confused with leukocytes and their relative abundance varies among individuals, which can be a source of errors during analysis.

To characterize the different types of immune cells and determine their relative abundance, flow cytometry analysis provides a rapid and informative assessment of a large number of cells within a heterogeneous cell population. The combination of light scatter, which provides information about cell size and granularity with specific cell surface antigens, is a widely used application of flow cytometry, giving valuable insight into cell compositions during various conditions. In addition, flow cytometry can be used to measure a long and constantly expanding list of parameters including intracellular antigens, total DNA content (cell cycle analysis, cell kinetics, ploidy, etc.), phagocytosis, apoptosis, proliferation, viability, and enzymatic activity. In addition, based on the same principles fluorescent activated cell sorting (FACS) is used to physically sort specific cell populations, so as to purify populations of interest.

We describe below protocols adapted to *X. laevis* for purifying immune cells from peripheral blood and for performing flow cytometry analysis. However, given that the collection of *Xenopus*-specific monoclonal antibodies (mAbs) and polyclonal antibodies (pAbs) is still limited, morphological characterization and differential counting of leukocytes using cytospin and Giemsa staining remains useful in many immunological studies. Therefore, we also describe the basic protocol for cytospin and Giemsa staining generally applicable to *X. laevis, X. tropicalis* and other amphibian species.

2 Materials

2.1 Isolation of Lymphocytes

1. Amphibian (A) PBS buffer: Dissolve 8 g NaCl, 0.2 g KCL, 1.44 g Na_2HPO_4 and 0.24 g KH_2PO_4 in 800 mL of H_2O. Adjust the pH to 7.4 with HCl, and then add H_2O to 1 L. Adjust PBS to amphibian osmolarity (225 ± 5 mOsm/L) by the addition of sterile H_2O. Add 250 mL H_2O to 1 L PBS.
2. Heparin at 1000 IU/mL from suppliers.
3. Percoll.
4. Hemocytometer.
5. 0.4% Trypan Blue solution.
6. 15 mL conical tubes.

2.2 Flow Cytometry

1. Flow cytometry staining buffer (FCSB): Add 10 g BSA to 700 mL APBS. Stir well until dissolved. Add 50 μL of 1 M NaN_3 solution. Adjust volume to 1000 mL by adding additional APBS. Filter through 0.22 μm filter and store at 4 °C.
2. Centrifuge.
3. Microcentrifuge tubes (1.5 mL, flip-top).
4. FACS tubes (5 mL round-bottom polystyrene tubes).
5. Flow cytometer.
6. Anti-*Xenopus* mAbs ([5]; *Available anti-Xenopus antibodies can be found at the Xenopus laevis Research Resource for Immunobiology* https://www.urmc.rochester.edu/microbiology-immunology/xenopus-laevis.aspx) (*see* **Note 1**).
7. Fluorescent-conjugated secondary antibodies (*see* **Note 2**).
8. Fluorescent-conjugated streptavidine.
9. Propidium iodide: 1 mg/mL stock solution, 1–2 μg/mL final concentration (*see* **Note 3**).

2.3 Cytospin

1. Benchtop cytocentrifuge (e.g., Shandon Southern or Thermo Fischer Scientific).
2. Microscope.
3. Superfrost Micro slides (no special coating necessary).
4. Absorbing paper.
5. Plastic reservoir.
6. Diamond pen.
7. Coplin jars.
8. Squirt bottle.
9. 95% Ethanol.

10. APBS.
11. Permount mounting medium.
12. Giemsa Stain in buffered methanol solution.

3 Methods

3.1 Isolation of Lymphocytes from X. laevis Adult Frog's Blood

1. Collect blood in ice-cold APBS containing 100 μL sodium heparin (1000 IU/mL) as anticoagulant and keep on ice. If needed add APBS/heparin to the blood to make a final volume of 4 mL (*see* **Note 4**).
2. In a 15 mL centrifuge tube add 4 mL 51% Percoll (diluted in APBS). Carefully layer 4 mL of blood/heparin/APBS mixture on the top without disturbing the percoll layer 1:1 volume).
3. Centrifuge at 450 × *g* for 25 min at 4 °C. Red blood cells will pellet at the bottom of the tube (Fig. 1).
4. Collect the cloudy interface on the top of the percoll—this fraction contains the leukocytes- using a 1 mL pipet and transfer to a 1.5 mL microcentrifuge tube.
5. Pellet cells at 13,000 × *g* for 1 min at 4 °C. Resuspend the pellet in 1 mL APBS. Mix gently.
6. Repeat **step 5** to remove residual Percoll.
7. Perform a cell count and viability analysis using a hemocytometer and Trypan blue exclusion. Mix 1 volume (e.g., 50 μL) cell suspension with 1 volume of 0.4% Trypan blue. Load 10 μL of this mixture on an assembled hemacytometer. Count unstained and blue (dead) cells separately. Cell density (cell/mL) is determined by: Cell count × dilution factor × 10^4 (conversion of 0.1 μL into mL). Percent of cell death is determined

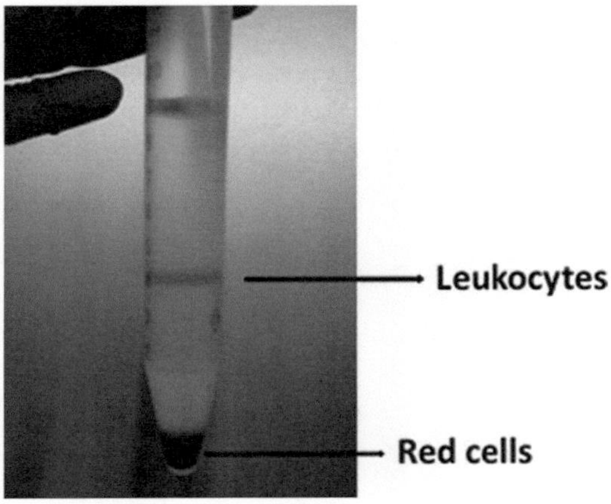

Fig. 1 Conical tube after centrifugation of *X. laevis* blood showing the band containing peripheral blood leukocytes and at the bottom erythrocytes

by: number of blue cells/number of blue cells + number of unstained cells.

3.2 Cell Surface Staining and Flow Cytometry

1. In addition to the sample of interest also prepare the following control samples: (1) unstained cells; and (2) cells incubated with an isotype control secondary antibody alone.
2. Dilute cells in flow cytometry staining buffer (FCSB) so that the final cell concentration is between 1×10^6 and 1×10^7 cells/mL. Aliquot 100 μL of cell suspension into each FACS tube (*see* **Note 5**).
3. Add 100 μL hybridoma supernatant (50 to 10 μg/mL of Ab) containing the primary antibody. Mix gently by pipetting.
4. Incubate on ice for 30 min.
5. Wash the cells by adding 1 mL FCSB to remove any unbound antibody (*see* **Note 6**). Pellet the cells by centrifugation at $13{,}000 \times g$ for 1 min at 4 °C. Discard the supernatant.
6. Resuspend the cells gently in 100 μL of FCSB by pipetting.
7. Dilute the appropriate fluorescent-conjugated secondary antibody 1:100 (0.1 to 1 μg/mL) in 100 μL of FCSB. Add to the cell suspension. Incubate for 30 min on ice. Protect from light.
8. Wash the cells by adding 1 mL FCSB to remove any unbound antibody. Centrifuge the suspended cells at $13{,}000 \times g$ for 1 min at 4 °C and discard the supernatant.
9. Resuspend the cells in 100 μL of FCSB.
   - If single staining is performed proceed to **step 19**
   - If double staining is performed proceed to **step 10**
10. Dilute the biotinylated primary antibody to 1 μg/mL in FCSB. Add 100 μL of the diluted primary antibody to the cells in **step 9** (*see* **Note** 7).
11. Incubate on ice for 30 min.
12. Wash the cells by adding 1 mL FCSB to remove any unbound antibody. Centrifuge the suspended cells at $13{,}000 \times g$ for 1 min at 4 °C and discard the supernatant. Resuspend the cells in 100 μL of FCSB.
13. Dilute fluorescent-conjugated streptavidine 1/100 in FCSB. Add 100 μL of diluted fluorescent-conjugated streptavidin to the cells in **step 12**.
14. Incubate on ice for 30 min.
15. Wash the cells by adding 1 mL FCSB to remove any unbound antibody. Centrifuge the suspended cells at $13{,}000 \times g$ for 1 min at 4 °C and discard the supernatant.
16. Resuspend the cells in 200–400 μL of FCSB.

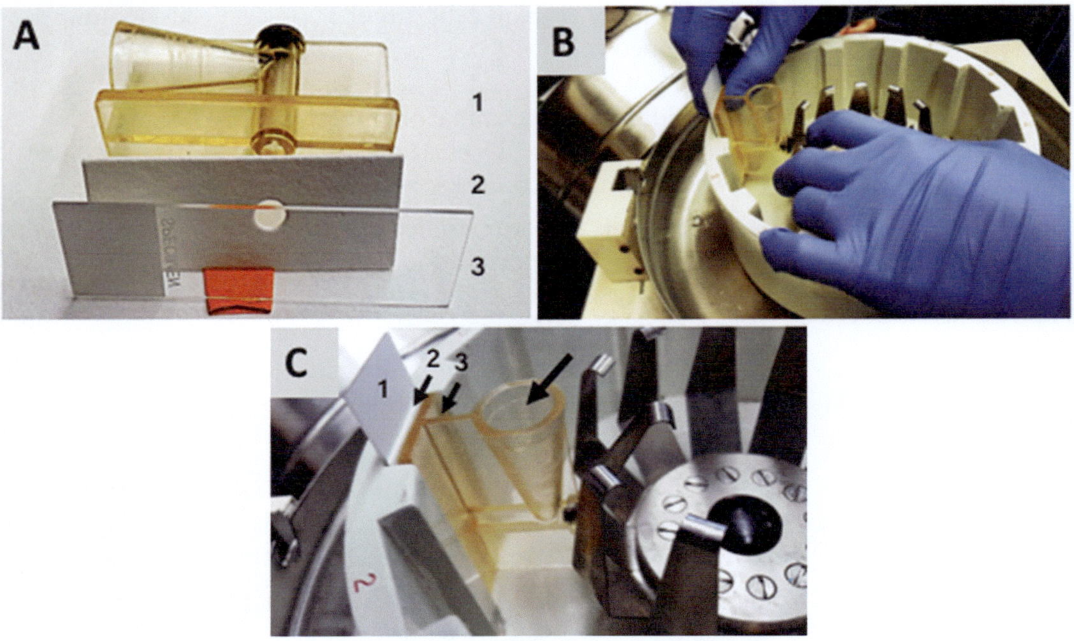

Fig. 2 View of the different parts of a cytocentrifuge. (**a**) The cytospin device is made by assembling a plastic reservoir (1) and an adsorbing paper glossy side up (2) to a microscope slide (3). (**b**) The cytospin design is inserted in the cytocentrifuge (check the holes of the adsorbing paper and plastic reservoir are aligned). (**c**) View of the installed cytopsin device. When all the devices are added, cell suspensions (200 μL) are loaded with a P200 pipetman (large arrow)

17. Just prior to analysis add 0.5 μL of propidium iodide stock solution (1 mg/mL) to each sample.
18. Analyze samples by flow cytometry (*see* **Note 8**).

3.3 Cytospin

1. For each slide use 200,000–400,000 total cells in a volume of 200 μL of APBS containing 10% BSA. Less cells can be used (20,000–40,000 cells), but the low density on the slide will be challenging for differential counting.
2. Assemble the cytospin device (microscope slide, adsorbing paper glossy side up, and plastic reservoir; Fig. 2). Use a pencil or an alcohol resistant marker to label the slides. Check that the holes of the absorbing paper and the plastic reservoir are aligned.
3. Centrifuge at 500–600 rpm (25–30 × *g*) for 5 min.
4. Disassemble the device, circle the cell spot on the microscope slide using a diamond pen.
5. Dry the cells for 20 min.
6. Fix and stain by adding a few drops of Giemsa solution directly on the cytospin and incubate for 1 min (cells are fixed by the methanol contained in the solution).

7. Differentiate by dipping the slides in a coplin jar filled with APBS diluted 2× with distilled water (pH = 7.0) and incubate for 5–10 min.
8. Quickly wash with distilled water, using a squirt bottle (avoid to flush directly on the cells).
9. Destain rapidly (a few seconds) with 95% EtOH, using a squirt bottle (avoid to flush directly on the cells).
10. Stop differentiation by washing with distilled water.
11. Check rapidly that cells are still stained under the microscope (do not let it dry). If the staining is too weak, repeat from point 6.
12. Let it dry for 5–10 min, then mount with Permount (*see* **Note 9**). In case the staining is not adequate (either saturated blue staining without differentiated red or to weak staining), it is possible to repeat by dissolving the Permount in Xylene, removing gently the coverslip, then transfer in ethanol (to remove the Xylene). The preparation can then be stained again following the steps from point 6.

4 Notes

1. Very few commercially available Abs cross-react with *Xenopus* (e.g., anti-human CD3 epsilon ABCAM). Fortunately, there is a collection of *X. laevis*-specific monoclonal recognizing surface receptors including CD5, CD8, CD45, IgM, IgY, MHC class I and II as well as less defined leukocyte markers. These mAbs are made available by the *Xenopus* Research Resource in Immunobiology. More information can be found online at https://www.urmc.rochester.edu/microbiology-immunology/xenopus-laevis.aspx).
2. As secondary reagents, fluorescently labeled goat- or rabbit anti-mouse isotype-specific secondary antibody is commercially available. It is to note that many secondary Ab exhibit some level of nonspecific cross-reaction with *Xenopus*. Therefore, it is recommended to pre-adsorb these reagents on *Xenopus* total blood cells (erythrocyte and leukocytes) or even better total splenocytes obtained by dilacerations of an adult spleen. This is done by mixing 1 volume of blood or splenic cells washed in APBS with 1 volume Ab for 1 h. on ice with occasional gentle mixing. Ab diluted 2× is recovered after pelleting cells by centrifugation. It is also possible to use biotinylated mAbs and use fluorescently conjugated streptavidine.
3. Propidium iodine is used as a viability marker.
4. Adult frogs can be bled either at the posterior tarsal vein [6] or by cardiac puncture [7]. Between 0.5 up to 1 mL can be easily

collected from medium size *X. laevis* adults (8–10 cm). For tadpoles, blood (0.1 mL) can be collected by heart puncture. Blood can also be collected from *X. tropicalis* by cardiac puncture [8].

5. 0.5 to 1 × 10^6 cells per sample is ideal for analyses but as little as 50,000 cells can be used.
6. When using a biotinylated primary antibody for the first time it is recommended to conduct a titration curve to determine which dilution allows for the strongest specific signal with the least background for your sample.
7. Background caused by nonspecific antibody binding can be minimized by increasing the concentration of BSA in the staining buffer to 2–5% during the step when the nonspecific antibody is added. One can also incubate cells with 0.5% *X. laevis* serum for 30 min on ice prior to staining.
8. In case the background caused by autofluorescence is too high and overlaps between two different fluorescent signals, it is recommended to use fluorescent probes that emit light farther away. For example, a FITCS probe (excitation at 493 nm, emission at 525 nm) could be combined with an Alexfluor probe that requires excitation with a different laser and emit at 645 nm and detection at 660 nm.
9. Differential count on cytospin is more cumbersome than flow cytometry, since it requires the counting of morphologically different cell type in randomly chosen quadrants and applying statistical evaluation. It is also possible to stain cytospins with other dyes including May Grünwald and hematoxylin eosine. It is also possible to use cytospins to stain with specific antibodies. In such a case, after drying, cells are fixed for 1 min with 3.7% formaldehyde, permeabilized with 100% cold methanol, and blocked with 1% bovine serum albumin (BSA) in PBS, before incubation with antibodies [9, 10].

Acknowledgments

The authors are supported by R24-AI-059830 and IOB-074271 from the NSF.

References

1. Robert J, Ohta Y (2009) Comparative and developmental study of the immune system in Xenopus. Dev Dyn 238(6):1249–1270. https://doi.org/10.1002/dvdy.21891
2. Iwasaki H, Akashi K (2007) Hematopoietic developmental pathways: on cellular basis. Oncogene 26(47):6687–6696. https://doi.org/10.1038/sj.onc.1210754

3. Rosenbauer F, Tenen DG (2007) Transcription factors in myeloid development: balancing differentiation with transformation. Nat Rev Immunol 7(2):105–117. https://doi.org/10.1038/nri2024
4. Hadji-Azimi I, Coosmans V, Canicatti C, Perrenot N (1987) Atals of adult Xenopus laevis hematology. Dev Comp Immunol 11:807–874
5. Robert J, Gantress J, Cohen N, Maniero GD (2004) Xenopus as an experimental model for studying evolution of hsp--immune system interactions. Methods 32(1):42–53
6. Nedelkovska H, Cruz-Luna T, McPherson P, Robert J (2010) Comparative in vivo study of gp96 adjuvanticity in the frog Xenopus laevis. J Vis Exp (43). doi:https://doi.org/10.3791/2026
7. Coosemans V, Hadji-Azimi I (1988) Immunoglobulin Fc receptor molecules on Xenopus laevis splenocytes. Immunology 65(4):641–645
8. Maxham LA, Forzan MJ, Hogan NS, Vanderstichel RV, Gilroy CV (2016) Hematologic reference intervals for Xenopus tropicalis with partial use of automatic counting methods and reliability of long-term stored samples. Vet Clin Pathol 45(2):291–299. https://doi.org/10.1111/vcp.12362
9. Chen G, Ward BM, Yu KH, Chinchar VG, Robert J (2011) Improved knockout methodology reveals that frog virus 3 mutants lacking either the 18K immediate-early gene or the truncated vIF-2alpha gene are defective for replication and growth in vivo. J Virol 85(21):11131–11138. https://doi.org/10.1128/JVI.05589-11
10. De Jesús Andino F, Chen G, Li Z, Grayfer L, Robert J (2012) Susceptibility of Xenopus laevis tadpoles to infection by the ranavirus Frog-Virus 3 correlates with a reduced and delayed innate immune response in comparison with adult frogs. Virology 432:435–443

Chapter 20

Isolation and Culture of Amphibian (*Xenopus laevis*) Sub-Capsular Liver and Bone Marrow Cells

Amulya Yaparla and Leon Grayfer

Abstract

The *X. laevis* sub-capsular liver is thought to be the principal hematopoietic site of *Xenopodinae* species from early development and, in case of certain species, into adulthood. The *Xenopus* bone marrow appears to be comprised of precursor cells committed to myeloid lineages, such as macrophage- and granulocyte-progenitor cells. With alarming increases in the contribution of pathogenic infections to the global amphibian declines, now more than ever a better understanding of the mechanisms controlling amphibian immune cell ontogeny and functionality is warranted. Accordingly, here we detail the isolation and culture of the *X. laevis* hematopoietic cells from the sub-capsular liver and bone marrow. Considering the immunological roles attributed to these amphibian organs, the respective cell isolation protocols described here will be pertinent to garnering further insights into the coordinated regulation of amphibian hematopoiesis and immune defense mechanisms.

Key words Bone marrow, Sub-capsular liver, Frog immunity, Amphibian hematopoiesis, Myelopoiesis, Leukocytes

1 Introduction

Different amphibian species appear to use distinct organs and tissues for hematopoiesis [1–4]. The amphibian sub-capsular (peripheral) region of the liver is considered to be the principal site for hematopoiesis from early stages of development and in some cases into adult life [5, 6]. Generally speaking, urodelian amphibians (newts and salamanders) and some of the more primordial anurans (frogs and toads, such as the *Xenopodinae* frog species) utilize their sub-capsular liver toward both larval and adult hematopoiesis, whereas more recently diverged anuran species appear to utilize their liver and/or kidney tissues for tadpole blood cell development, and their bone marrow for adult hematopoiesis [7]. Moreover, amphibian peripheral liver-mediated blood cell development may also be subject to seasonal variations, at least in some

Kris Vleminckx (ed.), *Xenopus: Methods and Protocols*, Methods in Molecular Biology, vol. 1865,
https://doi.org/10.1007/978-1-4939-8784-9_20,

species [1], adding another layer of complexity to the blood cell ontogeny of these complex organisms.

In more recently diverged species such as reptiles, birds, and mammals, the bone marrow is a soft and fatty tissue where blood cell development takes place [1–4]. Interestingly and as mentioned above, some amphibian species also utilize their bone marrow for hematopoiesis and thereby represent the earliest evolutionary origins of this site for blood cell development [1, 8]. Notably, the amphibians that tend to utilize their bone marrow rather than their peripheral liver toward hematopoiesis tend to be more terrestrial [2] and this evolutionary shift from hepatic- to bone marrow-hosted hematopoiesis coincides with frog development of more vascular innervation of their bone marrow [9]. In this respect, it is fascinating that caecilians (limbless amphibians), urodelians, and primitive anurans (frogs of the *Bombina* and *Xenopus* genera) have considerably more limited or no bone marrow hematopoietic activity and concomitantly possess very rudimentary vascularization of their bone marrow [9]. By contrast, more terrestrial anurans like the American bullfrog (*Rana catesbiana*) possess much more developed femoral vascularization, coinciding with more pronounced bone marrow-mediated hematopoietic activity [9].

With the many hematopoietic and immunological roles attributed to the sub-capsular liver of amphibian species, and considerably less know about the roles of the amphibian bone marrow, there is much left to be learned about the specific and coordinated regulation of these processes. Indeed, the amphibian liver serves as a model organ for several immunological and regenerative studies [10]. Owing to its roles in hematopoiesis, isolation of *X. laevis* sub-capsular liver cells provides a means to study various types of blood cells that belong to an array of cell lineages. Concurrently, the isolation and study of the *Xenopus* bone marrow cells may shed new light into the evolution of vertebrate blood cell development, and in particular, myelopoiesis.

Here, we described the isolation and culture of sub-capsular liver and bone marrow cells from the amphibian, *X. laevis*, which may be espoused toward an array of research purposes.

2 Materials

Prepare all solutions using ultrapure water and analytical grade reagents. Prepare and store APBS at room temperature and keep the media refrigerated until use. These and other related reagents and resources are available via the *Xenopus laevis* research resource for immunology: https://www.urmc.rochester.edu/microbiology-immunology/xenopus-laevis/protocols.

1. Amphibian Phosphate-Buffered Saline (APBS): 100 mM NaCl, 13 mM Na_2HPO_2, 1.5 mM KH_2PO_4 pH 7.7 (with NaOH). Sterilize by passage through a 0.2 μm filter.
2. Primatone stock: 10% (w/v) in water; aliquoted and frozen at −20 °C until use.
3. Penicillin-Streptomycin stock: 10,000 units in H_2O.
4. Gentamycin stock: 50 mg/mL in H_2O.
5. Mammalian Serum-Free Medium Recipe (MSF) (1 L): Iscove's Modified Dulbecco's Medium (powder), 100 units/mL Penicillin-Streptomycin, 0.1 mM Non-Essential Amino Acids Solution, 3μL/L *2*-Mercaptoethanol, 0.1% (w/v) Primatone, × 35 mM $NaHCO_3$, pH 7.7 (with NaOH). Sterilize over a 0.2-μm filter and store at 4°C for up to 3 months.
6. Complete-Amphibian Medium (C-AM): for 270 mL combine 200 mL MSF, 60 mL ultrapure water, 5 mL Fetal Bovine Serum, 5 mL Penicillin-Streptomycin stock, 80 μL Gentamycin stock. Mix well, sterilize by passage through a 0.2 μm filter, and store at 4°C for up to 3 months.
7. Tricane mesylate solution; 5% (w/v) buffered with 0.5 g/L Sodium bicarbonate.
8. 51% sterile Percoll solution (diluted with sterile APBS).
9. 70 μm cell strainer.

3 Methods

3.1 Dissection of X. laevis

1. Before beginning your frog dissection, make sure that you have thoroughly sterilized all dissection tools in 70% ethanol.
2. Euthanize the adult frogs by immersion in tricaine mesylate solution (*see* **Note 1**).
3. Place the euthanized frog on the dissection tray with ventral side up. Sterilize the animal with 70% ethanol.
4. Use forceps to lift the skin and abdominal muscles away from the body cavity and make an incision in the lower abdominal area (Fig. 1a).
5. Continue the incision laterally from both sides, around the center and proceed all the way up to generate a semi-circular opening (*see* **Note 2**).
6. Lift the skin flap to open up the body cavity and locate the liver lobes (Fig. 1a).

3.2 Isolation and Culture of X. laevis Peripheral Liver Cells

1. Lift one of the liver lobes with forceps and make an incision underneath. Collect the lobe into a 15 mL tube with 2 mL of sterile APBS over ice. Repeat the same procedure with the second lobe located on the other side (*see* **Note 3**).

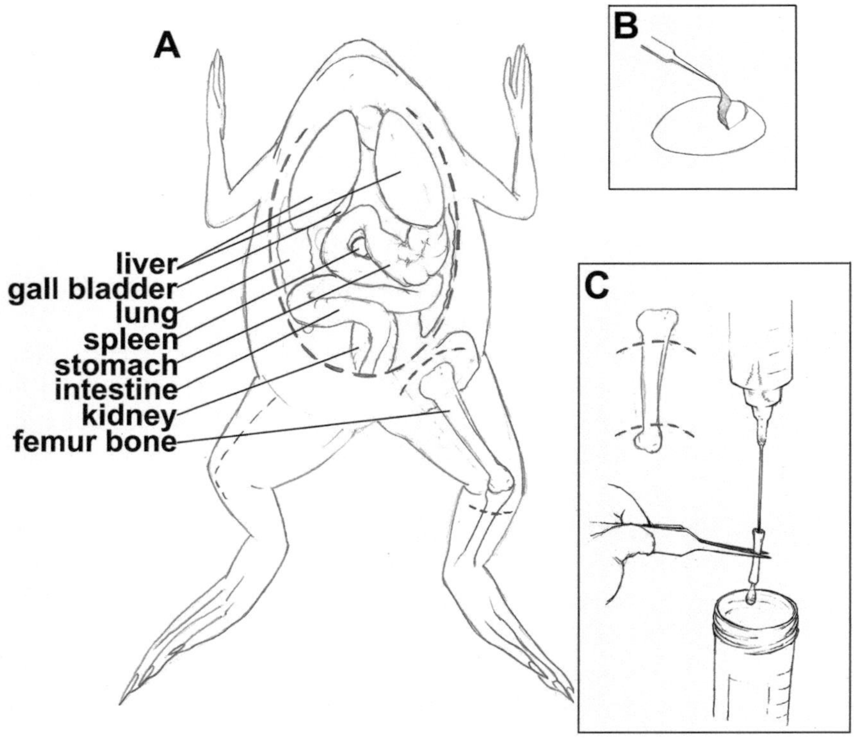

Fig. 1 Isolation of *Xenopus laevis* sub-capsular liver and bone marrow. (**a**) Cut along the frog abdomen (dotted line), keeping scissors pointing up, to reveal underlining tissues and organs, labeled. (**b**) Locate and remove the liver lobes. In a biosafety cabinet gently peel the peripheral liver. (**c**) Dissect out the femur bones and in a biosafety cabinet, cut the caps and flush the femurs with APBS

2. Proceed to the laminar flow hood to continue with the isolation procedure. Transfer the liver lobes, one at a time from the 15 mL tube onto a sterile six well plate or a petri dish.
3. Place a 70 μm cell strainer onto a 50 mL conical tube. With the help of sterile tweezers, carefully peel the outer layer of the liver and place the peeled fragments onto the cell strainer (Fig. 1b).
4. Using a sterile plunger from a 3 mL syringe, gently macerate the pealed layer of the peripheral liver against the cell strainer. Simultaneously, add sterile APBS to the strainer to drain the peripheral liver cells into the 50 mL tube, to a final volume of 10 mL. Keep the cells on ice until you are ready to proceed onto the next step.
5. Add 4 mL of 51% percoll to sterile 15 mL conical tubes. Gently layer the liver peripheral cell suspension from **step 4** onto the percoll using a sterile serological pipette. Be careful not to disturb the interface between the percoll and the cell suspension layers.

6. Centrifuge percoll/liver cell suspension at 600 × *g* for 25 min at 4°C. This will separate the hematopoietic cells at the interface with the percoll while pelleting the red blood cells (RBCs).
7. While being careful not to disturb the interface cell layer, remove the majority of the liquid (4–6 mL) up to the interface. Collect the cells at the percoll interface into a fresh 15 mL conical tube.
8. Wash the cells with 10 mL of cold APBS by centrifugation at 600 × *g* for 10 min at 4°C.
9. Gently decant the supernatant without disturbing the resulting cell pellets. Resuspend the cell pellets in 200–400 μL of C-AM medium and enumerate the cells by hemocytometer counts, preferably using a cell viability/exclusion stain such as trypan blue (*see* **Note 4**).
10. Culture the enumerated liver periphery cells in C-AM medium with treatment(s) of choice at 27 °C with 5% CO_2.

3.3 Isolation and Culture of *X. laevis* Bone Marrow Cells

1. Before beginning your frog dissection, make sure that you have thoroughly sterilized all dissection tools in 70% ethanol and have sterile, autoclaved paper towels or gauze on hand (sub-section 3.1, **step 4**).
2. To isolate the femur bones, cut the skin sideways along the upper thigh of the frog (Fig. 1a). Using a pair of tweezers and scissors, cut into the thigh until you locate the bone. Firmly grasp the femur bone with tweezers and cut alongside the bone to remove excess flesh surrounding it.
3. Separate the femur bone from the rest of the frog leg by cutting at the tibia, beneath the tibeofemoral joints (the frog knee) connecting the femur (Fig. 1a). Make sure that the femur cap remains intact.
4. Place the femur bone into a sterile (autoclaved) paper towel or gauze and gently rub the femur bones to remove remaining frog muscle, making sure not to directly touch the bone.
5. Once clean of flesh, transfer the femur into a sterile 15 mL conical tube containing 3 mL of APBS and place the tube on ice. Repeat **steps 1–3** for the other leg and collect both the femur bones from the same animal into the same conical tube.
6. Transfer the femur bones into a sterile six well plate or a petri dish in the laminar flow hood or biosafety cabinet (*see* **Note 5**).
7. Using a pair of tweezers, firmly hold the femur bone and cut the caps on either side (*see* **Note 6**).
8. Hold the cut femur bone vertically with tweezers over an opened 50 mL conical tube. Using a 20 mL syringe and a 24 gage needle, flush each bone with 10 mL of APBS (Fig. 1c).

9. Wash the cells with cold APBS by centrifugation at 600 × *g* for 10 min at 4°C.
10. Gently decant the supernatants without disturbing the pellets and resuspend the pellets in 200–400 μL of C-AM (*see* **Note 4**).
11. Enumerate the cells by hemocytometer counts, preferably using a cell viability/exclusion stain such as trypan blue.
12. Culture the enumerated bone marrow cells in C-AM at a desired cell density with treatment(s) of choice at 27 °C and 5% CO_2 incubator.

4 Notes

1. It will take 5–7 min to euthanize the frogs. You may reuse the euthanization solution for up to 6 months. When reusing this solution, make sure to adjust this solution to room temperature before euthanizing animals to prevent animal discomfort.
2. Take care to keep your scissors pointing upward to prevent rupture of animal organs.
3. Make sure to carefully remove the bile duct from the right lobe of the liver.
4. Ahead of establishing the cell cultures, adjust the culture medium to 22–27 °C.
5. Make sure that you have a set of sterile dissection tools ready in the hood in 70% ethanol.
6. Excess ethanol may be removed from dissection tools by immersion in sterile APBS.

Acknowledgment

AY thanks the GWU Department of Biological Sciences of Teaching and Research Stipends and thanks the Harlan Research Fund for summer research support. LG thanks the GWU Department of Biological Sciences for research support at the new investigator level. AY and LG thank Yaprak Ozakman for her assistance with generating the figures presented here.

References

1. Akulenko N (2012) Haemopoietic system of the anurans: the role of bone marrow and liver. Vestn Zool 46:28–35
2. Carver FJ, Meints RH (1977) Studies of the development of frog hemopoietic tissue in vitro. I. Spleen culture assay of an erythropoietic factor in anemic frog blood. J Exp Zool 201:37–46
3. Golub R, André S, Hassanin A et al (2004) Early expression of two TdT isoforms in the

hematopoietic system of the Mexican axolotl. Implications for the evolutionary origin of the N-nucleotide addition. Immunogenetics 56:204–213

4. Lane MC, Sheets MD (2002) Primitive and definitive blood share a common origin in Xenopus: a comparison of lineage techniques used to construct fate maps. Dev Biol 248:52–67
5. Nogawa-Kosaka N, Sugai T, Nagasawa K et al (2011) Identification of erythroid progenitors induced by erythropoietic activity in Xenopus laevis. J Exp Biol 214:921–927
6. Chen XD, Turpen JB (1995) Intraembryonic origin of hepatic hematopoiesis in Xenopus laevis. J Immunol 154:2557–6257
7. Lopez D, Lin L, Monaghan JR et al (2014) Mapping hematopoiesis in a fully regenerative vertebrate: the axolotl. Blood 124: 1232–1241
8. de Abreu Manso PP, de Brito-Gitirana L, Pelajo-Machado M (2009) Localization of hematopoietic cells in the bullfrog (Lithobates catesbeianus). Cell Tissue Res 337:301–312
9. Tanaka Y (1976) Architecture of the marrow vasculature in three amphibian species and its significance in hematopoietic development. Am J Anat 145:485–497
10. Manning MJ, Horton JD (1969) Histogenesis of lymphoid organs in larvae of the south African clawed toad, Xenopus laevis (Daudin). J Embryol Exp Morphol 22:265–277

Chapter 21

Isolation and Primary Culture Methods of Adult and Larval Myogenic Cells from *Xenopus laevis*

Kazi Taheruzzaman and Akio Nishikawa

Abstract

During amphibian metamorphosis, larval-to-adult conversion of the myogenic system occurs and there are two distinct types of muscle stem cells; larval myogenic cells have a death-fate by apoptosis in the presence of thyroid hormone T_3, and adult myogenic cells have a life-fate under the same conditions. Here, we describe isolation and culture methods for adult and larval myogenic cells from the frog, *Xenopus laevis*. Both types of cultured myogenic cells undergo cell division and cell differentiation, i.e., formation of multinucleated myotubes in serum-containing medium. Insulin-like growth factor-1 promotes cell division and differentiation in both cells. Basic properties and some applications of primary culture are also described.

Key words *Xenopus laevis*, Myogenic cells, Primary cell culture, Cell division, Cell differentiation, Thyroid hormone, Insulin-like growth factor-1, Death-fate, Muscle stem cells

1 Introduction

In our previous study [1], we first established a method for primary culture of two muscle stem cells with different developmental fates, i.e., adult and larval stem cells of *Xenopus laevis*. The reason for focusing on the two different stages in muscle, i.e., larval and adult muscles, is that during metamorphosis, the myogenic system converts from the larval type to the adult type [2–4]. During metamorphosis, the larval-type myogenic precursor cells and muscle fibers die under the influence of T_3, and adult muscle precursor cells respond to the same T_3 stimulus and proliferate, differentiating into adult-type muscles as an adult (frog) organ. By establishing the primary culture methods for larval and adult-type myogenic precursor cells, we can clarify the regulatory mechanism for muscle conversion from larval-type cells (with a death-fate) to adult-type cells (without such a fate).

The primary culture method for myogenic cells from many vertebrates usually involves enzymatic digestion and mechanical dissociation of muscle tissue fragments [5–10]. In the present

Kris Vleminckx (ed.), *Xenopus: Methods and Protocols*, Methods in Molecular Biology, vol. 1865,
https://doi.org/10.1007/978-1-4939-8784-9_21, © Springer Science+Business Media, LLC, part of Springer Nature 2018

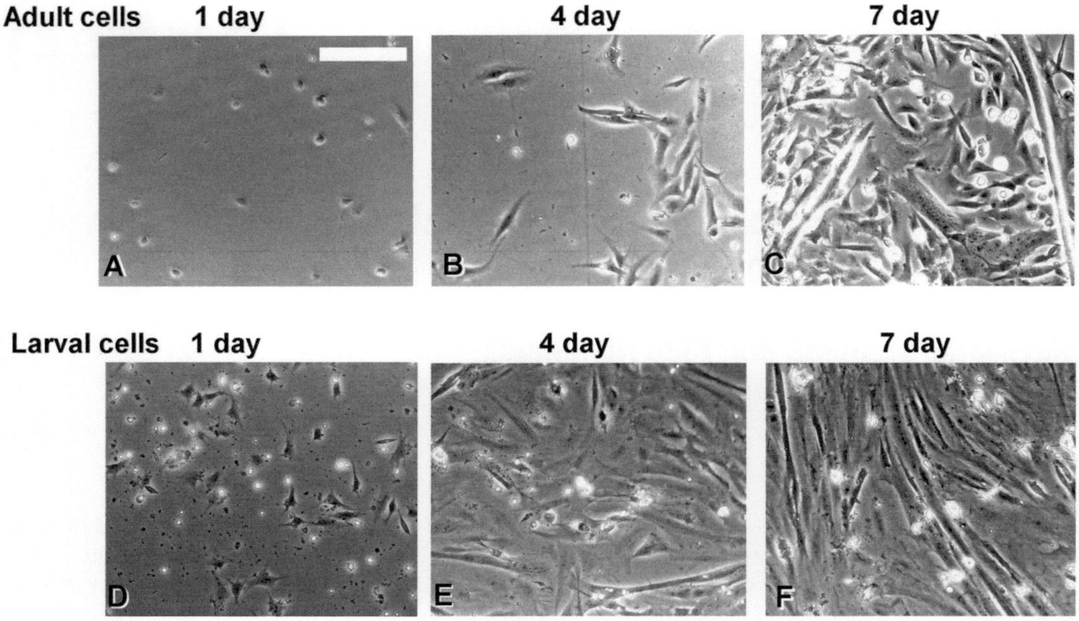

Fig. 1 Phase contrast photomicrographs of adult (**A–C**) and larval (**D–F**) myogenic precursor cells. They were cultured for 1 day (**A**, **D**), 4 days (**B**, **E**), and 7 days (**C**, **F**) on gelatin-coated well plates. The scale bar in (**A**) shows 100 μm

study, a similar principle was used for larval and adult myogenic cell cultures from the frog, *Xenopus laevis*. Similar to primary muscle cell cultures of mammals [5, 6], birds, [7], and fish [8–10], proteolytic enzymes such as collagenase and trypsin, and mechanical dissociation (up and down pipetting) were effective for isolating muscle stem cells from *Xenopus laevis* muscle tissues. Based on anti-desmin immunostaining results, the purity of *Xenopus laevis* myogenic cells was 88% for larval cells and 81% for adult cells [1]. The adhesion ratio of these cells to a collagen-coated culture dish on the first day was 30–50% for the larval type and 50–60% for the adult type. These adherent cells divide in the culture medium containing serum and cell fusion occurs; the resulting cells are differentiated into multinucleated myotubes (Fig. 1). Myotube differentiation was suppressed in adult-type cell cultures with a high serum content (40%) in the medium (Fig. 2), but such suppression did not occur in larval-type cells. Additionally, cultured adult-type cells were able to continue cell division and differentiation by detaching them from culture dishes using trypsin and subsequent subculturing, whereas there was no cell division and differentiation in larval-type cells in subculture. Therefore, it is considered that there is an essential difference between adult and larval muscle stem cells, and comparing the properties of both can clarify the developmental process of adult muscle stem cells.

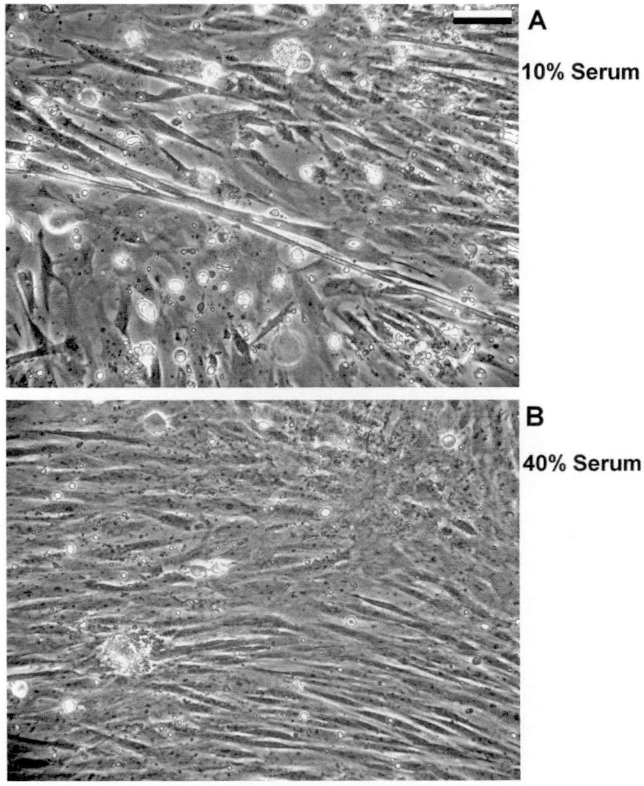

Fig. 2 High-serum suppression of myotube differentiation in adult myogenic cell cultures. Adult myogenic cells were cultured for 11 days in control medium (**A**) (70%DE + 10%RTS) or 40% serum medium (**B**) (70%DE + 40%RTS) and were photographed through a phase contrast micrograph. Myotube differentiation was suppressed in high serum condition (**B**). The scale bar in (**A**) shows 100 μm

Using the established culture methods for both myogenic cells, we characterized their basic and T_3 responses in vitro [1]. Thus, the behavior of both cell types in vitro was found to reflect exactly the difference in the "life-or-death fate," as described below.

1. The size of the cells attached to the culture plate is different between adult and larval myogenic cells. The diameter of the adult cells (5 μm) is smaller than that of the larval cell type (15 μm).
2. Timing of the initiation of differentiation in the culture period is different between adult and larval cells. Adult cells begin differentiation in vitro later (on day 4) than larval cells (on day 2).
3. Growth activity of myogenic cells is different between adult and larval cells. The adult cell number increases fourfold to fivefold during 1 week of culture period, while larval cell number increases twofold during the same period.

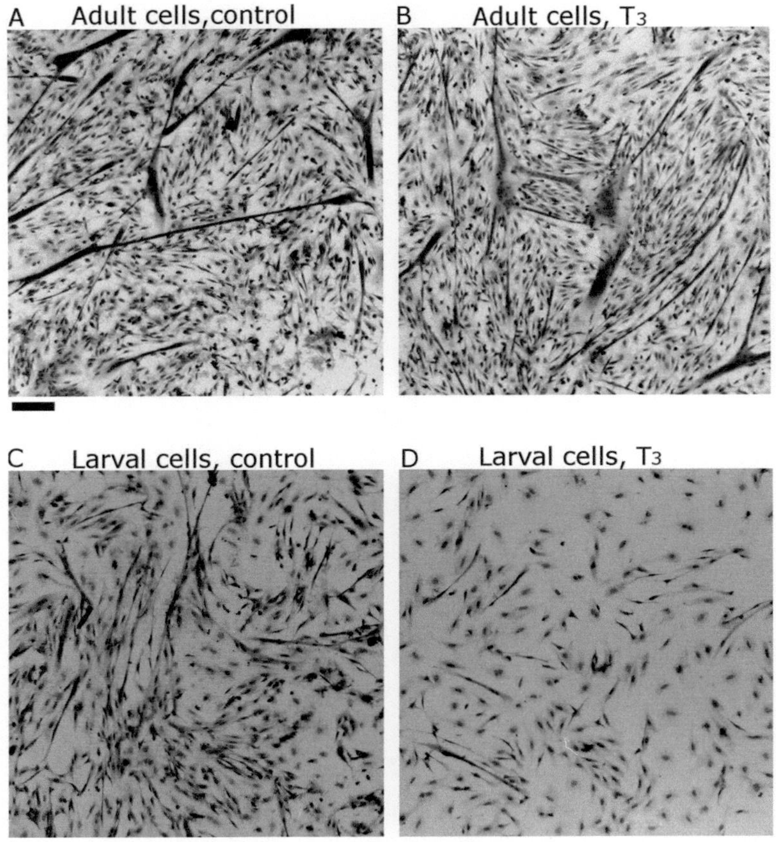

Fig. 3 Effect of thyroid hormone (T_3) on *Xenopus laevis* myogenic cells. (**A–D**) Giemsa-stained photomicrographs of cultured adult (**A**, **B**) and larval (**C**, **D**) myogenic precursor cells. They were cultured for 7 days in control medium (**A**, **C**) or T_3 (10^{-8} M) medium (**B**, **D**) and were fixed, stained, and taken photomicrographs. In T_3 medium the number of larval myogenic cells decreased in large scale but that of adult cells did not

4. Larval myogenic cells (myoblasts) respond to T_3 and die in culture unlike adult myogenic cells (Fig. 3).
5. Myotube death is promoted by T_3 only in the larval cell culture, but not in adult cell cultures.
6. T_3 increases the differentiation activity (accelerated myotube formation) in adult myogenic cells but decreases myotube formation activity in larval cells.

Previously, we have shown that the dorsal muscle during metamorphosis should also be examined in addition to tail and leg muscles. Dorsal muscles were found to convert the larval cell type to the adult cell type in a progressive anterior to posterior gradient of cell replacement, which consists of both larval muscle cell death and adult muscle differentiation [2]. Thus, it is reasonable to think that there should be two different types of myogenic stem cells within tadpole dorsal muscles (a larval type with a death-fate in response to T_3 and an adult type with a life-fate in response to T_3). Subsequent studies clarified that two types of muscle stem cells exist within the same region (i.e., dorsal muscles in the metamorphosing tadpole) [11]. In this study, cells were dissociated from tadpole dorsal muscles using the present method and then cell separation was performed using both a unit albumin-gravity sedimentation and Percoll-density gradient centrifugation [11]. The result showed that two different cell types, small and large types, were isolated from dorsal muscles at 1:1 (small: large) ratio. Cultivation of each of two types of dorsal muscle cells clarified that large cells could proliferate under T_3 but the number of small cells decreased under the same condition. These results suggested that small dorsal stem cells are a larval type with a "death fate" and large dorsal stem cells are a pre-adult type without a "death fate."

The present primary myogenic cell culture system used 10% fetal calf serum (FCS) in the culture medium because the serum contains many growth factors that are required for cell growth and differentiation. The percentage of serum in the medium seems to be critical for cell division and differentiation of adult and larval myogenic cells in vitro; no cell division and myotube differentiation occurred in low-serum medium (1% FCS). Detailed studies on mitotic suppression experiments using AraC (a DNA synthesis inhibitor) [12] showed that muscle differentiation occurs when cell division occurs, and thus, the dependency of differentiation on cell division is revealed. Insulin-like growth factor-1 (IGF-1), a major growth factor in the blood, was shown to promote the cell cycle and cell differentiation in the frog muscle stem cell culture system (Fig. 4). This differentiation-promoting action of IGF-1 was observed both in adult and larval cell types, but in the presence of thyroid hormone (T_3), the differentiation-promoting effect was specifically abolished in larval-type stem cells. Therefore, this is a useful approach to elucidate the detailed mechanism of IGF-1's action on muscle stem cells by comparing the difference in IGF responses in adult and larval-type muscle stem cells.

2 Materials

Culture medium and reagent solutions were prepared using deionized distilled water (DDW), i.e., ultra-pure water, and all chemicals were of analytical grade. Glassware should be washed carefully with detergent and running water and rinsed three times with DDW. Culture medium (Dulbecco's modified Eagle Medium) is used at 70% of the original concentration used for mammalian cells.

1. Antibiotic stock solution: Penicillin and streptomycin mixture (×1000). Add 10×10^6 units penicillin and 10 g streptomycin

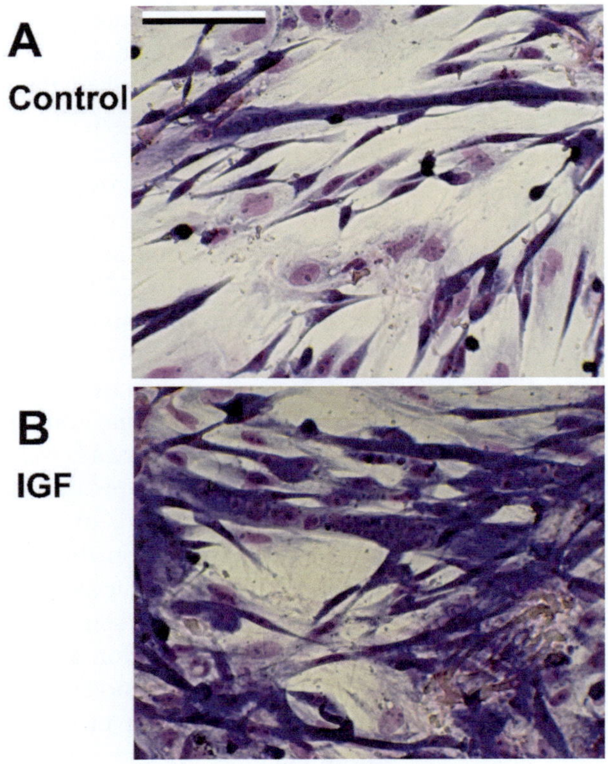

Fig. 4 Effect of IFG-1 on cell differentiation activity. Adult myogenic cells were cultured in control medium (**A**) (70% DE + 10%RTS) or in IGF-1 medium (**B**) (70% DE + 10%RTS supplemented with 100 ng/mL IGF-1) for 6 days, fixed and stained with Giemsa solution, and taken photomicrographs. The scale bar shows 100 μm. In IGF-1 medium, enhanced myotube differentiation occurred (**B**)

to 10 mL DDW. Sterilize with a 0.2 μm filter, aliquot, and store at −20 °C.

2. Basic culture medium (70% DE): 70% Dulbecco's modified Eagle medium at pH 7.4. Add 7 g powder of Dulbecco's modified Eagle medium (DMEM-low glucose) and 1 g sodium bicarbonate to a 1 L beaker containing 950 mL DDW. Mix and adjust pH with NaCl (1 N). Make up to 1000 mL with DDW. Sterilize by filtration with a 0.2 μm filter, add 1 mL sterilized antibiotic stock solution, and store at 4 °C, in an autoclaved glass bottle (*see* **Note 1**).

3. Resin-treated fetal calf serum (RTS): Heat-inactivate fetal calf serum (FCS; *see* **Note 2**) by incubating a bottle containing FCS at 56 °C for 30 min. FCS is subsequently treated with an anion exchange resin (AG-8, Bio-Rad, Hercules, CA) to remove thyroid and steroid hormones in the serum [13] . The resin (1.5 g) is washed three times with DDW and added to 30 mL heat-inactivated FCS in a 50 mL plastic tube. Gently

rotate the tube (resin-FCS) on a rotator at room temperature for 6 h. Centrifuge the tubes at 1500 × *g* for 10 min and the supernatant (FCS) should be transferred to a new 50 mL plastic tube with 1.5 g fresh washed-resin. After mixing for 18 h, the tube should be centrifuged at 1500 × *g* for 10 min and the supernatant transferred to a new 50 mL plastic tube. The resin-treated FCS (RTS) is then sterilized by filtration with a 0.2 μm filter and store at −20 °C until use.

4. Culture medium (70% DE + 10% RTS): Aseptically combine 90 mL sterilized 70% DE and 10 mL sterilized serum (RTS) in 100 mL glass bottle.
5. Anesthesia solution: 0.03% MS222 (Ethyl 3-aminobenzoate methanesulfonate salt). Weigh 0.03 g MS222 powder in 100 mL DDW and dissolve. Prepare just before use.
6. Amphibian phosphate-buffered saline (APBS): 10 mM Na_2HPO_4-NaH_2PO_4 buffer (pH 7.2), 0.1 M NaCl, 1 mM KCl, 1 mM $CaCl_2$, 1 mM $MgCl_2$. To a 1 L bottle, add 879 mL DDW, 20 mL 5 M NaCl, 40.5 mL 0.2 M Na_2HPO_4, 9.5 mL 0.2 M NaH_2PO_4, and 1 mL 2 M KCl. Sterilize by autoclaving. When the solution is cooled, add 50 mL autoclaved $CaCl_2$ and $MgCl_2$ mixture (20 mM each) under a laminar flow hood.
7. Ca^{2+}, Mg^{2+}-free Amphibian phosphate-buffered saline (APBS(−)): 10 mM Na_2HPO_4-NaH_2PO_4 buffer (pH 7.2), 0.1 M NaCl, 1 mM KCl. To a 1 L bottle, add 929 mL DDW, 20 mL 5 M NaCl, 40.5 mL 0.2 M Na_2HPO_4, 9.5 mL 0.2 M NaH_2PO_4, and 1 mL 2 M KCl. Sterilize by autoclaving and store at 4 °C.
8. Trypsin solution (0.25%): add 0.25 g trypsin powder (e.g., Trypsin 250, Difco) to 100 mL APBS(−) (pH 7.4) in a 100 mL beaker and mix using a magnetic stirrer at 4 °C. Sterilize with 0.2 μm filter, aliquot, and store at −20 °C.
9. Collagenase solution (0.3%): add 0.3 g collagenase powder (bacterial collagenase, e.g., Wako Pure Chemical Industries) to 100 mL 70% DE serum-free basic culture medium in a 100 mL beaker and mix with a magnetic stirrer at 4 °C. Sterilize with a 0.2 μm filter, aliquot, and store at −20 °C.
10. EDTA (200 mM): ethylenediaminetetraacetate (EDTA, pH 7). Add 200 mL 0.2 M EDTA·4Na and 100 mL 0.2 M EDTA·2Na to a glass bottle. Sterilize by autoclaving and store at room temperature.
11. Gelatin solution (0.01%): dilute a sterile tissue-culture grade 2% gelatin solution derived from bovine skin (e.g., from Sigma) with sterilized DDW. Store at 4 °C.
12. Insulin-like growth factor-1 (IGF-1) stock solutions: IGF-1 stock solution-1 (500 μg/mL). Add 0.1 mL sterilized DDW

to 50 μg IGF-1 powder (e.g., human recombinant from Sigma) in an original tube on ice. Let it stand for 20 min on ice, mix gently with a plastic 200 μL tip, and aliquot into plastic sterile tubes (10 μL/tube). Store at −20 °C until further dilution. IGF-1 stock solution-2 (5 μg/mL). Add 990 μL filter (0.2 μm)-sterilized 0.1% bovine serum albumin in APBS(−) to 10 μL IGF-1 stock solution-1, mix gently with a plastic 1000 μL tip, and aliquot into plastic sterile tubes (100 μL/tube). Store at −20 °C until use.

13. Nylon mesh filter (200 mesh): Cut a nylon mesh sheet (0.075 mm sieve opening, i.e., 200 mesh) into 6 cm squares, boil for 3 min, and wash them several times with DDW. Further cut the washed sheets into small squares (1.8 cm), wrap with aluminum foil, place them into a glass petri dish, and sterilize by autoclaving (121 °C, 20 min).
14. 70% EtOH: Ethanol is diluted with DDW to 70% (v/v).
15. Glassware: 10 cm glass Petri dishes and beakers (50 mL and 100 mL). Sterilize these glassware by dry-heating (180 °C, 2 h) before operation.
16. Microfuge tubes (1.5 mL) and 50 mL plastic tubes.
17. Forceps (No. 5) and surgical scissors (Mayo scissors and iris scissors).
18. Strainer (stainless steel).
19. Cell counting chamber (Thoma): Any type can be used.
20. Micropipettes.
21. Binocular dissection microscope.
22. Incubator (low temperature incubator) and shaker.
23. CO_2 incubator (low temperature incubator).
24. Cell culture plate: any size of cell culture plates (96-well, 24-well, 12-well, and 6-well) can be used.
25. Copper sulfate solution (10%): add 10 g copper sulfate pentahydrate to 90 mL DDW in beaker and mix using a magnetic stirrer. Stock in a glass bottle.

3 Methods

3.1 Dissection of Adult and Larval Muscles

All procedures after anesthetizing the frogs (or tadpoles) are to be performed in a clean room at low temperature (on ice or 4 °C). Leg muscles of small frogs are used as a source of adult-type myogenic cells, and tail muscles of tadpoles at pre-metamorphic stage are used as a source of larval-type myogenic cells.

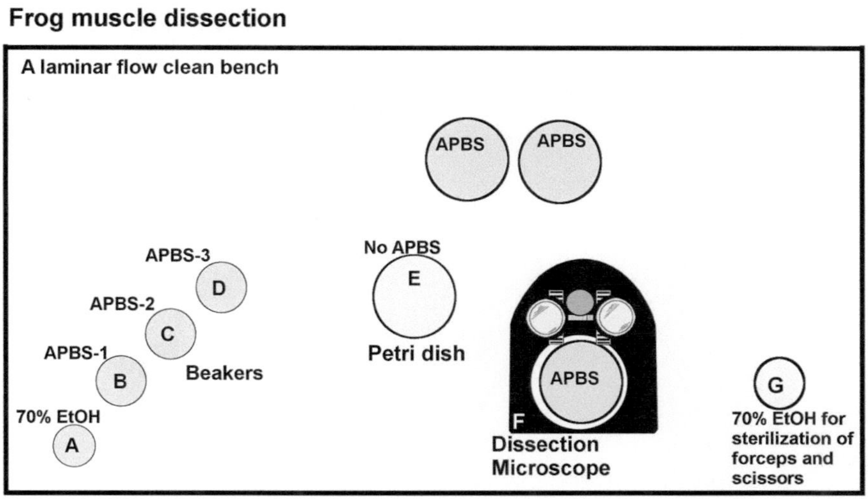

Fig. 5 The position of glasswares, solutions, and a dissection microscope on a laminar flow clean bench: An example for adult muscles dissection. Note that a Petri dish (E) does not contain APBS (No APBS)

3.1.1 Dissection of Froglet Hind–Limb Muscles

1. Wash 1–3 frogs (*see* **Note 3**) with running water for 5 min in a plastic case and then with deionized water for 2 min in a stainless steel strainer.
2. Put the washed frogs in a container with 100 mL 0.03% MS222 until they are anesthetized (about 10 min). Transfer the anesthetized frogs into a container with 500 mL cold water and then put these frogs in a Petri dish containing cold water on ice.
3. In a laminar flow hood, place two 50 mL beaker (containing 70% EtOH; Fig. 5A, G), three 100 mL beakers (containing cold PBS; Fig. 5B-D), four 10 cm Petri dishes (one without PBS and the three containing cold APBS), and a binocular dissection microscope (Fig. 5F). Carefully wipe the binocular microscope using cotton wool soaked in 70% EtOH.
4. Using forceps, soak anesthetized frogs in a 50 mL beaker containing 70% EtOH for 7 s, and then wash three times for 7 s each in 100 mL beakers containing APBS. Transfer the frogs onto a 10 cm Petri dish (without APBS; Fig. 5E). Cut off the frog's body-leg boundaries (Fig. 6A, B), and the ankles (Fig. 6C, D) using Mayo scissors and remove the leg skin with forceps (*see* **Note 4**), in such a way as to "take their pants off." Transfer the de-skinned legs into 10 cm Petri dish containing cold APBS, and dissect all leg muscles from the bones using iris scissors and forceps under a binocular microscope. Transfer the muscles into another 10 cm Petri dish containing cold APBS. Remove the nerve cord (white) and blood vessels (black) from muscle fragments, and mince the muscles into 1–2 mm pieces using iris scissors and forceps under a binocular microscope.

Fig. 6 Parts of frog legs which can be used for adult muscle dissection. Leg fragments between A (or B) and C (or D) can be used

5. After dissection, put all muscle fragments into three to six 1.5 mL plastic microfuge tubes containing cold APBS on ice using a micropipette with the tip end precut to 2–3 mm in diameter (*see* **Note 5**).

3.1.2 Dissection of Larval Tail Muscles

1. Judge the stage of tadpoles on the ice-cooled Petri dish under a binocular microscope and carefully select tadpoles at stage 53–54 (*see* **Note 6**). Wash the appropriate number of tadpoles with running water for 5 min in a plastic case and then with deionized water for 2 min in a stainless steel strainer.
2. Put the washed tadpoles in a container with 100 mL 0.03% MS222 until they are anesthetized (about 3 min). Transfer the anesthetized tadpoles into a container with 500 mL cold water and then put these frogs in a Petri dish containing cold water on ice.
3. In a laminar flow hood, place two 50 mL beakers (containing 70% EtOH; Fig. 7A, G), three 100 mL beakers (containing cold APBS; Fig. 7B-D), four 10 cm Petri dishes (containing cold APBS), and a binocular dissection microscope (Fig. 7F). Carefully wipe the binocular microscope with cotton wool soaked in 70% EtOH.
4. Use forceps to soak anesthetized tadpoles in a 50 mL beaker containing 70% EtOH for 7 s, and then wash three times for 7 s in 100 mL beakers containing APBS. Transfer the tadpoles into a 10 cm Petri dish (containing cold APBS). Cut the spinal cord (Fig. 8A), and cut off the body-tail boundaries (Fig. 8B) and the center part of the tail (Fig. 8C) using scissors and forceps. Transfer tail fragments (between B and C) into another

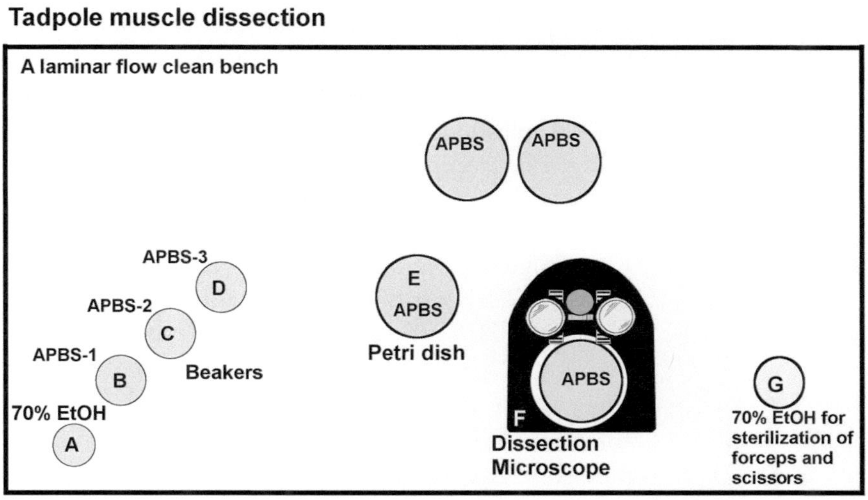

Fig. 7 The position of glasswares, solutions, and a dissection microscope on a laminar flow clean bench: An example for larval muscles dissection. Note that a Petri dish (E) contains APBS

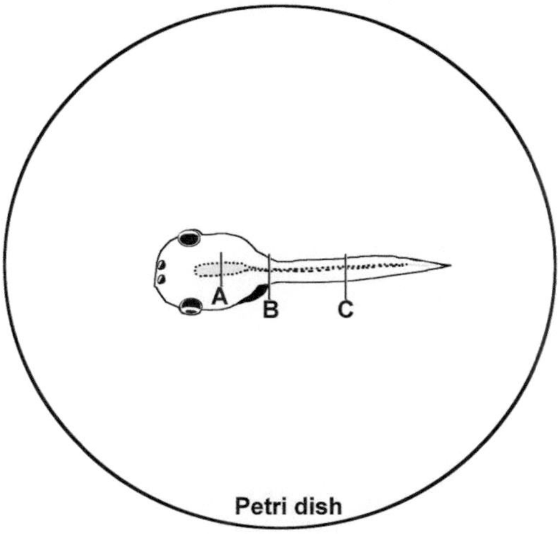

Fig. 8 Parts of tadpole tail which can be used for larval muscle dissection. A tail fragment between B and C can be used. Before cutting off the tail, the spinal cord should be cut at position A

Petri dish (containing cold APBS) and remove the tail skin using forceps. Transfer the de-skinned tails into another Petri dish (containing cold APBS), and dissect off all tail muscles from the central elements (notochord and spinal cord) using scissors and forceps under a binocular microscope. Transfer the muscles to another 10 cm Petri dish containing cold APBS. Remove the nerve cord (white) and blood vessels (black) from muscle fragments using forceps and micro-scissors.

Mince the muscles into 1–2 mm pieces using micro dissecting scissors under a binocular microscope.

5. After dissection, place all muscle fragments into one to three 1.5 mL plastic microfuge tubes containing cold APBS on ice using a micropipette with the tip end precut to 1–2 mm in diameter (*see* **Note 7**).

3.2 Collagenase Treatment of Dissected Muscles

1. Centrifuge the microfuge tubes containing minced muscle pieces at 240 × *g* for 30 s, discard the supernatant (APBS), add 1 mL cold 70% DE, and suspend the muscle pieces.
2. Centrifuge the microfuge tubes for 30 s at 240 × *g*, discard the supernatant (70% DE), add 0.6–0.8 mL 0.3% collagenase solution, and suspend the muscle fragments using a Pasteur pipette. Close the tube caps, place the tubes horizontally on a shaker, which is placed into an incubator, and incubate the tubes for 90 min at 26–27 °C with gentle shaking (*see* **Note 8**).

3.3 Gelatin–Coating the Culture Plates

During collagenase treatment, culture plates should be coated with gelatin using the following steps:

1. Pour 50 μL 0.01% gelatin solution into a 6 mm culture well ("96-well plate") in a laminar flow hood. The volume of the collagen solution depends on the size of the plate. For example, for a 35 mm well ("6-well plate"), the volume is 1.5 mL.
2. Close the culture plate cover and leave in a refrigerator (4 °C) for at least 1 h.
3. In a laminar flow hood, discard the collagen solution using a micropipette, add 50 μL basal medium (70% DE), and discard the medium before cell inoculation (*see* **Note 9**). In this manner, the cells can be inoculated onto the gelatin-coated plate (*see* Subheading 3.8, **step 2**).

3.4 Trypsin Treatment of Collagenase-Treated Materials

1. Centrifuge the microfuge tubes containing collagenase-digested materials for 2 min at 360 × *g*. Discard the supernatant (collagenase solution) using a Pasteur pipette.
2. Pour 0.8 mL trypsin solution (0.25%) into the microfuge tubes, and briefly suspend the partially digested tissues using a Pasteur pipette. Close the tube cap, place the tubes on a shaker, which is placed into an incubator, and incubate for 15 min at 26–27 °C. After incubation, the tubes containing trypsin-digested tissues are to be placed on ice.

3.5 Filtration of Digested Muscle Tissue Through Nylon Mesh

1. Place a nylon mesh (200 mesh) onto new 1.5 mL plastic microfuge tubes (*see* **Note 10**). Place one drop of culture medium (70%DE + 10% RTS) onto the nylon mesh surface, and carefully reverse the mesh (Fig. 9A) (*see* **Note 11**).

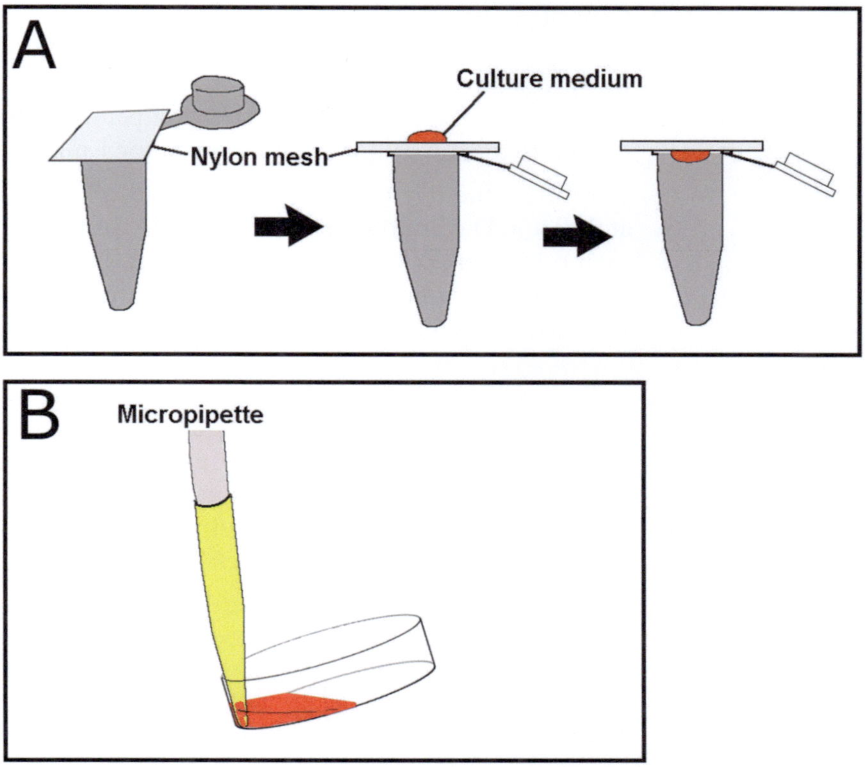

Fig. 9 How to set the nylon mesh on a 1.5 mL tube (**A**) and how to change medium with a micropipette (**B**)

2. Add about 0.5 mL culture medium to the microfuge tubes containing trypsin-digested tissues (*see* **Note 12**) (diluted sample volume becomes about 1.4 mL) and dissociate tissues by gently pipetting 30 times with a Pasteur pipette (diameter, 1 mm). Put several drops of dissociated tissue-materials on the nylon filter surface with a Pasteur pipette, wait for seconds the tissue drops passively to pass down through the nylon mesh filter. Repeat several times of these steps until the half volume (about 0.7 mL) of tissue-materials are passed down through a filter. Use a new filter-tube for the rest half volume of tissue-digested materials. Discard the nylon mesh which contains residual unfiltered materials, add 0.7 mL culture medium (70%DE + 10% RTS) into each tube, and gently mix the cell suspension. Place these tubes on ice.

3.6 Washing Dissociated Cells

1. Centrifuge the microfuge tubes containing filtered cells at 440 × *g* for 3 min and resuspend the cell pellet in 0.6 mL culture medium (70%DE + 10% RTS) ("washed cells-1"). If there are many tubes, combine cell suspensions of two tubes each into one tube. Place these tubes on ice for 1 min.

2. Centrifuge the microfuge tubes ("washed cells-1") for 2 min at 340 × *g*. Discard the supernatants and resuspend the cells using 0.6 mL culture medium (70%DE + 10%RTS) ("washed cells-2"). If there are four tubes, combine the cell suspension from the four tubes into two tubes. Let the tubes stand for 1 min on ice.
3. Centrifuge the microfuge tubes ("washed cells-2") for 2 min at 320 × *g*. Discard the supernatants and resuspend the cells in each tube using 0.52 mL culture medium ("washed cells-3"). Combine two tubes of "washed cells-3" into one tube (total volume is 1.04 mL). Place the tube on ice for 1 min.

3.7 Cell Counting

1. Cell sampling: disperse cells (washed cells-3) within the microfuge tube using a 1000 μL micropipette by pipetting three times, immediately take 40 μL of the cell suspension using a 200 μL micropipette, and put it into a new microfuge tube (*see* **Note 13**). Keep the rest of cell suspension (about 1 mL) on ice until it is used for cell inoculation.
2. Add 10 μL 200 mM EDTA to the cells (40 μL cell suspension in the tube) using a 200 μL micropipette. Mix the cell suspension well enough to dissociate cell aggregates into single cells by repeated pipetting (40 times up-and-down) (*see* **Note 14**). Be careful not to make bubbles in the tubes when pipetting.
3. Take 8 μL cell suspension from the tube after mixing 3–4 times using a micropipette. Pour the cell suspension (8 μL) onto a cell counting chamber (i.e., hemocytometer) and count the cell number under a phase-contrast microscope (*see* **Note 15**). Do not count ovoid cells (red blood cells) and black cells (pigment cells). Count round cells that have good integrity. Take 8 μL cell suspension from the same tube after mixing 3–4 times using a micropipette and count the cell number as above. Repeat the cell count again with a third 8 μL aliquot of cell suspension. Calculate the mean value of the three cell counts.

3.8 Inoculation and Culture of Adult and Larval Myogenic Cells

1. Transfer an appropriate number of cells (adult or larval myogenic cells; inoculum number of 2×10^4 cells for one 6 mm well) into a new 1.5 mL microfuge tube (or a 50 mL plastic tube). Add culture medium (70%DE + 10% RTS) into the microfuge tube to adjust final volume of culture medium (100 μL/2×10^4 cells/well) (*see* **Note 16**). Otherwise, when the cell density of the original cell suspension is lower than 2×10^4/100 μL, pellet the cells by centrifugation for 2 min at 290 × *g* and suspend them with an appropriate volume of culture medium (100 μL/2×10^4 cells).
2. Disperse the cells using a 200 μL micropipette and immediately inoculate 100 μL of cell suspension (2×10^4 cells) into

each 6 mm well coated with a gelatin solution (Subheading 3.3). Resuspend the cells before each-time inoculation and repeat the inoculation.

3. After inoculation, close the culture plates with the cover, slowly transfer the plates into a humidified incubator with 4% CO_2 and 96% air, and incubate them at 25 °C for 1–2 weeks.
4. Change used culture medium for fresh medium at 1 day (18 h), 4 days, and 7 days after inoculation. If the culture period lasts more than 1 week, change the medium every 3 days throughout the period (*see* below).

3.9 Culture of Adult and Larval Myogenic Cells

The first medium change (18–20 h after inoculation) should be done according to the following procedures:

1. Observe the cells under a phase contrast microscope to check for contamination. If bacteria or fungi are found on several culture wells, add 100 μL copper sulfate solution into the wells to kill the bacteria (or fungi). This process is important to prevent contamination from spreading to other clean cell culture wells.
2. Discard all culture medium in a culture plate using a 200 μL micropipette (or a Pasteur pipette; *see* Fig. 9B), and immediately pour 100 μL of basic medium (70% DE) along the inner surface of the outer wall of the culture plate.
3. Discard basic medium (70% DE) from the culture plate and immediately pour 100 μL of culture medium (70% DE + 10% RTS) along the inner surface of the culture plate outer wall.
4. If growth factors or hormones are to be added, add 1–2 μL stock solution (100× or 50×) of the factors (or hormones) into a well containing 100 μL culture medium. These factors should be added each time the culture medium is changed.
5. For second (or later) medium changes start again from **step 1**, but without washing with 70% DE.

4 Notes

1. Freshly prepared culture medium can be used for up to 3 months if stored at 4 °C.
2. Checking the serum lot is an important point. When starting the experiment, examine cell adhesiveness, cell toxicity, and cell growth activity using several samples of different FCS lots and select the best FCS lots for myogenic cell cultures.
3. Young frogs (*Xenopus laevis*; 2.5–3 cm in body length) are used for primary culture of adult-type myogenic cells. Feed the frogs every day for 1 week before dissecting the muscles.

4. During the procedure, scissors and forceps should be cleaned occasionally by dipping them in 70% EtOH (Fig. 5G) and then drying the residual alcohol.
5. Usually muscle fragments from one frog (two legs) are collected into one to two 1.5 mL microfuge tubes.
6. Tadpoles at stage 53–54 of *Xenopus laevis* were grown from fertilized eggs in a laboratory room at 23 °C and used for primary culture of larval-type myogenic cells. The tadpole stage was judged by Nieuwkoop and Farber's criteria [14]. Tadpoles must be fed every day for 2 weeks before the procedure. The fastest growing group of tadpoles should be used (about 4–5 weeks after fertilization) to get good results in the cellular response (e.g., response to T_3, adhesiveness, cell growth, and differentiation activities). If 7–30 tadpoles are used, the cell yield should be at least $30–120 \times 10^4$.
7. Usually muscle fragments from tails of 10 tadpoles are collected into one 1.5 mL microfuge tube.
8. The microfuge tubes should be placed horizontally.
9. These gelatin-coated plates can be prepared 1 day before use.
10. Use two new microfuge tubes for one tube containing trypsin-digested material.
11. These steps (wetting a filter with culture medium and up-side-down reversion of the filter) are important steps for the digested materials easily to pass down through the filter.
12. Addition of 0.5 mL culture medium is needed to decrease the viscosity of trypsin-digested materials.
13. The important point for cell sampling is that sampling must be done quickly when cell density distribution becomes equalized all over the cell suspension, during pipetting. For the quick manipulation, therefore, two types of pipet tips (200 μL and 1000 μL) should be set on two micropipetters, before starting sampling.
14. Dissociation of cells into a single cell is important for exact estimating the cell number. Because sampled cells usually contain cell aggregates, complete separation of cells must occur by repeated pipetting in medium containing EDTA (final 40 mM). Cell aggregates can be seen through the tube in transmitted light. So, repeat pipetting until you cannot see cell aggregates.
15. If a hemocytometer chamber is too thick to adjust focus, invert the hemocytometer to make it possible to adjust the focus through the apparatus' cover glass side.
16. The recommended appropriate inoculum numbers and culture medium volumes are summarized in Table 1.

Table 1

Recommended appropriate inoculum cell numbers and volumes of culture medium

Type of culture plate	Diameter (mm)	Plate plane area (cm^2)	Relative area (96-well = 1)	Inoculum cell number (per well)	Volume of culture medium (mL/well)
96-well	6	28.27	1	$1–2 \times 10^4$	0.1
24-well	15	176.71	6.25	$6–12.5 \times 10^4$	0.6
12-well	21	346.36	12.25	$12–25 \times 10^4$	1.2
6-well	35	962.11	34.03	$34–70 \times 10^4$	3

Acknowledgment

This work was supported by a Shimane University Strategic Grant for Innovative Research (Functional Healthy Rice Project).

References

1. Shibota Y, Kaneko Y, Kuroda M et al (2000) Larval-to-adult conversion of a myogenic system in the frog, *Xenopus laevis*, by larval-type myoblast-specific control of cell division, cell differentiation, and programmed cell death by triiodo-L-thyronine. Differentiation 66:227–238
2. Nishikawa A, Hayashi H (1994) Isoform transition of contractile proteins related to muscle remodeling with an axial gradient during metamorphosis in *Xenopus laevis*. Dev Biol 165:86–94
3. Nishikawa A, Hayashi H (1995) Spatial, temporal and hormonal regulation of programmed muscle cell death during metamorphosis of the frog *Xenopus laevis*. Differentiation 59:207–214
4. Nishikawa A, Murata E, Akita M et al (1998) Roles of macrophages in programmed cell death and remodeling of tail and body muscle of *Xenopus laevis* during metamorphosis. Histochem Cell Biol 109:11–17
5. Soriano-Arroquia A, Clegg PD, Molloy AP et al (2017) Preparation and culture of myogenic precursor cells/primary myoblasts from skeletal muscle of adult and aged humans. J Vis Exp (120):55047. https://doi.org/10.3791/55047
6. Danoviz ME Yablonka-Reuveni Z (2012) Skeletal muscle satellite cells: background and methods for isolation and analysis in a primary culture system. Methods Mol Biol 798:21–52
7. Schmid C, Steiner T, Froesch ER (1983) Preferential enhancement of myoblast differentiation by insulin-like growth factors (IGF I and IGF II) in primary cultures of chicken embryonic cells. FEBS Lett 161:117–121
8. Froehlich JM, Seiliez I, Gabillard J-C, et al (2014) Preparation of primary myogenic precursor cell/myoblast cultures from basal vertebrate lineages. J Vis Exp (86):e51354. doi:https://doi.org/10.3791/51354
9. Gabillard JC, Sabin N, Paboeuf G (2010) In vitro characterization of proliferation and differentiation of trout satellite cells. Cell Tissue Res 342:471–477
10. Mulvaney DR, Cyrino JEP (1995) Establishment of channel catfish satellite cell cultures. Bas Appl Myo 5:65–70
11. Shimizu-Nishikawa K, Shibota Y, Takei A et al (2002) Regulation of specific developmental fates of larval- and adult-type muscles during metamorphosis of the frog *Xenopus*. Dev Biol 251:91–104
12. Miyata S, Yada T, Ishikawa N et al (2017) Insulin-like growth factor 1 regulation of proliferation and differentiation of *Xenopus laevis* myogenic cells in vitro. In Vitro Cell Dev Biol-Anim 53:231–247
13. Samuels HH, Stanley F, Casanova J (1979) Depletion of L-3, 5, 3′-triiodothyronine and L-thyroxine in euthyroid calf serum for use in cell culture studies of the action of thyroid hormone. Endocrinol 105:80–85
14. Nieuwkoop PD, Faber J (1967) Normal table of *Xenopus laevis* (Daudin). North-Holland Publisher, Amsterdam

Index

A

Amphibian 133, 148, 164, 175, 195, 244, 252, 265–267, 275–280, 289
Amphibian hematopoiesis 275, 276
Aquatic recirculating systems 2, 4, 5, 7, 10
Aquatic vertebrates 265
Axon navigation 120, 126

B

BATCH-GE 68, 79, 80, 83–90, 153, 158
Bone marrow 275–280
Brain 120–122, 124–127, 129, 130, 157, 158, 217, 218, 220–222, 229, 233–235, 237–240, 243–249

C

Calcium imaging 218, 219, 225–227
Cancer .... v, vi, 34, 35, 49, 147, 243
Cell ablation 134, 141–144
Cell cycle .... v, 134, 243–249, 266, 287
Cell differentiation 243, 260, 284–288, 298
Cell division 35, 56, 119, 284, 287
Cell imaging 252
Cilia 254, 256, 258, 261
Clustered Regularly Interspaced Short Palindromic Repeats (CRISPR) .... v, 34, 39, 43, 47, 55, 56, 67–81, 83–85, 91, 92, 94, 96, 97, 105–116, 147, 164, 252, 255
Comparative immunology 265
Congenital heart disease (CHD) .... vi, 34, 164
CRISPR/Cas9 .... v, vi, 33–51, 55, 56, 67–81, 83–85, 91, 92, 94, 96, 97, 105–116, 147, 164, 252, 255
Cutsite 84–85, 87–89

D

Data-analysis 83–90, 199, 246
Development .... v, vi, 1, 9, 16, 19, 22, 26, 29, 33–35, 56, 63, 92, 119, 120, 133, 144, 150, 151, 153, 164, 175–177, 196–199, 205, 207–211, 220, 233, 243, 244, 251, 252, 255, 256, 258, 261, 275, 276, 283, 284
Developmental immunology 265
Disease model .... v–vi, 34, 147

E

Electroporation .... vi, 119, 120, 122, 126, 217
Epidermis 22, 125, 251–262

F

F0 .... 34, 35, 47–49, 56, 67–80, 105–116, 148, 151, 159, 164
False discovery rate (FDR) 196, 199, 204, 205
Flow cytometry 243–249, 265–272
Fluorescence 94–96, 101, 140, 156, 219, 220, 226, 227, 229, 239
Fluorescent dyes 50, 220, 225, 229
Frog immunity 275
Fructose-based high-refractive index solution (fbHRI) 234, 236, 238, 239

G

Gene set enrichment 195
Gene symbols 198, 199, 208–209, 211
Genome editing .... v, vi, 33–35, 37–40, 47, 48, 56, 62, 67–75, 80, 83–92, 105, 106, 147, 150–152, 173, 252
Genotyping .... v, 35, 47, 49, 67–81, 94, 99–100, 108, 113–114, 148, 152–153
Goblet cell 254, 257, 258
Golden Gate cloning 57, 60
Guide RNA (gRNA) 34–51, 70, 150

H

Heteroduplex mobility assay (HMA) 67, 70–72, 80, 83
High resolution melting analysis (HRMA) 67, 68, 75–78, 80
Homology directed repair (HDR) 84, 88, 89, 105, 106, 110, 115
Human chorionic gonadotropin (hCG) 20, 23, 58, 62, 111, 139
Human disease modeling .... v, 1, 33, 34, 147
Husbandry .... v, 1–16

I

In vitro fertilization (IVF) 19, 22, 24–26, 28, 31, 48, 176, 195, 255
In vivo imaging 126
Ionocyte 253

Kris Vleminckx (ed.), *Xenopus: Methods and Protocols*, Methods in Molecular Biology, vol. 1865,
https://doi.org/10.1007/978-1-4939-8784-9, © Springer Science+Business Media, LLC, part of Springer Nature 2018

K

k-means clustering............197, 209–211
Knock-in............v, 34, 57, 91–101, 105–116

L

Leukocytes............265, 266, 268, 271
Linux............84, 86, 89
Living imaging............119

M

Mass spectrometry............49, 176, 180, 186, 189, 198, 212
Mass spectrum............196, 204
MaxQuant............198–205, 207–208, 211, 212
Microhomology mediated end joining repair (MMEJ)............38, 40, 91–101
Microinjection............26, 27, 30, 35–37, 42–49, 51, 56, 58, 62, 119, 150, 151, 164, 217, 252–253, 255
Mitral-tufted cells............218, 220, 223–229
Mucociliary............251–262
Multiciliated cell (MCC)............251, 253, 254, 256, 258, 260
Muscle stem cells............283, 284, 287
Mutagenesis efficiency............87, 89
Myelopoiesis............276
Myogenic cells............283

N

Neural wiring............120
Neurodevelopment............233
Neuronal tracing............217
Next-generation sequencing (NGS)............80, 83–90
Nitroreductase/metronidazole (NTR/MTZ)............134

O

Olfactory receptor neurons............218, 221, 226, 227, 229
Olfactory system............218, 221, 225
Oocyte............23, 25, 31, 106–108, 110–113, 115, 116, 175, 176, 196, 255, 260
Optical coherence tomography (OCT)............164, 165, 167–173
Optic tract............120, 126
Ovine luteinizing hormone (oLH)............20, 23

P

Photoreceptors............133, 138, 140, 142
Primary cell culture............vi, 284, 287
Protein dynamics............177, 196, 199
Proteomics............175, 195, 252, 259

Q

Quantitative multiplexed proteomics............197, 199

R

Regeneration............77, 133, 134, 144, 243
Retina............120, 122, 124–126, 129, 133, 138, 140, 142
Ribonucleoprotein (RNP)............37, 46–49, 91, 150

S

Sample preparation............69, 70, 84–85, 177, 195, 197, 200, 202, 205, 207
Single cell morphology............225
Single guide RNA (sgRNA)............94, 96–100, 105, 107, 109, 111, 112, 114, 115, 164–167, 173, 255
Small secretory cell............251
Sub-capsular liver............275–280

T

TALEN design............v, 57–60, 94
Tandem mass tags (TMT)............177, 179, 184–185, 190, 196, 197, 199–201, 203, 205, 207–208, 211–213
3D-Imaging............234
Thyroid hormone............34, 234–237, 286, 287
TIDE............67, 68, 74–76, 79, 114, 153, 157
Tissue clearing............234, 236, 238, 239
Tissue-specific............33–51, 56, 94
Transcription activator-like effector nucleases (TALEN)............v, vi, 33, 55–63, 67, 83, 84, 92, 94–99, 106
Transgenesis............93, 101
Tumor suppressor gene............40, 148, 150, 151

W

Web tools............198

X

Xenopus laevis............v, vi, 1–16, 19–31, 34, 55, 56, 91, 92, 94–95, 98, 99, 108, 110, 111, 113, 115, 134, 136, 139, 164, 175, 195–200, 202, 205, 207–211, 213, 218, 219, 221, 226, 234, 236, 239, 240, 252, 255, 258, 259, 261, 262, 265–270, 272, 275–280, 284
Xenopus tropicalis............v, vi, 1–16, 19–31, 33–51, 55–57, 60, 62, 67–81, 84, 86, 87, 91, 92, 94, 96, 98, 99, 108, 110, 111, 113, 115, 148, 164, 243–249, 259, 266, 272

Y

Yolk............69, 70, 80, 178, 180–182

MIX
Papier aus verantwortungsvollen Quellen
Paper from responsible sources
FSC® C105338

If you have any concerns about our products,
you can contact us on
ProductSafety@springernature.com

In case Publisher is established outside the EU,
the EU authorized representative is:
Springer Nature Customer Service Center GmbH
Europaplatz 3, 69115 Heidelberg, Germany

Printed by Libri Plureos GmbH
in Hamburg, Germany